Advances in Intelligent Systems and Computing

Volume 1170

AF880279

The series "Advances in Intelligent Systems and Computing" contains publications on theory, applications, and design methods of Intelligent Systems and Intelligent Computing. Virtually all disciplines such as engineering, natural sciences, computer and information science, ICT, economics, business, e-commerce, environment, healthcare, life science are covered. The list of topics spans all the areas of modern intelligent systems and computing such as: computational intelligence, soft computing including neural networks, fuzzy systems, evolutionary computing and the fusion of these paradigms, social intelligence, ambient intelligence, computational neuroscience, artificial life, virtual worlds and society, cognitive science and systems, Perception and Vision, DNA and immune based systems, self-organizing and adaptive systems, e-Learning and teaching, human-centered and human-centric computing, recommender systems, intelligent control, robotics and mechatronics including human-machine teaming, knowledge-based paradigms, learning paradigms, machine ethics, intelligent data analysis, knowledge management, intelligent agents, intelligent decision making and support, intelligent network security, trust management, interactive entertainment, Web intelligence and multimedia.

The publications within "Advances in Intelligent Systems and Computing" are primarily proceedings of important conferences, symposia and congresses. They cover significant recent developments in the field, both of a foundational and applicable character. An important characteristic feature of the series is the short publication time and world-wide distribution. This permits a rapid and broad dissemination of research results.

Indexed by SCOPUS, DBLP, EI Compendex, INSPEC, WTI Frankfurt eG, zbMATH, Japanese Science and Technology Agency (JST), SCImago.

All books published in the series are submitted for consideration in Web of Science.

More information about this series at http://www.springer.com/series/11156

Debasis Giri · Anthony T. S. Ho ·
S. Ponnusamy · Nai-Wei Lo
Editors

Proceedings of the Fifth International Conference on Mathematics and Computing

ICMC 2019

Springer

Editors
Debasis Giri
Department of Information Technology
Maulana Abul Kalam Azad University of Technology
Haringhata, West Bengal, India

Anthony T. S. Ho
Faculty of Engineering and Physical Sciences
University of Surrey
Guildford, UK

S. Ponnusamy
Indian Institute of Technology Madras
Chennai, India

Nai-Wei Lo
Department of Information Management
National Taiwan University of Science and Technology
Taipei City, Taiwan

ISSN 2194-5357 ISSN 2194-5365 (electronic)
Advances in Intelligent Systems and Computing
ISBN 978-981-15-5410-0 ISBN 978-981-15-5411-7 (eBook)
https://doi.org/10.1007/978-981-15-5411-7

This Springer imprint is published by the registered company Springer Nature Singapore Pte Ltd.
The registered company address is: 152 Beach Road, #21-01/04 Gateway East, Singapore 189721, Singapore

Committee

Chief Patron
Achyuta Samanta, Founder KIIT & KISS

Patrons
Ved Prakash, Chancellor, KIIT
Hrushikesha Mohanty, Vice Chancellor, KIIT
Sasmita Samanta, Registrar, KIIT

General Co-Chairs
P. K. Saxena, Former Director, DRDO, SAG, Delhi, India
P. D. Srivastava, IIT Bhilai, India
Sudarshan Nanda, Eminence & Research Chair, KIIT, India

Program Co-Chairs
S. Ponnusamy, IIT Madras, India
Anthony TS Ho, University of Surrey, UK
Nai-Wei Lo, National Taiwan University of Science and Technology, Taiwan
Debasis Giri, Maulana Abul Kalam Azad University of Technology, West Bengal, India

Organizing Chairs
Monideepa Roy, School of Computer Engineering, KIIT
Tanmoy Maitra, School of Computer Engineering, KIIT

Organizing Co-Chair
Sujoy Datta, KIIT, India

Finance Chairs
Sujoy Datta, KIIT, India
Tanmoy Maitra, KIIT, India

Public Relations and Publicity Committee
Hrudaya Kumar Tripathy, KIIT, India
Niladri Bihari Puhan, IIT Bhubaneswar

Jitendra Kumar Rout, KIIT, India
Rajdeep Chatterjee, KIIT, India
Sagar Patnaik, KIIT, India

Transport and Hospitality Committee
Pinaki Sankar Chatterjee, KIIT, India
Manas Ranjan Lenka, KIIT, India
Arup Sarkar, KIIT, India

Registration Committee
Saurabh Bilgaiyan, KIIT, India
Bhaswati Sahoo, KIIT, India
Roshni Pradhan, KIIT, India

Event Management Committee
Ajay Kumar Jena, KIIT, India
Sital Dash, KIIT, India

Food and Refreshment Committee
Mahendra Kumar Gourisaria, KIIT, India
Ajay Kumar Jena, KIIT, India

Local Advisory Committee
Samaresh Mishra, KIIT, India
Ganga Bishnu Mund, KIIT, India
Madhabananda Das, KIIT, India
Biswajit Sahoo, KIIT, India
Prasant Kumar Pattnaik, KIIT, India
Alok Jagdev, KIIT, India
Bhabani Shankar Prasad Mishra, KIIT, India

Technical Program Committee

TPC for Computing
Takeshi Koshiba, Waseda University, Japan
Gerardo Pelosi, Politecnico di Milano, Leonardo da Vinci, Italy
Bart Mennink, Radboud University, The Netherlands
Michal Choras, University of Technology and Life Sciences, Poland
Sherali Zeadally, University of Kentucky, USA
Marko Holbl, University of Maribor Smetanova 17, Maribor, Slovenia
Chung-Huang Yang, National Kaohsiung Normal University, Taiwan
Junwei ZHOU, Wuhan University of Technology, China
Kouichi Sakurai, Kyushu University, Japan
Fagen Li, University of Electronic Science and Technology, China
Bimal Roy, ISI Kolkata, India
Saru Kumari, Agra College, India
Weizhi Meng, Technical University of Denmark, Denmark
Debasis Giri, Maulana Abul Kalam Azad University of Technology, WB, India

Christina Boura, Université de Versailles Saint-Quentin-en-Yvelines, France
Debiao He, Wuhan University, China
Saibal K. Pal, DRDO, Delhi, India
P. K. Saxena, SAG, DRDO, Delhi, India
Athanasios V. Vasilakos, Luleå University of Technology, Sweden
Noboru Kunihiro, The University of Tokyo, Japan
Niladri Puhan, IIT Bhubanewar, India
Sarmistha Neogy, Jadavpur University, India
Dhananjoy Dey, SAG, DRDO, India
Cheng-Chi Lee, Fu Jen Catholic University, Taiwan
Keshav Dahal, University of the West of Scotland, Scotland
Anthony TS Ho, University of Surrey, UK
Jiqiang Lu, Beihang University, China
Subrata Dutta, NIT Jamshedpur, India
Yoshinori Aono, National Institute of Information and Communications Technology, Japan
Chandrashekhar Meshram, Rani Durgavati University, India
Rajat Subhra Chakraborty, IIT Kharagpur, India
Sokratis Katsikas, NTNU, Norway
Josep Domingo-Ferrer, Universitat Rovira i Virgili, Spain
Jun Peng, University of Texas—Rio Grande Valley, Texas
Shay Gueron, University of Haifa and Intel Corporation, Israel
Shan-Hsiang She, National Taiwan University of Science and Technology, Taiwan
Sarbani Roy, Jadavpur University, India
Dipanwita Roychowdhury, IIT Kharagpur, India
Sujoy Mistry, Maulana Abul Kalam Azad University of Technology, India
Shih-Hao (Henry) Chang, Tamkang University, Taiwan
Chi-Yuan Chen (Justin), National Ilan University, Taiwan
Chan Yeob Yeun, Khalfia University, Abu Dhabi, UAE
Indivar Gupta, SAG, DRDO, Delhi, India
Samiran Chattopadhyay, Jadavpur University, India
Olivier Blazy, Université de Limoges, France

TPC for Mathematics

Margareta Heilmann, University of Wuppertal, Germany
Ashok Kumar Das, IIIT Hyderabad, India
Abdalah Rababah, Jordan University of Science and Technology, Jordan
Emel Aşıcı, Karadeniz Technical University, Turkey
Maria A. Navascues, Escuela de Ingenieríay Arquitectura, Tfno
Alip Mohammed, The Petroleum Institute, Abu Dhabi
Mujahid Abbas, University of Pretoria (UP), Pretoria, South Africa
Madhumangal Pal, Vidyasagar University, India
Fahreddin Abdullayev, Mersin University, Turkey
Ljubisa Kocinac, University of Nis, Serbia
Nita H. Shah, Gujarat University, Navrangpura, Ahmedabad, India

Preface

The Fifth International Conference on Mathematics and Computing (ICMC 2019) was held at the Kalinga Institute of Industrial Technology (KIIT) Deemed to be University, Bhubaneswar from February 6 to 9, 2019. Bhubaneswar is a smart city and capital of the Indian state of Odisha and KIIT is a premier institution training engineers and computer scientists for the past several years.

In response to the call for papers for ICMC 2019, 81 papers were submitted in the conference. The papers were evaluated and ranked on the basis of their significance, novelty, and technical quality by at least two reviewers per paper. After a careful blind refereeing process, 20 papers were selected for inclusion in the conference proceedings. The papers cover current research in Cryptography, Security, Abstract Algebra, Functional Analysis, Fluid Dynamics, Fuzzy Modeling and Optimization, etc. ICMC 2019 will have eminent personalities both from India and abroad (UK, Taiwan, Japan, Bangladesh, and Turkey), who will deliver invited addresses and tutorial talks. The speakers from India are recognized leaders in government, industry and academic institutions like Indian Statistical Institute Kolkata, Indian Statistical Institute Chennai, IIT Kharagpur, Maulana Abul Kalam Azad University of Technology, West Bengal, Jadavpur University, etc. All of them are involved in research dealing with the current issues of interest related to the theme of the conference. The conference was two invited talks on February 6, 2019 in the 1-day pre-conference workshop addressed by Dr. Debarshi Sanyal (Principal Project Officer (Research), National Digital Library of India, Indian Institute of Technology Kharagpur), and Dr. Kamalesh Acharya (School of Computer Sciences, National Institute of Science Education and Research, India). In addition to these there were the invited talks on the main conference during February 7–9, 2019 by Prof. T. S. Ho (University of Surrey, UK), Prof. Feng Hao (University of Warwick, UK), Prof. S. Ponnusamy (IIT Madras, India), Prof. Al-Sakib Khan Pathan (Southeast University, Bangladesh), Prof. Sarbani Roy (Jadavpur University, India), Prof. Goutam Kumar Paul (Indian Statistical Institute, Kolkata, India), Prof. Debasis Giri (Maulana Abul Kalam Azad University of Technology, West Bengal, India).

A conference of this kind would not be possible to organize without the full support from different people across different committees. All logistics and general organizational aspects are looked after by the organizing committee members, who spent their time and energy in making the conference a reality. We also thank all the technical program committee members and external reviewers for thoroughly reviewing the papers submitted for the conference and sending their constructive suggestions within the deadlines. Our hearty thanks to Springer for agreeing to publish the proceedings in its "Advances in Intelligent Systems and Computing" series.

We are indebted to the Kalinga Institute of Industrial Technology (KIIT) Deemed to be University, and Defense Research and Development Organization (DRDO) for sponsoring the event. Their support has significantly helped in raising the profile of the conference.

Last but not the least, our sincere thanks go to all authors who submitted papers to ICMC 2019 and to all speakers and participants. We sincerely hope that the readers will find the proceedings stimulating and inspiring.

Haringhata, India Debasis Giri
Chennai, India S. Ponnusamy
Guildford, UK Anthony T. S. Ho
Taipei, Taiwan Nai-Wei Lo

Message from Program Co-chairs

It was a great pleasure for us to organize the Fifth International Conference on Mathematics and Computing 2019 (ICMC 2019) to be held during February 6–9, 2019 at Kalinga Institute of Industrial Technology (KIIT), Deemed to be University, Bhubaneswar, Odisha, India. Our main goal in this conference was to provide an opportunity to the participants to learn about contemporary research in Cryptography, Security, Modeling, and different areas of Mathematics and Computing. In addition, we aim to promote the exchange of ideas among themselves and with experts participating in the conference as the keynote speakers. It is our sincere hope that the conference will help the participants in their research and training and opened new avenues for work for those who are either starting their research or are looking for extending their area of research to a new area of current research Mathematics and Computing.

The inauguration ceremony of the conference was held on February 8, 2019. There were the keynote talks of one hour by **Prof. T. S. Ho** (University of Surrey, UK), **Prof. Feng Hao** (University of Warwick, UK), **Prof. S. Ponnusamy** (IIT Madras, India), **Prof. Al-Sakib Khan Pathan** (Southeast University, Bangladesh), **Prof. Sarbani Roy** (Jadavpur University, India), **Prof. Goutam Kumar Paul** (Indian Statistical Institute, Kolkata, India), **Prof. Debasis Giri** (Maulana Abul Kalam Azad University of Technology, West Bengal, India), **Dr. Debarshi Sanyal** (Principal Project Officer (Research), National Digital Library of India, Indian Institute of Technology Kharagpur), and **Dr. Kamalesh Acharya** (School of Computer Sciences, National Institute of Science Education and Research, India).

After an initial call for papers, 81 papers were submitted at the conference. All the submitted papers were sent to external referees. After a thorough referring process 20 papers were recommended for publication for the conference proceedings.

We are truly thankful to the speakers, participants, referees, organizers, and student volunteers for their support and help without which it would have been impossible to organize the conference. Our special thanks go to Prof. P. K. Saxena,

Prof. P. D. Srivastava, Prof. Sudarshan Nanda, Prof. Tanmoy Maitra and Prof. Monideepa Roy for their hard work and support. In addition, we would like to thanks the sponsors and host university, i.e., KIIT Deemed to be University and funding agency i.e., DRDO for supporting Fifth ICMC 2019.

Debasis Giri
S. Ponnusamy
Anthony T. S. Ho
Nai-Wei Lo

Message from the General Co-chairs

As we all are aware, Mathematics has always been a discipline of interest not only to theoreticians but also to all practitioners irrespective of their specific profession. Be it Science, Technology, Economics, Commerce or even Sociology, new Mathematical principles and models have been emerging and helping in new research and in drawing inferences from practical data as well as through logic. The past few decades have seen enormous growth in applications of Mathematics in different areas, which are multidisciplinary in nature. Cryptography, Security, and Signal processing are such areas which have got more focus recently due to need for secure communication while connecting with others. With emerging computing facilities and speeds, a phenomenal growth has happened in the problem-solving area. Earlier, some observations were made and conjectures were drawn, which remained conjectures till somebody either could prove it theoretically of found counterexamples. But today, we can write algorithms and use computers for long calculations, verifications, or generation of huge amount of data. With available computing capabilities, we can find factors of very large integers of the size of hundreds of digits; we can find inverses of very large size matrices and solve a large set of linear equations and so on. Thus, Mathematics and Computations have become more integrated areas of research these days and it was thought to organize an event where thoughts may be shared by researchers and new challenging problems could be deliberated for solving these.

Apart from many other interdisciplinary areas of research, cryptography and security have emerged one of the most important areas of research with discrete mathematics as a base. Several research groups are actively pursuing the research on different aspects of cryptology not only in terms of new crypto-primitives and algorithms but a whole lot of concepts related to authentication, integrity, and security proofs/protocols are being developed, many times with open and competitive evaluation mechanism to evolve standards.

As Conferences, Seminars, and Workshops are the mechanisms to share knowledge and new research results giving us a chance to get new innovative ideas for futuristic needs as threats and computational capabilities of adversaries are ever increasing, it was thought appropriate to organize the present Conference focusing on Mathematics and Computations covering theoretical as well as practical aspects of research, Cryptography and Security being ones of these.

Eminent personalities working in Mathematical Sciences and related areas have been invited from abroad as well as from within the country, who would deliver invited talks and tutorials for participants. The talks by these speakers intend to cover a wide spectrum, viz., Number Theoretic Concepts, Cryptography, Algebraic Concepts like Quasi-Groups and applications, etc. The Conference is spread over 5 days (February 6–9, 2019), where the first day was dedicated to a 1-day workshop. The main conference was planned with special talks by experts and paper presentations in each session.

I hope that the conference would meet the aspirations of the participants and meet its objective of ideas and current research being shared and new targets/problems identified; more so the young researchers and students would get new directions to pursue their future research.

The conference would not have been possible without the enthusiasm and hard work of a number of colleagues. We wish to express my sincere appreciation to the extended Technical Program Committee. We would like to thank the host university KIIT Deemed to be University, Bhubaneswar, Odisha, India for funding and supporting this conference Last, but certainly not least, our thanks go to all the authors who submitted papers and all the attendees.

P. K. Saxena
P. D. Srivastava
Sudarshan Nanda

Contents

About the Editors

Debasis Giri is currently working as an Associate Professor in the Department of Information Technology, Maulana Abul Kalam Azad University of Technology (Formerly known as West Bengal University of Technology), West Bengal, India. Prior to this, he also held academic positions as Professor in the Department of Computer Science and Engineering and Dean in the School of Electronics, Computer Science and Informatics, Haldia Institute of Technology, Haldia, India. He did his masters (M.Tech. and M.Sc.) both from IIT Kharagpur, India, and also completed his Ph.D. from IIT Kharagpur, India. He is tenth all India rank holder in Graduate Aptitude Test in Engineering (GATE) in 1999. He received a certificate from All India Science Teachers' Association in Science Aptitude & Talent Search Test in 1988. His current research interests include Cryptography, Information Security, E-commerce security and Design & Analysis of Algorithms. He has authored more than 75 research papers in reputed international journals and conference proceedings. He is editorial board member and reviewer of many International journals. He is the founder of the International Conference on Mathematics and Computing. He is also program committee member of several international conferences. He is a member of IEEE, and a life member of Cryptology Research Society of India, and the International Society for Analysis, its Applications and Computation (ISAAC).

Prof. Anthony T. S. Ho holds a Personal Chair in Multimedia Security and served as Head of Department of Computer Science, University of Surrey from 2010 to 2015. He is also a Tianjin Distinguished Professor, Guest Professor of Tianjin University of Science and Technology and of Wuhan University of Technology, China, as well as Visiting Professor of University of Malaya, Malaysia. He was the recipient of the prestigious Institution of Engineering and Technology (IET) Innovation in Engineering Award under the Security category for his research and commercialization work on digital watermarking in 2006.

Professor Ho obtained his M.Sc. in Applied Optics from Imperial College London in 1980 and his Ph.D. in Digital Image Processing from King's College London, University of London in 1983. After graduation, he worked in technical management positions in industry for 11 years in the UK and Canada. From 1994 to 2005, He was a Senior Lecturer and then Associate Professor at Nanyang Technological University (NTU), Singapore. He has published more than 160 articles in international journals and conference proceedings as well as 8 international patents granted related to watermarking and steganography.

Professor Ho is Founding Editor-in-Chief of the International Journal of Information Security and Applications (JISA) and an Area Editor for Signal Processing: Image Communication, both published by Elsevier. He serves as an Associate Editor for IEEE Transactions on Information Forensics and Security (TIFS) and Associate Editor for IEEE Signal Processing Letters (SPL) (2014–2016), as well as an Associate Editor for EURASIP Journal of Image and Video Processing published by Springer. He is a Fellow of Institution of Engineering and Technology (FIET), Fellow of Institute of Physics (FInstP) and Fellow of British Computer Society (FBCS).

Dr. S. Ponnusamy is a Senior Professor in the Department of Mathematics of IIT Madras. He has earned his B.Sc. (1980) and M.Sc. (1992) from the University of Madras. He completed his Ph.D. (1989) at IIT Kanpur. His current research interest includes Complex Analysis, Quasiconformal and Harmonic Mappings, Special Functions and Functions Spaces.

Dr. S. Ponnusamy is the Founding "Fellow of the Forum de Analystes, Chennai, India", 1992. He was elected as a "Fellow of The National Academy of Sciences, India" in the year 2002. He served five years as Head of the Indian Statistical Institute, Chennai Centre (Oct. 2012 to Oct. 2017). He is the currently the President of the Ramanujan Mathematical Society, India.

Dr. S. Ponnusamy has been associating with a number of organizations for popularization of science and serves on the editorial boards of several peer-reviewed international journals. He is often invited to give plenary talks at international conferences and to give lectures in universities all over the world. He took part as a researcher in many funded projects in India and Abroad. He has been refereeing for more than 100 journals.

Dr. Ponnusamy has written four books and another book with Herb Sivlerman. One of his books won the Best Selling Author award in 2002. He has edited several volumes, and international conference proceedings. He has published more than 250 technical articles in reputed international journals (such as Advances in Mathematics, Annales Academia Scientiarum Fennica Mathematica, Applied Mathematics and Computation, Archiv der Mathematik, Bulletin des Sciences Mathematiques, Bulletin/Journal of the Australian Mathematical Society, Complex Variables and Elliptic Equations, Complex Analysis and Operator Theory, Computational Methods and Function Theory, Indagationes Mathematicae, Journal of London Mathematical Society, Integral Equations and Operator Theory, Journal Computational and Applied Mathematics, Journal of Mathematical Analysis and Applications, Mathematika, Mathematishe Annalen, Mathematische Nachrichten, Mathematische Zeitschrift, Monatshefte fuer Mathematik, Nonlinear Analysis, Potential Analysis, Proceedings of American Mathematical Society,

Results in Mathematics, Rocky Mountain Journal of Mathematics The Journal of Geometric Analysis) and has solved several long standing open problems and conjectures.

Dr. Ponnusamy has been a visiting professor of number of universities in abroad (eg. Hengyang Normal University and Hunan Normal University, China; Kazan Federal University and Petrozavodsk State University, Russia; University Sains Malaysia, Malaysia; University of Aalto, University of Turku, and University of Helsinki, Finland; Texas Tech University, Lubbock, USA).

Nai-Wei Lo received the B.S. degree in engineering science from National Cheng Kung University, Tainan, Taiwan, in 1988 and the M.S. and Ph.D. degrees in computer science and electrical engineering from the State University of New York at Stony Brook, Stony Brook, NY, USA, in 1992 and 1998, respectively. He is currently a Professor and the Chairman of Department of Information Management, National Taiwan University of Science and Technology, Taipei, Taiwan. His research interests include application and system security, IoT/IoV security, cloud security and Web technology. He currently serves as an associate editor of Journal of Information Security and Applications. He is a senior member of IEEE.

On the Design of a Secure Anonymous User Authentication Scheme for Multiple Base Stations Based Wireless Sensor Networks

Soumya Banerjee, Vanga Odelu, Ashok Kumar Das, Samiran Chattopadhyay, and Debasis Giri

Abstract In recent years, security and data privacy in Wireless Sensor Networks (WSNs) have gained great thrust. Several anonymity-preserving user authentication and key establishment mechanisms for WSNs with single base station have been suggested in the literature. However, sometimes a single base station may be insufficient to maintain the proper quality of services or it may be a single point of failure in WSNs. Very recently Amin et al. suggested a multiple base stations based anonymous user authentication and key establishment approach for WSNs to address the aforementioned problem. In this paper, we first demonstrate that Amin et al.'s scheme has several security pitfalls. To address the security issues in Amin et al.'s scheme, we design a more secure and anonymous user authentication and key establishment mechanism for WSNs that is also based on multiple base stations concept. An exhaustive comparative study demonstrates that the proposed mechanism

S. Banerjee (✉) · S. Chattopadhyay
Department of Information Technology, Jadavpur University, Salt Lake City, Kolkata 700 098, India
e-mail: soumyaBanerjee@outlook.in

S. Chattopadhyay
e-mail: samirancju@gmail.com

V. Odelu
Department of Computer Science and Information Systems, Birla Institute of Technology and Science, Pilani Hyderabad Campus, Hyderabad 500 078, India
e-mail: odelu.vanga@hyderabad.bits-pilani.ac.in

A. K. Das (✉)
Center for Security, Theory and Algorithmic Research, International Institute of Information Technology, Hyderabad 500 032, India
e-mail: iitkgp.akdas@gmail.com; ashok.das@iiit.ac.in

D. Giri
Department of Information Technology, Maulana Abul Kalam Azad University of Technology, Kolkata 700064, India
e-mail: debasis_giri@hotmail.com

D. Giri et al. (eds.), *Proceedings of the Fifth International Conference on Mathematics and Computing*, Advances in Intelligent Systems and Computing 1170, https://doi.org/10.1007/978-981-15-5411-7_1

gives significantly better security and more functionality attributes with comparable commutation as well as computation overheads while those are compared with Amin et al's. approach.

1 Introduction

The utilization of WSNs has exploded in several domains ranging from military to civilian applications, including disaster management and remote monitoring [1]. A WSN comprises of many inexpensive sensing devices (sensors or nodes), which are restricted in available power, processor, and memory, which are dispersed over a target or deployment field. The sensors are accessed through a base station (also called the gateway node), which is not resource constrained in its capabilities. The sensors can be deployed randomly or manually depending on various application scenarios.

Recent advances in WSN research advocates a heterogeneous WSN architecture [17], where limited capability sensors are supplemented with relatively more capable nodes acting as cluster heads (group heads). The sensing information from local sensors are aggregated in cluster heads and forwarded to the base station. However, sometimes a single base station may be insufficient for a large geographically covered WSN or it may be a single-point failure in WSN. Hence, the use of multiple base stations is very much desirable [2].

1.1 Motivation

In WSNs, the collected data at the base stations may be not always real time as the base stations collect the sensing information from the sensors periodically. There are some critical applications, such as battlefield and healthcare applications, which require real-time data analysis to take appropriate actions by the users (e.g., a general or major in case of battlefield application and a doctor in case of healthcare application). This is then possible when an external authorized party, called a user, is given direct access to the services from the sensors in WSNs. This necessitates an anonymity-preserving authentication scheme for the user and the accessed sensor node to mutually authenticate each other and establish a session key among themselves. However, the possibility of physical sensor node captures attack and other potential attacks such as replay, impersonation, man-in-the-middle, online/off-line guessing, and privileged-insider attacks in addition to resource limitations of sensor nodes makes designing such a scheme while maintaining security and efficiency, challenging.

Very recently, Amin et al. [2] presented "A robust mutual authentication protocol for WSN with multiple base stations". Unfortunately, in this paper, we demonstrate that Amin et al's. scheme has several security vulnerabilities. To address the security limitations identified in Amin et al's. scheme, we propose a more secure user authentication and key agreement mechanism for multiple base stations based WSNs.

1.2 Network and Threat Models

We adhere to the same network model as presented in [2] for modeling WSNs with multiple base stations shown in Fig. 1. It describes the network connection between the different principles in the authentication scheme. The system administrator deploys the sensors and the cluster heads in the target field. The cluster heads (and the sensors) are associated with their nearest base station. The users must register with any one of the available nearby base stations, henceforth designated as the home base station for those users. However, the users can communicate with all other bases stations over the Internet, which are designated as the foreign base stations.

This work adheres to the de-facto Dolev-Yao (DY) threat model [9]. Under the DY threat model, it is assumed that the adversary $\mathcal{A}$ has ultimate authority over the open channel. Therefore, it is within $\mathcal{A}$'s capability to eavesdrop upon all transmitted messages and also to modify, delete, or forge them. It is further assumed that $\mathcal{A}$ can acquire a user's stolen/lost smart card and can then extract all secret credentials stored in its memory through the power analysis attacks as discussed in [14, 16]. Finally,

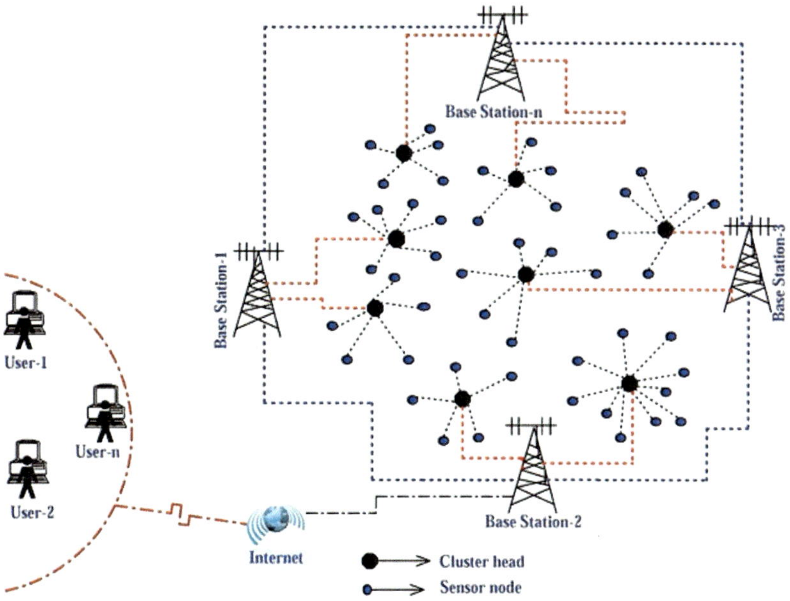

Fig. 1 Communication model for the multiple base stations based WSN [2]

the sensor nodes and cluster heads are considered susceptible to physically capture by $\mathcal{A}$ due to hostile environmental nature of WSNs. However, the base stations are considered trusted entities.

1.3 Research Contributions

The principal contributions are briefed below.

- We first analyzed the recent work by Amin et al. [2], which is an anonymity-preserving authentication scheme for WSNs having multiple base stations, and discovered that their scheme fails to defend several crucial attacks, like off-line password guessing attack through stolen smart card attack, smart card impersonation, parallel session, and password change attacks. Furthermore, their scheme also fails in preserving anonymity.
- We then design a more secure user authentication scheme resistant to the security limitations found in Amin et al's. scheme.
- Finally, through a detailed comparative study, we demonstrate that the proposed scheme is superior to the scheme of Amin et al. [2], in terms of security and functionality features.

1.4 Related Work

In this section, through a brief study we review authentication schemes for the following categories: (1) homogeneous (distributed) uni-base station WSN [11], (2) heterogeneous uni-base station WSN [5], and (3) heterogeneous multiple base stations WSN, i.e., WSN with multiple base stations [17].

Wong et al. [23] introduced a lightweight scheme for user authentication. Tseng et al. [18] then demonstrated several security limitations in Wong et al's. scheme such as replay and forgery attacks. Vaidya et al. [21] identified that the scheme proposed in [18] was vulnerable to replay and man-in-the-middle attacks. They also presented an improved variation of the scheme [23] to make it resilient to the mentioned attacks.

Das [7] presented an efficient scheme in WSNs. The schemes presented in [7, 23] lack password change facility. Khan et al. [13] presented that if a password change mechanism is not provided, it allows the adversary with unlimited guessing of the user's password. They showed that the scheme [7] couldn't resist a bypassing attack on the gateway node and it also lacks mutual authentication. Later, Vaidya et al. [20] found that the scheme [7] was further vulnerable under other known attacks (e.g., stolen smart card, password guessing and gateway node impersonation attacks). Furthermore, they demonstrated that the scheme [13] too was unsafe to several known attacks, like stolen smart card, privileged-insider, forgery, and node capture attacks.

Turkanović et al. [19] presented Internet of Things (IoT) in context of WSNs and presented a user authentication scheme for ad hoc WSNs. However, the authors in [17] described several vulnerabilities in the scheme [19]. In addition, some of the state-of-art user authentication protocols [3, 4, 6, 15, 22, 24] have been suggested for WSNs.

Amin and Biswas [17] presented the idea of multiple gateway nodes based scheme to address the coverage issue. Amin et al. [2] presented a user authentication and key agreement schemes for multiple base stations based WSNs. However, we demonstrate that the scheme is defenceless against several known security issues that are detailed in Sect. 3 .

2 Review of Amin et al's. Scheme

A short overview of Amin et al's. scheme [2] has been presented in the following for various described phases.

2.1 System Setup

During network setup, administrator of the system (SA) selects a shared secret key X_{SH}. For m number of sensors (SN_k) and n number of cluster heads (CH_j), SA selects unique identities ID_{SN_k} and ID_{CH_j}, and computes $X_{SN_k} = h(ID_{SN_k} || X_{SH})$, $X_{CH_j} = h(ID_{CH_j} || X_{SH})$, respectively, for $k \in [1, m]$ and $j \in [1, n]$. After embedding these credentials into the sensors and cluster heads, the SA deploys them in a target field. It is assumed that there are t base stations. Each BS_l, for $l \in [1, t]$, maintains a table that contains the identities corresponding to all the deployed sensors and cluster heads administered by it. BS_l also saves its private key X_l. Additionally, X_{SH} is known only among the base stations.

2.2 User Registration Phase

Before availing services, a user U must register with any $BS_l, l \in [1, t]$ through the steps described below:

Step 1. U requests BS_l for smart card with its chosen identity ID_u and "user credentials" (issued by government approved agency) over a secure channel (e.g., in person).

Step 2. If ID_u is not already present in BS_l's database, it generates a temporary identity TID_u, calculates $A_u = h(ID_u || SCN_u || X_l)$ and saves the information TID_u, $E_{X_l}(ID_u)$, user credentials, and SCN_u in its database. Finally, BS_l securely issues a smart card accommodating $\{TID_u, A_u, ID_{BS_l}, h(\cdot), SCN_u\}$, where SCN_u

is the U's smart card's unique serial number and $h(\cdot)$ is a 'cryptographically secure collision-resistant one-way hash function'.

Step 3. On receiving the smart card, U provides the password PW_u and calculates $B_u = h(ID_u \oplus PW_u)$, $C_u = A_u \oplus B_u$, $D_u = h(TID_u \| ID_u \| PW_u)$ and replaces A_u with C_u and D_u in the smart card. The smart card then contains the information $\{TID_u, C_u, D_u, ID_{BS_l}, h(\cdot), SCN_u\}$.

2.3 Login and Authentication Phases

A registered user U first inputs ID_u and PW_u to login in WSNs with his/her smart card, which then calculates $D_u^* = h(TID_u \| ID_u \| PW_u)$. If $D_u^* = D_u$, login is successful; otherwise, this phase is terminated.

After logging in successfully, through mutual authentication session key is established. This procedure comprises three sub-phases, namely (1) authentication among a user and base station, (2) authentication among the base station and sensor nodes, and (3) authentication among two base stations. In the following, we will only discuss the authentication mechanism among user and base station. The details of the other two authentication procedures are provided in [2].

Step 1. After a successful login, the smart card picks a random number $R_u \in \mathbb{Z}_p^*$, where $Z_p^* = \{1, 2, \ldots, p-1\}$ and p is a tolerably large prime, computes $B_u^* = h(ID_u \oplus PW_u)$, $A_u^* = C_u \oplus B_u^*$, $E_u = h(ID_u \| R_u \| A_u^*)$, $F_u = R_u \oplus h(ID_{BS_l} \| A_u^*)$ and sends the message $M_1 = \langle TID_u, E_u, F_u, ID_{SN_k} \rangle$ to BS_l, where ID_{SN_k} is the identity of the target sensor node.

Step 2. After receiving M_1, BS_l retrieves the tuple $\langle TID_u, E_{X_l}(ID_u)$, "user credentials", $SCN_u \rangle$ from its database and calculates $ID_u = D_{X_l}(E_{X_l}(ID_u))$, $A_u^* = h(ID_u \| SCN_u \| X_l)$, $R_u = F_u \oplus h(ID_{BS_l} \| A_u^*)$, $E_u^* = h(ID_u \| R_u \| A_u^*)$. If $E_u^* = E_u$, BS_l selects $R_l, TID_u' \in \mathbb{Z}_p$ and calculates $G_1 = R_u \oplus R_l$, $G_2 = h(A_u^* \| R_u \| R_l)$, $G_3 = TID_u' \oplus h(A_u^* \| R_u)$ and sends the message $M_2 = \langle G_1, G_2, G_3 \rangle$ to U.

Step 3. After receiving M_2, U recovers $R_l = R_u \oplus G_1$ and calculates $G_2^* = h(A_u^* \| R_u \| R_l)$. If $G_2^* = G_2$, U recovers $TID_u' = G_3 \oplus h(A_u^* \| R_u)$ and computes the session key with BS_l as $SK_{ul} = h(ID_u \| R_u \| R_l)$ and $G_4 = h(SK_{ul} \| TID_u')$. User U then communicates with BS_l by sending the message $M_3 = \langle G_4 \rangle$ to BS_l.

Step 4. On receiving M_3, BS_l calculates $SK_{lu} = h(ID_u \| R_u \| R_l)$ and $G_4^* = h(SK_{lu} \| TID_u')$. If $G_4^* = G_4$, BS_l accepts SK_{lu} as the session key with U and sends the acknowledgment to user U.

Step 5. On receiving the acknowledgment, U calculates $D_u^* = h(TID_u' \| ID_u \| PW_u)$ and updates D_u and TID_u in smart card with D_u^* and TID_u', respectively. Finally, U sends the acknowledgment BS_l.

Step 6. On receiving the acknowledgment, BS_l replaces TID_u with TID_u' in the tuple $\langle TID_u, E_{X_l}(ID_u)$, "user credentials", $SCN_u \rangle$ in its database.

2.4 Password Renewal Phase

The steps to update the password of a valid user, say U are described in this section using the following steps:

Step 1. U first enters his/her identity ID_u and password PW_u to login in WSNs through the smart card. Which then calculates $D_u^* = h(TID_u || ID_u || PW_u)$. If $D_u^* = D_u$, login is successful; if not the process is terminated. On successful login, U inputs new password PW_u^{new}.

Step 2. The smart card then calculates $A_u^* = C_u \oplus B_u$, $B_u^{new} = h(ID_u || PW_u^{new})$, $C_u^{new} = A_u^* \oplus B_u^{new}$, and $D_u^{new} = h(TID_u || ID_u || PW_u^{new})$.

The smart card finally replaces $\langle C_u, D_u \rangle$ with $\langle C_u^{new}, D_u^{new} \rangle$ in its memory.

3 Security Vulnerabilities in Amin et al's Scheme

In this section, we show that in the presence of a passive/active adversary, the following potential attacks are possible.

- **Privileged-insider attack through off-line password guessing and lost/stolen smart card attacks**: Assume an adversary $\mathcal{A}$, who is a privileged- insider, knows ID_u from during registration. Once the user registration process is over, $\mathcal{A}$ can gain access the user's smart card and through power analysis attacks [14, 16], $\mathcal{A}$ extract all the credentials $\{TID_u, C_u, ID_{BS_l}, h(\cdot), SCN_u, D_u\}$. Now, $\mathcal{A}$ can guess a password, say $PW_u^{guessed}$ and calculate $D_u^* = h(TID_u || ID_u || PW_u^{guessed})$. $\mathcal{A}$ can verify the guessed password by checking if $D_u^* == D_u$. Hence, it is apparent that low-entropy passwords are easily verified in Amin et al's. scheme.
- **User impersonation attack**: Suppose a privileged adversary $\mathcal{A}$ having the correct password PW_u is using the above attacks, and credentials TID_u and C_u from the compromised smart card, can generate $R_u' \in \mathbb{Z}_p$ and calculate $B_u^* = h(ID_u \oplus PW_u)$, $A_u^* = C_u \oplus B_u^*$, $E_u = h(ID_u || R_u' || A_u^*)$, and $F_u = R_u' \oplus h(ID_{BS_l} || A_u^*)$. Thus, $\mathcal{A}$ can send the message $M_1 = \langle TID_u, E_u, F_u, ID_{SN_k} \rangle$ to BS_l on behalf of the smart card of the user U. Hence, Amin et al's. scheme is susceptible to user impersonation attack.
- **Parallel session attack**: A privileged adversary $\mathcal{A}$ can eavesdrop the messages $M_1 = \langle TID_u, E_u, F_u, ID_{SN_k} \rangle$ and $M_2 = \langle G_1, G_2, G_3 \rangle$. $\mathcal{A}$ having guessed the correct password PW_u can calculate $B_u^* = h(ID_u \oplus PW_u)$, $A_u^* = C_u \oplus B_u^*$, $R_u^* = F_u \oplus h(ID_{BS_l} || A_u^*)$, $R_l^* = G_1 \oplus R_u^*$. Thus, $\mathcal{A}$ can easily independently calculate the session key $SK = h(ID_u || R_u^* || R_l^*)$ making feasibility of parallel session attack on Amin et al's. scheme.
- **Anonymity and untraceability**: A privileged adversary $\mathcal{A}$ is aware of the mapping of TID_u to ID_u. Thus, by eavesdropping on an authentication message $M_1 = \langle TID_u, E_u, F_u, ID_{SN_k} \rangle$, $\mathcal{A}$ can easily identify and trace the user U. Furthermore, from the eavesdropped message $M_2 = \langle G_1, G_2, G_3 \rangle$, $\mathcal{A}$ can also calculate $TID_u' =$

$G_3 \oplus h(A_u^* || R_u^*)$ and update TID_u to identify and trace the user U through subsequent authentications. As a result, Amin et al's. scheme does not safeguard anonymity & untraceability properties.
- **Password change attack**: A privileged adversary $\mathcal{A}$ can simply execute the password update phase to alter the user's password PW_u with a different password PW_u' having already guessed the user password PW_u extracted from the stolen or lost smart card. Hence, the password change attack is also easily mounted on Amin et al's. scheme.

Remark 1 In Amin et al's. scheme, in order to access a sensor node SN_k, which is not administered by the home base station BS_l but by some other base station, say BS_y, three sessions are established: (1) between user U and BS_l, (2) between BS_l and BS_y, and (3) between BS_y and SN_k or CH_j. However, this puts an unnecessary burden on BS_l for both communication and computational overheads. With assistance from BS_l, U and BS_y can agree upon a session key after completion of mutual authentication. For further communications, it is not essential for BS_l involvement.

4 The Proposed Scheme

This section presents a more secure anonymous authentication and session key establishment scheme for WSNs with support of multiple base stations, which is free from all the mentioned security vulnerabilities discussed in Sect. 3.

4.1 Setup Phase

This phase is similar to that presented in Sect. 2.1. The system administrator SA embeds ID_{SN_k} and X_{SN_k} to a sensor node SN_k, and ID_{CH_j} and X_{CH_j} to a cluster head CH_j before their deployment. The sensors and cluster heads are accessible through base stations BS_l, $l \in [1, t]$. Each BS_l maintains a table that contains ID_{CH_j} and ID_{SN_k} corresponding to all sensors and cluster heads accessible through it. BS_l also saves its private key X_l. Additionally, X_{SH} is only known among the base stations.

4.2 User Registration Phase

A user U registers with any BS_l, $l \in [1, t]$ through the described steps:

Step 1. User U chooses identity ID_u and requests BS_l for a smart card SC_u with chosen ID_u and UC_u over a secure channel, where UC_u contains the user credentials issued by some government approved agency.

Step 2. If ID_u is already present in BS_l's database, it requests U to select another new identity. Otherwise, BS_l generates a secret key $R_{sc} \in \mathbb{Z}_p$, and calculates $TID_u = E_{X_l}(ID_u||R_{sc}||TS_l)$ and $EIDR_{sc} = E_{X_l}(ID_u||R_{sc})$. TS_l represents the current time stamp of BS_l. BS_l commits the tuple $\langle TID_u, EIDR_{sc}, h(UC_u), SCN_u\rangle$ to memory and securely issues a smart card with the values $\{TID_u, R_{sc}, ID_{BS_l}, h(\cdot), SCN_u\}$ where SCN_u is its unique serial number.

Step 3. After U receives the smart card, he/she provides password PW_u, imprints biometrics Bio_u, and calculates $(\sigma_u, \tau_u) = Gen(Bio_u)$, $ER_{sc} = R_{sc} \oplus h(ID_u||PW_u||\sigma_u)$, $D_u = h(TID_u||\sigma_u||PW_u)$, $ETID_u = TID_u \oplus h(R_{sc}||\sigma_u||PW_u)$. U then replaces TID_u and R_{sc} with $ETID_u$ and ER_{sc}, respectively, and inserts D_u, τ_u into the smart card SC_u. Here, we utilize user biometric verification through fuzzy extractor method [8]. A fuzzy extractor consists of two procedures:

Gen: A randomized generation method which takes user personal biometrics and outputs a pair (σ_u, τ_u), where σ_u and τ_u are the biometric secret key and the public reproduction parameters associated with biometrics Bio_u. Formally, $(\sigma_u, \tau_u) = Gen(Bio_u)$.

Rep: A deterministic reproduction method which from a noisy biometrics, say Bio'_u and τ_i constructs the original biometric secret key σ_i. Formally, $\sigma_i = Rep(Bio'_u, \tau_u)$ as long as the Hamming distance between Bio_u and Bio'_u is less than or equal to the threshold for error tolerance.

4.3 Login and Authentication Phases

A registered user U first inputs ID_u, PW_u, and Bio_u to login in the system for accessing data in real time from the WSNs. The smart card SC_u then calculates $\sigma_u = Rep(Bio_u, \tau_u)$, $R^*_{sc} = ER_{sc} \oplus h(ID_u||PW_u||\sigma_u)$, $TID_u = TID_u \oplus h(R^*_{sc}||\sigma_u||PW_u)$, $D^*_u = h(TID_u||\sigma_u||PW_u)$. If $D^*_u = D_u$, login is successful.

A successful login is followed by authentication and session key establishment phase, which is composed of the sub-phases: (1) authentication between U and home base station, (2) authentication between U and non-home base station, and (3) authentication among base station & sensors. The description of these sub-phases is discussed below.

(1) Authentication between the user and home base station

The steps for negotiating a session key between U and home base station, when the requested sensor node SN_k is accessible through the home base station, are described below and summarized in Fig. 2.

Step 1. After successful login, the smart card SC_u selects $R_u \in \mathbb{Z}_p$, generates the current timestamp TS_u, computes $E_u = h(ID_u||R_u||R^*_{sc}||TS_u)$, $F_u = R_u \oplus h(ID_{BS_l}||R^*_{sc})$ and $RID_{SN_k} = ID_{SN_k} \oplus h(R^*_{sc}||ID_u)$, and then sends the message $M_1 = \langle TID_u, E_u, F_u, RID_{SN_k}, TS_u\rangle$ to BS_l, where ID_{SN_k} is the identity of the targeted sensor node SN_k.

User U	BS_l
$\{SC_u = \langle ETID_u, ER_{sc}, D_u, \tau_u, ID_{BS_l}, h(\cdot)\rangle\}$	$\{X_l\}$
Enter ID_u, PW_u and Bio_u.	
Compute $\sigma_u = Rep(Bio_u, \tau_u)$,	
$R_{sc}^* = ER_{sc} \oplus h(ID_u\|\|PW_u\|\|\sigma_u)$,	
$TID_u = ETID_u \oplus h(R_{sc}^*\|\|\sigma_u\|\|PW_u)$,	If $TS_l - TS_u > \Delta T$, reject.
$D_u^* = h(TID_u\|\|\sigma_u\|\|PW_u)$.	Lookup $\langle TID_u, EIDR_{sc}, h(UC_u), SCN_u\rangle$.
If $D_u^* \neq D_u$, reject.	Retrieve $(ID_u\|\|R_{sc}) = D_{X_l}(EIDR_{sc})$.
Generate $R_u \in \mathbb{Z}_p$, TS_u.	Calculate $R_u^* = F_u \oplus h(ID_{BS_l}\|\|R_{sc})$,
Compute $E_u = h(ID_u\|\|R_u\|\|R_{sc}^*\|\|TS_u)$,	$ID_{SN_k} = RID_{SN_k} \oplus h(R_{sc}^*\|\|ID_u)$,
$F_u = R_u \oplus h(ID_{BS_l}\|\|R_{sc}^*)$,	$E_u^* = h(ID_u\|\|R_u^*\|\|R_{sc}\|\|TS_u)$.
$RID_{SN_k} = ID_{SN_k} \oplus h(R_{sc}^*\|\|ID_u)$.	If $E_u^* \neq E_u$, reject.
$M_1 = \langle TID_u, E_u, F_u, RID_{SN_k}, TS_u\rangle$ $\longrightarrow$	Generate TID_u', $R_l \in \mathbb{Z}_p$.
	Update $\langle TID_u', EIDR_{sc}, h(UC_u), SCN_u\rangle$.
	Compute $SK_{lu} = h(ID_u \|\|R_{sc} \|\|R_u^* \|\|R_l)$,
If $TS_l - TS_u^{new} > \Delta T$, reject.	$G_1 = TID_u' \oplus h(R_{sc}^*\|\|R_u)$,
Calculate $R_l^* = R_u \oplus G_2$,	$G_2 = R_u^* \oplus R_l$,
$TID_u' = G_1 \oplus h(R_{sc}^*\|\|R_u)$,	$G_3 = h(R_{sc}^*\|\|TID_u'\|\|SK_{lu}\|\|TS_l)$.
$SK_{ul} = h(ID_u\|\|R_{sc}^*\|\|R_u\|\|R_l^*)$	$M_2 = \langle G_1, G_2, G_3, TS_l\rangle$ $\longleftarrow$
$G_3^* = h(R_{sc}\|\|\|TID_u'\|\|SK_{ul}\|\|TS_l)$.	
If $G_3^* \neq G_3$, reject.	
Acknowledgement $\longrightarrow$	Accept SK_{lu} as the session key
	Acknowledgement $\longleftarrow$
Accept SK_{ul} as the session key.	
Compute $D_u' = h(TID_u'\|\|\sigma_u\|\|PW_u)$,	
$ETID_u' = TID_u' \oplus h(R_{sc}^*\|\|PW_u)$.	
Update $ETID_u, D_u$ with $ETID_u', D_u'$ in SC_u.	

Fig. 2 Summary of authentication between user and home base station

Step 2. On receiving M_1, BS_l verifies the freshness of TS_u by the verifying if $TS_l - TS_u > \Delta T$, where ΔT and TS_l are the maximum transmission delay and received timestamp of M_1, respectively. BS_l then looks up $\langle TID_u$, $EIDR_{sc}$, $h(UC_u)$, $SCN_u\rangle$, retrieves $(ID_u\|\|R_{sc}) = D_{X_l}(TID_u)$, and computes $R_u^* = F_u \oplus h(ID_{BS_l} \|\|R_{sc})$, $ID_{SN_k} = RID_{SN_k} \oplus h(R_{sc}^* \|\|ID_u)$, $E_u^* = h(ID_u \|\|R_u^* \|\|R_{sc} \|\|TS_u)$. If $E_u^* = E_u$, BS_l generates TID_u', $R_l \in \mathbb{Z}_p$ and updates $\langle TID_u'$, $EIDR_{sc}$, $h(UC_u)$, $SCN_u\rangle$. BS_l then calculates $SK_{lu} = h(ID_u \|\|R_{sc} \|\|R_u^* \|\|R_l)$, $G_1 = TID_u' \oplus h(R_{sc}^* \|\|R_u)$, $G_2 = R_u^* \oplus R_l$, $G_3 = h(R_{sc}^* \|\|TID_u' \|\|SK_{lu} \|\|TS_l)$, and sends the message $M_2 = \langle G_1, G_2, G_3, TS_l\rangle$ to U.

Step 3. On receiving M_2, SC_u checks the freshness of TS_l. SC_u then calculates $R_l^* = R_u \oplus G_2$, $TID_u' = G_1 \oplus h(R_{sc}^* \|\|R_u)$, $SK_{ul} = h(ID_u \|\|R_{sc}^* \|\|R_u \|\|R_l^*)$ and $G_3^* = h(R_{sc} \|\|TID_u' \|\|SK_{ul} \|\|TS_l)$. If $G_3^* = G_3$, SC_u sends an acknowledgment to BS_l.

Step 4. On receiving acknowledgment, BS_l accepts SK_{lu} as the session key and sends an acknowledgment to U. Finally, on receiving acknowledgment, SC_u accepts SK_{ul} as the session key, calculates $D_u' = h(TID_u' \|\|\sigma_u \|\|PW_u)$, $ETID_u = TID_u' \oplus h(R_{sc}^* \|\|PW_u)$, and updates D_u and $ETID_u$ with D_u' and $ETID_u'$ in its memory, respectively.

(2) Authentication between the user and non-home base station

When the requested sensor node SN_k is not accessible through the home base station, U and the non-home base station can mutually authenticate and establish a session key in presence of the home base station. The steps required in this procedure are described below.

Step 1. After successful login, the smart card SC_u of U selects a random nonce $R_u \in \mathbb{Z}_p$, computes $E_u = h(ID_u || R_u || R^*_{sc} || TS_u)$, $F_u = R_u \oplus h(ID_{BS_l} || R^*_{sc})$ and $RID_{SN_k} = ID_{SN_k} \oplus h(R^*_{sc} || ID_u)$, and then sends the message $M_1 = \langle TID_u, E_u, F_u, RID_{SN_k}, TS_u \rangle$ to BS_l, where ID_{SN_k} is the identity of the targeted sensor node.

Step 2. On receiving M_1, BS_l verifies the freshness of TS_u. If it is valid, BS_l looks up $\langle TID_u, EIDR_{sc}, h(UC_u), SCN_u \rangle$, retrieves $(ID_u || R_{sc}) = D_{X_l}(TID_u)$, and computes $R^*_u = F_u \oplus h(ID_{BS_l} || R_{sc})$ and $E^*_u = h(ID_u || R^*_u || R_{sc} || TS_u)$. If $E^*_u = E_u$, BS_l generates $TID'_u \in \mathbb{Z}_p$ and current timestamp TS_l, and updates $\langle TID'_u, EIDR_{sc}, h(UC_u), SCN_u \rangle$. BS_l also calculates $G_l = TID'_u \oplus h(R_{sc} || R^*_u) \oplus h(X_{SH})$, $IDR_u = h(ID_u || R_{sc} || R^*_u)$, $F_l = IDR_u \oplus h(ID_{BS_y} || X_{SH})$, $E_l = h(IDR_u || X_{SH} || R^*_u)$ and $ID_{SN_k} = RID_{SN_k} \oplus h(R_{sc} || ID_u)$ and $RID^*_{SN_k} = ID_{SN_k} \oplus h(IDR_u || TS_l)$, and sends the message $M_2 = \langle E_l, F_l, G_l, RID^*_{SN_k}, TS_l \rangle$ to BS_y.

Step 3. On receiving M_2, BS_y checks the freshness of TS_l. After that it calculates $IDR_u = F_l \oplus h(ID_{BS_y} || X_{SH})$, $E^*_l = h(IDR_u || X_{SH} || TS_l)$. If $E^*_l = E_l$, BS_y computes $ID_{SN_k} = RID^*_{SN_k} \oplus h(IDR_u || TS_l)$, picks a random number $R_l \in \mathbb{Z}_p$ and current timestamp TS_y, and calculates $SK_{yu} = h(IDR_u || R_y)$, $G_1 = G_l \oplus h(X_{SH})$, $G_2 = IDR_u \oplus R_y$, $G_3 = h(G_1 || SK_{yu} || TS_y)$, and sends the message $M_3 = \langle G_1, G_2, G_3, TS_y \rangle$ to U.

Step 4. On receiving M_3, U checks the freshness of TS_l and then calculates $R^*_y = R_u \oplus G_2$, $TID'_u = G_1 \oplus h(R^*_{sc} || R_u)$, $SK_{uy} = h(h(ID_u || R^*_{sc} || R_u) || R^*_y)$ and $G^*_3 = h(G_1 || SK_{ul} || TS_l)$. If $G^*_3 = G_3$, U sends an acknowledgment to BS_y.

Step 5. On receiving acknowledgment, BS_y accepts SK_{yu} as the session key and sends an acknowledgment to U. U then accepts SK_{uy} as the session key, calculates $D'_u = h(TID'_u || \sigma_u || PW_u)$, $ETID'_u = TID'_u \oplus h(R^*_{sc} || PW_u)$ and updates D_u and $ETID_u$ with D'_u and $ETID'_u$, respectively, in SC_u.

(3) Authentication between the base station and sensor nodes

This procedure remains similar to that in Amin et al's. scheme [2]. However, instead of sending the identities ID_{SN_k}, ID_{CH_j} and ID_{BS_l} of accessed sensor node SN_k, cluster head CH_j and home base station BS_l, respectively, to CH_j and SN_k by BS_l and CH_j, respectively, it is essential to send the pseudo-identities RID_{SN_k}, RID_{CH_j} and RID_{BS_l}.

4.4 Password and Biometric Update Phase

The user U inputs ID_u, PW_u, and Bio_u to change the credentials through the steps described in Sect. 4.3. If the login is successful, U provides the updated password PW'_u, imprints the updated biometric Bio'_u and calculates $(\sigma'_u, \tau'_u) = Gen(Bio'_u)$,

$ER'_{sc} = R^*_{sc} \oplus h(ID_u || PW'_u || \sigma'_u)$, $D'_u = h(TID_u || \sigma'_u || PW'_u)$, $ETID'_u = TID_u \oplus h(R_{sc} || \sigma_u || PW_u)$. Finally, U replaces $ETID_u$, ER_{sc} and τ_u with $ETID'_u$, ER'_{sc} and τ'_u, respectively, in the smart card SC_u.

4.5 Smart Card Revocation Phase

A smart card, if it is lost or stolen, can be revoked by user U through a request for a new smart card to his/her respective home base station BS_l. Hence, the steps are identical to those for the registration phase discussed in Sect. 4.2.

5 Security Analysis

We analyze our scheme in this section and demonstrate that it resists several known attacks.

(1) **Replay attack**: Assume that message M_1 is replayed to the home base station BS_l by an adversary $\mathcal{A}$. However, BS_l will reject the message based on the timestamp validation. Similarly, our scheme resists replay attacks for all other messages.

(2) **Impersonation attacks**: An adversary $\mathcal{A}$ may attempt to impersonate the message M_1 to the base station BS_l on behalf of a user U, where $M_1 = \{TID_u, E_u, F_u, ID_{SN_k}, TS_u\}$. However, TID_u is encrypted with the secret X_l, and both E_u and F_u are composed from the hash of secret values such as R_{sc} and R_u. Thus, $\mathcal{A}$ cannot impersonate the message M_1 and user impersonation attack is protected in the proposed scheme. Similarly, our scheme is also secure against impersonation of base stations, cluster head, and sensor nodes.

(3) **Man-in-the-middle attack**: Assume that an adversary $\mathcal{A}$ tries to attempt to execute man-in-the-middle attack by intercepting and then modifying the messages. But as we previously demonstrate above, the messages cannot be modified without knowledge of secret credentials. Thus, the proposed scheme is secure against man-in-the-middle attack.

(4) **Lost/stolen smart card and off-line guessing attacks**: Suppose $\mathcal{A}$ recovers a lost/stolen smart card SC_u of a registered authorized user U. Through the power analysis attacks [14, 16], $\mathcal{A}$ can extract all the secret credentials $ETID_u$, ER_{sc}, D_u, τ_i, and ID_{BS_l} stored in SC_u. Out of these, except for ID_{BS_l} and τ_i, none is stored in plaint text and hence, it requires combinations of user identity, password, and biometrics to derive other secret credentials. For $\mathcal{A}$ to subvert the proposed scheme through off-line guessing attack, he/she will have to simultaneously guess ID_u, PW_u and σ_u, which is computationally infeasible. Thus, the proposed method is resistant to such kinds of attacks.

(5) **Privileged-insider attack**: Assume that $\mathcal{A}$ is an adversary with privileged access to a base station and can learn the user identity ID_u and UC_u from user's registration. Further, assume that he/she can extract the credentials such as $ETID_u = TID_u \oplus h(R_{sc} ||\sigma_u ||PW_u)$, $ER_{sc} = R_{sc} \oplus h(ID_u ||PW_u ||\sigma_u)$, $D_u = h(TID_u ||\sigma_u ||PW_u)$, from the compromised smart card SC_u of the same user U. Even if ID_u is known, with the available information $\mathcal{A}$ must simultaneously guess PW_u and σ_u, which is still computationally infeasible. Therefore, the proposed scheme is resistant to privileged-insider attack.

(6) **Temporary session information attack**: Assume that an adversary $\mathcal{A}$ somehow learns one or both of the session specific secrets such as R_u, R_l or R_y. But, since the session key $h(ID_u ||R_{sc} ||R_u ||R_l)$ or $h(h(ID_u ||R_{sc} ||R_u) ||R_y)$ depends on ID_u and R_{sc} in addition to R_u, R_l or R_y. Thus, $\mathcal{A}$ neither subvert the session key nor glean any useful information about any long term secret values. This shows that the proposed scheme is resistant to such an attack.

(7) **Parallel session attack**: For an adversary $\mathcal{A}$ to successfully execute a parallel session attack, he/she needs to compose the session key $h(ID_u ||R_{sc} ||R_u ||R_l)$ or $h(h(ID_u ||R_{sc} ||R_u) ||R_y)$ by eavesdropping the messages. However, no secrets are compromised regardless of a lost/stolen smart card attack or a privileged-insider attack. Hence, the proposed scheme is resistant to parallel session attack.

(8) **Smart card impersonation attack**: Smart card impersonation attack can only be executed if adversary $\mathcal{A}$ can learn the $ETID_u$, ER_{sc} and D_u from a user U's compromised smart card SC_u. But, the secret credentials are not exposed through a compromised smart card even in the presence of a privileged-insider attack. Thus, the proposed scheme is resistant to smart card impersonation attack.

(9) **Password/biometric change attack**: For an adversary $\mathcal{A}$, to change the password or biometric information of a registered user U using a stolen smart card SC_u, $\mathcal{A}$ needs to successfully log in the system first. As this requires simultaneously guessing of the secret credentials ID_u, PW_u, and σ_u, it is computationally infeasible. Hence, the proposed scheme is also secure against password or biometric change attack.

(10) **Resilience against sensor node/cluster head capture attack**: As discussed in the threat model, some sensor nodes can be physically captured by an adversary $\mathcal{A}$. Assume that $\mathcal{A}$ physically captures a sensor node SN_k. $\mathcal{A}$ can then extract all the credentials such as identity ID_{SN_k} and secret key $X_{SN_k} = h(ID_{SN_k} ||X_{SH})$ from SN_k's memory. Similarly, physical capture of a cluster head CH_j leads to compromise credentials such as ID_{CH_j} and secret key $X_{CH_j} = h(ID_{CH_j} ||X_{SH})$ from it memory by $\mathcal{A}$. Since the identities ID_{SN_k} and ID_{CH_j} are unique, all the secret keys X_{SN_k} of sensors and X_{CH_j} of cluster heads are also distinct. Thus, $\mathcal{A}$ can only compromise the session key between SN_k and a user U residing in his/her home base station BS_l. However, other independent session keys established among the same user U and other uncompromised sensors remain secure as the secrets are distinct. This means that a compromised sensor node or cluster head does not affect the secure communications among the user

and other non-compromised sensors or cluster heads. As a result, the proposed scheme is unconditionally resistant to sensor/cluster head capture attack.

(11) **Anonymity and untractability**: Assume that an adversary $\mathcal{A}$ eavesdrops and monitors the messages $M_1 = \{TID_u, E_u, F_u, RID_{SN_k}, TS_u\}$ and $M_2 = \{G_1, G_2, G_3, TS_y\}$. The eavesdropped messages are devoid of any identifying information for the user, base station, cluster head and sensor node. Thus, anonymity property is maintained in our scheme. Furthermore, all of the eavesdropped parameters in the messages are either composed with random nonces or current timestamps and consequently, these messages are always unique across sessions. Therefore, untractability is also maintained in our scheme.

6 Comparative Study

In this section, we analyze the proposed scheme in terms of communication and computation overheads, and then for security and functionality features. After that, we benchmark the proposed scheme with Amin et al's. scheme [2]. For communication and computation costs comparison, we examine the following three cases:

- **Case I**: login and authentication between a user and home base station
- **Case II**: login and authentication between a user and a non-home base station (in the proposed scheme) or between a home base station and non-home base station (in Amin et al's. scheme [2])
- **Case III**: login and authentication between a base station and a sensor node via a cluster head

6.1 Communication Costs Comparison

For communication costs comparison, identity of user/base station, identity of sensor/cluster head, timestamp, hash output (if SHA-1 hashing algorithm is applied), random number and acknowledgment are assumed to be 160, 32, 32, 160, 160, and 8 bits, respectively. Table 1 summarizes the communications costs (in bits) of the messages exchanged for the schemes in three different cases (Case I-III). From this table, we observe that the proposed scheme requires more communication overhead than the Amin et al's. scheme. However, this acceptable owning to the superior security & functionality attributes of our scheme (see Table 3).

Table 1 Communication costs comparison

Scheme	Case I	Case II	Case III
Our	1200 bits	1872 bits	2720 bits
Amin et al. [2]	1168 bits	808 bits	2208 bits

Table 2 Computation costs comparison

Scheme	User	Home/non-home base station	Cluster head	Sensor	Total cost
Case I (Our)	$T_f + 11T_h$	$T_s + 5T_h$	–	–	$T_f + T_s + 16T_h \approx 79.775$ ms
Case II (Our)	$T_f + 10T_h$	$T_s + 17T_h$	–	–	$T_f + T_s + 27T_h \approx 85.275$ ms
Case III (Our)	–	$11T_h$	$14T_h$	$8T_h$	$33T_h \approx 16.5$ ms
Case I [2]	$9T_h$	$T_s + 16T_h$	–	–	$T_s + 25T_h \approx 16.7$ ms
Case II [2]	–	$8T_h$	–	–	$8T_h \approx 4$ ms
Case III [2]	–	$9T_h$	$10T_h$	$6T_h$	$25T_h \approx 12.5$ ms

6.2 Computation Costs Comparison

We denote T_f, T_s and T_h as the time needed for fuzzy extractor function operations, symmetric encryption/decryption and one-way cryptographic hashing operations, respectively. Table 2 compares the computation costs for the proposed scheme and Amin et al's. scheme in different cases. Based on the experimental results reported in [10, 12], we have $T_f \approx 63.075$ ms, $T_s \approx 8.7$ ms, and $T_h \approx 0.5$ ms, respectively. In the proposed scheme, we have utilized fuzzy extractor technique for local biometric verification of a registered user by the smart card to make the proposed scheme much stronger in terms of its security as compared to Amin et al's. scheme. Due to the fuzzy extractor, the computational cost at the user side increases. The greater computation cost is justified by the superior functionality & security attributes (see Table 3).

6.3 Functionality and Security Feature Comparison

Finally, in Table 3 we compare the functionality and security features of the proposed scheme with Amin et al's. scheme. We have used the following notations. ☑: a scheme resists the attack or it supports the feature; ☒: a scheme does not resist the attack or it does not support the feature; $\S_1$: user anonymity; $\S_2$: sensor

Table 3 Security & Functionality attributes comparison

Attribute	$\S_1$	$\S_2$	$\S_3$	$\S_4$	$\S_5$	$\S_6$	$\S_7$	$\S_8$	$\S_9$	$\S_{10}$	$\S_{11}$	$\S_{12}$	$\S_{13}$	$\S_{14}$
Amin et al. [2]	✗	✗	✗	✗	☑	☑	✗	✗	✗	✗	✗	✗	✗	☑
Our	☑	☑	☑	☑	☑	☑	☑	☑	☑	☑	☑	☑	☑	☑

node anonymity; $\S_3$: base station anonymity; $\S_4$: untraceability; $\S_5$: replay attack; $\S_6$: man-in-the-middle attack; $\S_7$: off-line password guessing attack; $\S_8$: impersonation attacks; $\S_9$: privileged-insider attack; $\S_{10}$: parallel session attack; $\S_{11}$: password change attack; $\S_{12}$: temporary session information attack; $\S_{13}$: three-factor authentication; $\S_{14}$: resilience against sensor node/cluster head capture attack. It is evident that Amin et al's. scheme [2] does not resist or support the majority of features listed in the table, except the features $\S_5$ and $\S_6$. On the other hand, our scheme offers greater security & functionality attributes as compared to the scheme of Amin et al. Additionally, only password and smart card are utilized in Amin et al's. scheme, whereas in the proposed scheme, password, smart card and biometrics are used. Thus, the proposed scheme operates under three-factor authentication.

7 Concluding Remarks

We discussed the issue of anonymous user authentication in multiple base stations based WSNs. We then provided several vulnerabilities and security issues in a recent user authentication mechanism suggested by Amin et al. We proposed a more secure scheme for anonymous user authentication and key agreement in multiple base stations based WSNs to erase the security pitfalls launched in Amin et al's. scheme. The security analysis and performance comparison show that the proposed scheme provides superior security and more functionality attributes as compared to those for Amin et al's. scheme.

References

1. Akyildiz, I.F., Su, W., Sankarasubramaniam, Y., Cayirci, E.: Wireless sensor networks: a survey. Comput. Netw. **38**(4), 393–422 (2002)
2. Amin, R., Islam, S.H., Biswas, G., Obaidat, M.S.: A robust mutual authentication protocol for WSN with multiple base-stations. Ad Hoc Netw. **75–76**, 1–18 (2018)
3. Challa, S., Das, A.K., Odelu, V., Kumar, N., Kumari, S., Khan, M.K., Vasilakos, A.V.: An efficient ECC-based provably secure three-factor user authentication and key agreement protocol for wireless healthcare sensor networks. Comput. Electr. Eng. **69**, 534–554 (2018)
4. Das, A.K., Kumari, S., Odelu, V., Li, X., Wu, F., Huang, X.: Provably secure user authentication and key agreement scheme for wireless sensor networks. Secur. Commun. Netw. **9**(16), 3670–3687 (2016)

5. Das, A.K., Sharma, P., Chatterjee, S., Sing, J.K.: A dynamic password-based user authentication scheme for hierarchical wireless sensor networks. J. Netw. Comput. Appl. **35**(5), 1646–1656 (2012)
6. Das, A.K., Sutrala, A.K., Odelu, V., Goswami, A.: A secure smartcard-based anonymous user authentication scheme for healthcare applications using wireless medical sensor networks. Wirel. Pers. Commun. **94**(3), 1899–1933 (2017)
7. Das, M.L.: Two-factor user authentication in wireless sensor networks. IEEE Trans. Wirel. Commun. **8**(3), 1086–1090 (2009)
8. Dodis, Y., Reyzin, L., Smith, A.: Fuzzy extractors: how to generate strong keys from biometrics and other noisy data. In: Annual International Conference on the Theory and Applications of Cryptographic Techniques- Advances in Cryptology (Eurocrypt'04). LNCS, vol. 3027, pp. 523–540. Springer, Interlaken, Switzerland (2004)
9. Dolev, D., Yao, A.: On the security of public key protocols. IEEE Trans. Inf. Theory **29**(2), 198–208 (1983)
10. He, D., Kumar, N., Khan, M.K., Lee, J.H.: Anonymous two-factor authentication for consumer roaming service in global mobility networks. IEEE Trans. Consum. Electron. **59**(4), 811–817 (2013)
11. He, D., Gao, Y., Chan, S., Chen, C., Bu, J.: An enhanced two-factor user authentication scheme in wireless sensor networks. Ad Hoc Sens. Wirel. Netw. **10**(4), 361–371 (2010)
12. Jiang, Q., Ma, J., Li, G., Yang, L.: An efficient ticket based authentication protocol with unlinkability for wireless access networks. Wirel. Pers. Commun. **77**(2), 1489–1506 (2014)
13. Khan, M.K., Alghathbar, K.: Cryptanalysis and security improvements of 'two-factor user authentication in wireless sensor networks'. Sensors **10**(3), 2450–2459 (2010)
14. Kocher, P., Jaffe, J., Jun, B.: Differential power analysis. In: Proceedings of Advances in Cryptology - CRYPTO'99, LNCS. vol. 1666, pp. 388–397. Springer, Santa Barbara, CA, USA (1999)
15. Kumari, S., Das, A.K., Wazid, M., Li, X., Wu, F., Choo, K.K.R., Khan, M.K.: On the design of a secure user authentication and key agreement scheme for wireless sensor networks. Concurr. Comput.: Pract. Exp. **29**(23), 1–24 (2017)
16. Messerges, T.S., Dabbish, E.A., Sloan, R.H.: Examining smart-card security under the threat of power analysis attacks. IEEE Trans. Comput. **51**(5), 541–552 (2002)
17. Sutrala, A.K., Das, A.K., Kumar, N., Reddy, A.G., Vasilakos, A.V., Rodrigues, J.J.P.C.: On the design of secure user authenticated key management scheme for multigateway-based wireless sensor networks using ECC. Int. J. Commun. Syst. **31**(8), 1–31 (2018)
18. Tseng, H.R., Jan, R.H., Yang, W.: An improved dynamic user authentication scheme for wireless sensor networks. In: IEEE Global Telecommunications Conference (GLOBECOM'07). pp. 986–990. Washington, DC, USA (2007)
19. Turkanović, M., Brumen, B., Hölbl, M.: A novel user authentication and key agreement scheme for heterogeneous ad hoc wireless sensor networks, based on the Internet of Things notion. Ad Hoc Netw. **20**, 96–112 (2014)
20. Vaidya, B., Makrakis, D., Mouftah, H.T.: Improved two-factor user authentication in wireless sensor networks. In: 6th IEEE International Conference on Wireless and Mobile Computing. Networking and Communications (WiMob'10), pp. 600–606. Niagara Falls, Canada (2010)
21. Vaidya, B., Sá Silva, J., Rodrigues, J.J.P.C.: Robust dynamic user authentication scheme for wireless sensor networks. In: 12th ACM International Conference on Modeling. Analysis and Simulation of Wireless and Mobile Systems (MSWiM'09), pp. 88–91. Tenerife, Canary Islands, Spain (2009)
22. Wazid, M., Das, A.K., Vasilakos, A.V.: Authenticated key management protocol for cloud-assisted body area sensor networks. J. Netw. Comput. Appl. **123**, 112–126 (2018)
23. Wong, K.H., Zheng, Y., Cao, J., Wang, S.: A dynamic user authentication scheme for wireless sensor networks. In: IEEE International Conference on Sensor Networks. Ubiquitous, and Trustworthy Computing (SUTC'06), vol. 1, pp. 1–8. Taichung, Taiwan (2006)
24. Wu, F., Xu, L., Kumari, S., Li, X., Shen, J., Choo, K.K.R., Wazid, M., Das, A.K.: An efficient authentication and key agreement scheme for multi-gateway wireless sensor networks in IoT deployment. J. Netw. Comput. Appl. **89**, 72–85 (2017)

Visual Secret Sharing Scheme with Distributed Levels of Importance of Shadows

Sabyasachi Dutta, Tamal Bhore, M. Kutubuddin Sardar, Avishek Adhikari, and Kouichi Sakurai

Abstract We put forward a construction of visual cryptographic scheme (VCS) to address the scenario, where there are several levels of importance of shadows. Traditional (k, n)-threshold VCS does not distinguish among the shadows—*all shadows are of equal importance*. Any k of them can recover the secret image whereas less than or equal to $k - 1$ shadows do not reveal any information about the secret image. We consider the scenario where there are several levels of essential participants (labeled according to their importance) and there must be an agreement across the levels to recover the secret image. This captures the distribution of importance of parties across several levels or compartments. We give a novel construction of visual cryptographic scheme realizing such access structures, where the parties are compartmented into different levels according to their importance. We come up with closed mathematical formulae of pixel expansion and relative contrast. We also consider more efficient

S. Dutta is grateful to the NICT, Japan for the financial support under the NICT International Exchange Program.
K. Sardar is financially supported by a Ph.D. fellowship (award no. 09/028(0975)/2016) CSIR, India.
A. Adhikari is thankful to Presidency University for providing FRPDF grant.

S. Dutta (✉) · K. Sakurai
Faculty of Information Science and Electrical Engineering, Kyushu University, Fukuoka, Japan
e-mail: saby.math@gmail.com

K. Sakurai
e-mail: sakurai@inf.kyushu-u.ac.jp

T. Bhore
Institute of Technical Education and Research, SOA University, Odisha, India
e-mail: tamaltml@gmail.com

M. K. Sardar
Department of Pure Mathematics, University of Calcutta, Kolkata 700019, India
e-mail: kutubpmath@gmail.com

A. Adhikari
Department of Mathematics, Presidency University, Kolkata, India
e-mail: avishek.adh@gmail.com

D. Giri et al. (eds.), *Proceedings of the Fifth International Conference on Mathematics and Computing*, Advances in Intelligent Systems and Computing 1170, https://doi.org/10.1007/978-981-15-5411-7_2

probabilistic construction and compare contrasts with several other probabilistic, non-expansible constructions. Numerical evidence shows that our scheme performs better than the available generic constructions.

Keywords Compartmented access structure · Monotone access structure · Pixel expansion · Relative contrast · Basis matrices · Non-expansible visual secret sharing

1 Introduction

In a visual secret sharing scheme, also known as visual cryptographic scheme (VCS), a secret image is encoded into n many noise-like shadows. The shadows are created in such a way so that only some predetermined "qualified" subsets of shadows reveal the secret image. No other subsets of shadows reveal information of any kind about the secret image.

Background
Naor and Shamir [12] proposed the first VCS for the (k, n)-threshold access structure. Greater than or equal to k shadows can visually recover the secret image. However, $k - 1$ or less many shadows do not reveal information of any sort about the secret image. Thus, the subsets of size greater than or equal to k are the qualified sets and rest of the subsets are "forbidden". This concept has been extended in [1, 2, 4] to general access structures. The underlying mathematical operation of a VCS is the OR operation, which enables the human visual system to visually recover the secret image from the shadows. The main problem with VCS is its huge share/shadow size (**pixel expansion**), which results in a very low contrast of the recovered secret image.

To solve the problem of reducing the pixel expansion keeping the contrast value of the recovered image same as the above-mentioned schemes, Yang [18] introduced a new model of visual secret sharing in which the shadows/shares have the same size of the secret image. That is, the schemes have no pixel expansion but the recovered image has the same contrast as before. However, this model of VCS allows an *error* during the reconstruction process, which makes the schemes *probabilistic*. In such probabilistic VCS, each pixel can be correctly reconstructed only with a probability given as a parameter of the scheme. A first attempt to provide VCS without pixel expansion was done by Ito et al. [10]. The fundamental works in this area of probabilistic visual cryptography were by Yang [18] and Cimato et al. [5]. Another convenient way to construct such *non-expansible* visual cryptographic scheme is via *Random Grids*. The method was proposed in [11], even before the visual secret sharing problem was considered. The idea was later rediscovered by [14, 15, 17] as an efficient tool to construct non-expansible visual cryptographic scheme.

Motivation
In a (k, n)-threshold access structure, all the shadows have the same importance, any k of them can recover the secret image. On the other hand, for general access structures,

the importance of shadows can vary arbitrarily. There are some interesting access structures which lie between the above two. Let us consider the scenario of a typical business organization consisting of the following main departments: Production, R&D, Purchasing, Human Resource, Finance, and Marketing. All the departments must agree in order to take up a brand new project for the company which is due to the distribution of responsibilities once the project ensues. If (at least) one of the departments does not have the necessary arrangements, the project must not take off. In a more technical language, the company has six departments/compartments and each of them may contain several representatives. From each compartment, a threshold number of representatives must take the responsibility in order to start a new project. This type of access structures is called *compartmented* access structure [16].

1.1 Related Work

Arumugam et al. [3] gave the construction of $(k, n)^*$-VCS to construct a scheme, where one participant is "essential" and he collaborates with any $k - 1$ parties from the rest to recover the secret image. Guo et al. [9] generalized $(k, n)^*$-VCS by extending to accommodate t many essential participants. We denote this by the notation t-$(k, n)^*$-VCS following the notations of [8]. In both the cases, there are two compartments—one contains the essential parties and the other contains the ordinary parties. The case when $t = 1$ is the work of Arumugam et al. [3], where one shadow is essentially important and $t = 0$ corresponds to the threshold access structure where all the shadows are of equal importance. Praveen et al. [13] considered the probabilistic model of t-$(k, n)^*$-VCS. Dutta et al. [7] studied t-$(k, n)^*$-VCS in the XOR model and gave an efficient construction to build basis matrices.

Our Contribution: In this paper, we consider a generalization in which there are several compartments of *essentiality*. We first give a construction of basis matrices that realizes the visual cryptographic scheme on the compartmented access structure. Our construction of basis matrices is in a recursive manner. Up to the best of our knowledge, the scheme is the first proposed scheme for general compartmented access structure in the literature of visual cryptography. We come up with a closed form of both pixel expansion as well as relative contrast. Numerical evidence shows that our scheme performs better in terms of both relative contrast as well as pixel expansion than the cumulative array based construction [4] and linear algebra based construction [1]. We also give a non-expansible (probabilistic) construction from our deterministic construction, using Yang's method [18]. We compare our results with several other non-expansible probabilistic constructions of VCS.

2 The Model and Preliminaries

We discuss some of the basic notations and definitions that are needed. Let $\mathcal{P} = \{1, 2, 3, \ldots, n\}$ denote a set of participants. Let $2^{\mathcal{P}}$ denote the set of all subsets of $\mathcal{P}$. Let $\mathcal{Q} \subset 2^{\mathcal{P}}$ and $\mathcal{F} \subset 2^{\mathcal{P}}$ such that $\mathcal{Q} \cap \mathcal{F} = \emptyset$, respectively, denote the set of all qualified sets and the set of all forbidden sets. The pair $(\mathcal{Q}, \mathcal{F})$ defines an access structure on the set $\mathcal{P}$ of participants.

Compartmented Access Structure and Notations: In a compartmented access structure, there are different compartments, say $\mathcal{C}_1, \mathcal{C}_2, \ldots, \mathcal{C}_u$ having the property that $\mathcal{C}_i \cap \mathcal{C}_j = \emptyset$ for all $i \neq j$ with $|\mathcal{C}_i| = n_i$ and positive integers $k, t_1, t_2, \ldots, t_u$ such that $k = t_1 + t_2 + \cdots + t_u$ and $n = n_1 + n_2 + \cdots + n_u$. The minimal qualified sets for this access structure are those subsets of $\mathcal{P}$ that contain t_1 many parties from $\mathcal{C}_1$, t_2, many parties from $\mathcal{C}_2, \ldots, t_u$, many parties from $\mathcal{C}_u$ so that the subsets contain exactly k many parties. We notice that if there are two compartments $\mathcal{C}_1, \mathcal{C}_2$ with $\mathcal{C}_1 = \{1, 2, \ldots, t\}$ and $\mathcal{C}_2 = \{t+1, t+2, \ldots, n\}$ and $t_1 = t$, $t_2 = k - t$ then we have the access structure of a t-$(k, n)^*$. Thus, compartmented access structure is a generalization of the t-$(k, n)^*$ access structure. To maintain the parity of notations, we denote a VCS for a compartmented access structure by $\boldsymbol{t}$-$(k, \boldsymbol{n})^*$-VCS, where the symbols $\boldsymbol{t}$ and $\boldsymbol{n}$ depict the vectors $(t_1, t_2, \ldots, t_u)$ and $(n_1, n_2, \ldots, n_u)$ respectively such that $|\mathcal{C}_i| = n_i$ for all $i = 1, \ldots, u$, $t_1 + t_2 + \cdots + t_u = k$ and $n_1 + n_2 + \cdots + n_u = n$.

We denote the collection of all minimal qualified sets of participants by $\mathcal{Q}_{min}$ such that $Q \in \mathcal{Q}_{min}$ means Q contains exactly t_1 many parties from $\mathcal{C}_1$, t_2 many parties from $\mathcal{C}_2, \ldots, t_u$ many parties from $\mathcal{C}_u$ so that total number of parties in Q is k. The collection of all maximal forbidden sets is denoted by $\mathcal{F}_{max}$ and is defined by $\mathcal{F}_{max} = \{F \in \mathcal{F} : \ for\ all\ i \notin F,\ F \cup \{i\} \in \mathcal{Q}\}$.

Example 1 If $\mathcal{P} = \{1, 2, 3, 4, 5, 6, 7\}$, $\mathcal{C}_1 = \{1, 2\}$, $t_1 = 1$ and $\mathcal{C}_2 = \{3, 4, 5\}$, $t_2 = 2$ and $\mathcal{C}_3 = \{6, 7\}$, $t_3 = 2$, then $\mathcal{Q}_{min}$ consists of the following minimal qualified subsets of participants $B_1 = \{1, 3, 4, 6, 7\}$, $B_2 = \{1, 3, 5, 6, 7\}$, $B_3 = \{1, 4, 5, 6, 7\}$, $B_4 = \{2, 3, 4, 6, 7\}$, $B_5 = \{2, 3, 5, 6, 7\}$, $B_6 = \{2, 4, 5, 6, 7\}$. Note that both $\{1, 3, 6\}$ and $\{1, 2, 3, 4, 5\}$ are some of the members of $\mathcal{F}$, while $\{1, 2, 3, 4, 5, 6\}$ is a member of $\mathcal{F}_{max}$.

Notations: Let S be an $n \times m$ binary matrix and let $X \subseteq \mathcal{P}$. The notation $S[X]$ denotes the matrix obtained by restricting the rows of S to the indices belonging to X. Moreover, for any $Y \subset \mathcal{P}$, the Boolean vector obtained by applying the OR operation "$\bigvee$", to the rows of $S[Y]$ is denoted by S_Y. The Hamming weight of the row vector S_Y is denoted by $w(S_Y)$.

Remark: In this paper, we are interested in monotone access structures only. As a result, in terms of defining our schemes, it is sufficient for us to restrict ourselves to the minimal qualified sets and maximal forbidden sets only.

We now define "basis matrices" with which $\boldsymbol{t}$-$(k, \boldsymbol{n})^*$-VCS can be realized in practice.

Definition 1 A pair of binary matrices (S^0, S^1) of dimension $n \times m$ is called "basis matrices" realizing a $\boldsymbol{t}$-$(k, \boldsymbol{n})^*$-VCS, if there exist two sets of nonnegative real numbers $\{\alpha_X\}_{X \in \mathcal{Q}_{min}}$ and $\{t_X\}_{X \in \mathcal{Q}_{min}}$ such that the following two conditions hold:

1. (*contrast condition*) If $B \in \mathcal{Q}_{min}$, then S^0_B satisfies $w(S^0_B) \leq t_B - \alpha_B \cdot m$; whereas, S^1 satisfies $w(S^1_B) \geq t_B$.
2. (*security condition*) If $F = \{i_1, i_2, \ldots, i_f\} \in \mathcal{F}_{max}$ then the two $f \times m$ matrices $S^0[F]$ and $S^1[F]$ obtained by restricting S^0 and S^1, respectively, to rows $i_1, i_2, \ldots, i_f$ are identical up to a column permutation.

m denotes the *pixel expansion* of the scheme. Moreover, α_X and $\alpha_X \cdot m$, respectively, denote the *relative contrast* and *contrast* of the reconstructed image by the collaboration of participants forming minimal qualified set X.

3 Construction of Basis Matrices

In this section we describe the method of constructing basis matrices for a monotone $\boldsymbol{t}$-$(k, \boldsymbol{n})^*$-VCS, where $\boldsymbol{t} = (t_1, t_2, \ldots, t_u)$, $\boldsymbol{n} = (n_1, n_2, \ldots, n_u)$ with $k = t_1 + t_2 + \cdots + t_u$.

Let $\mathcal{P}_1 = \{1, 2, \ldots, n_1\}$, $\mathcal{P}_2 = \{n_1 + 1, n_1 + 2, \ldots, n_1 + n_2\}$,...,$\mathcal{P}_u = \{n_1 + n_2 + \cdots + n_{u-1} + 1, \ldots, n_1 + n_2 + \cdots + n_{u-1} + n_u\}$ be the compartments in the access structure. We break the construction process into several simpler parts. Let on the participant set $\mathcal{P}_i$, for the (t_i, n_i)-VCS, the basis matrices be S^0_i and S^1_i and the corresponding pixel expansion be m_i, $i = 1, 2, \ldots, u$. Further, let $S^b_i = [C^b_{i,1}\ C^b_{i,2} \cdots C^b_{i,m_i}]$, $b \in \{0, 1\}$, where $C^b_{i,j}$ denotes the jth column of S^b_i, where $j = 1, 2, \ldots, m_i$.

We now give describe the steps toward constructing the VCS for compartmented access structure.

Step 1:
Here, we describe the construction method of the basis matrices for $\boldsymbol{t}_{ij}$-$(k_{ij}, \boldsymbol{n}_{ij})^*$-VCS, where $\boldsymbol{t}_{ij} = (t_i, t_j)$, $k_{ij} = t_i + t_j$ and $\boldsymbol{n}_{ij} = (n_i, n_j)$. For this construction, we consider the basis matrices (S^0_i, S^1_i) and (S^0_j, S^1_j) for the (t_i, n_i)-VCS and (t_j, n_j)-VCS, respectively.

Consider the two following matrices:

$$S^0_{ij} = \left[\begin{array}{cccc|cccc} m_j C^0_{i,1} & m_j C^0_{i,2} & \ldots & m_j C^0_{i,m_i} & m_j C^1_{i,1} & m_j C^1_{i,2} & \ldots & m_j C^1_{i,m_i} \\ S^0_j & S^0_j & \ldots & S^0_j & S^1_j & S^1_j & \ldots & S^1_j \end{array}\right] \text{ and}$$

$$S^1_{ij} = \left[\begin{array}{cccc|cccc} m_j C^0_{i,1} & m_j C^0_{i,2} & \ldots & m_j C^0_{i,m_i} & m_j C^1_{i,1} & m_j C^1_{i,2} & \ldots & m_j C^1_{i,m_i} \\ S^1_j & S^1_j & \ldots & S^1_j & S^0_j & S^o_j & \ldots & S^0_j \end{array}\right].$$

Through Lemma 1, we shall show that the above two matrices are indeed basis matrices for $\boldsymbol{t}_{ij}$-$(k_{ij}, \boldsymbol{n}_{ij})^*$-VCS. Here, the notation $m_j C^b_{i,k}$, $b \in \{0, 1\}$ stands for the repetition of the $C^b_{i,k}$ column m_j times.

Step 2:

Here, we describe the construction method of the basis matrices for $\boldsymbol{t}_{ijl}$-$(k_{ijl}, \boldsymbol{n}_{ijl})^*$-VCS where $\boldsymbol{t}_{ijl} = (t_i, t_j, t_l)$, $k_{ijl} = t_i + t_j + t_l$ and $\boldsymbol{n}_{ijl} = (n_i, n_j, n_l)$.

Consider the two following matrices:

$$S^0_{ijl} = \left[\begin{array}{cccc|cccc} m_l C^0_{ij,1} & m_l C^0_{ij,2} & \cdots & m_l C^0_{ij,m_{ij}} & m_l C^1_{ij,1} & m_l C^1_{ij,2} & \cdots & m_l C^1_{ij,m_{ij}} \\ S^0_l & S^0_l & \cdots & S^0_l & S^1_l & S^1_l & \cdots & S^1_l \end{array}\right] \text{ and}$$

$$S^1_{ijl} = \left[\begin{array}{cccc|cccc} m_l C^0_{ij,1} & m_l C^0_{ij,2} & \cdots & m_l C^0_{ij,m_{ij}} & m_l C^1_{ij,1} & m_l C^1_{ij,2} & \cdots & m_l C^1_{ij,m_{ij}} \\ S^1_l & S^1_l & \cdots & S^1_l & S^0_l & S^0_l & \cdots & S^0_l \end{array}\right].$$

Step $u-1$**:**

Construction of basis matrices for $\boldsymbol{t}_{12\ldots(u-1)u}$-$(k_{12\ldots(u-1)u}, \boldsymbol{n}_{12\ldots(u-1)u})^*$-VCS where $\boldsymbol{t}_{12\ldots u} = (t_1, t_2, \ldots, t_u)$, $k_{12\ldots u} = t_1 + t_2 + \cdots + t_u$ and $\boldsymbol{n}_{12\ldots u} = (n_1, n_2, \ldots, n_u)$. To construct the basis matrices, we consider the following matrices: $(S^0_{12\ldots(u-1)}, S^1_{12\ldots(u-1)})$ and (S^0_u, S^1_u). Note that the basis matrices $(S^0_{12\ldots(u-1)}, S^1_{12\ldots(u-1)})$ is obtained recursively.

Consider the matrices

$$S^0_{12\ldots u} = \left[\begin{array}{ccc|ccc} m_u C^0_{12\ldots(u-1),1} & \cdots & m_l C^0_{12\ldots(u-1),m_{12\ldots(u-1)}} & m_u C^1_{12\ldots(u-1),1} & \cdots & m_u C^1_{12\ldots(u-1),m_{12\ldots(u-1)}} \\ S^0_u & \cdots & S^0_u & S^1_u & \cdots & S^1_u \end{array}\right] \text{ and}$$

$$S^1_{12\ldots u} = \left[\begin{array}{ccc|ccc} m_u C^0_{12\ldots(u-1),1} & \cdots & m_l C^0_{12\ldots(u-1),m_{12\ldots(u-1)}} & m_u C^1_{12\ldots(u-1),1} & \cdots & m_u C^1_{12\ldots(u-1),m_{12\ldots(u-1)}} \\ S^1_u & \cdots & S^1_u & S^0_u & \cdots & S^0_u \end{array}\right].$$

Remark: Note that in the above constructions, some of the t_i's maybe 1. In that case, we consider $(1, n_i)$-VCS, although it does not make much sense in terms of practicality but it does mean that every participant alone can recover the secret. That means the parties must be given as their shares the secret itself! So the basis matrix S^0_i must be the $n \times 1$ column vector $[0, 0, 0, \ldots, 0]^t$ and S^1_i must be the $n \times 1$ column vector $[1, 1, 1, \ldots, 1]^t$. This also holds true if $n_i = 1$.

Let us first illustrate the above construction through the following examples.

Example 2 Let $\mathcal{P} = \{1, 2, 3, 4\}$ be the set of parties and let $C_1 = \{1, 2\}, C_2 = \{3, 4\}$ be a partition of $\mathcal{P}$ such that $t_1 = 1, t_2 = 2$ (notations have their usual meanings). That is at least one party from C_1 collaborating with at least two parties from C_2 can recover the secret image. Thus, $\mathcal{Q}_{min} = \{134, 234\}$. We consider two hypothetical schemes, viz., (1, 2)-VCS and (2, 2)-VCS. For these two imaginary schemes, the basis matrices would have been $S^0_1 = \begin{bmatrix} 0 \\ 0 \end{bmatrix}$, $S^1_1 = \begin{bmatrix} 1 \\ 1 \end{bmatrix}$ and $S^0_2 = \begin{bmatrix} 1\,0 \\ 1\,0 \end{bmatrix}$, $S^1_2 = \begin{bmatrix} 1\,0 \\ 0\,1 \end{bmatrix}$. Following Step1 of our construction method, we obtain the basis matrices of the actual scheme as

$$S^0 = \left[\begin{array}{cc|cc} 0 & 0 & 1 & 1 \\ 0 & 0 & 1 & 1 \\ 0 & 1 & 1 & 0 \\ 0 & 1 & 0 & 1 \end{array}\right] \text{ and } S^1 = \left[\begin{array}{cc|cc} 0 & 0 & 1 & 1 \\ 0 & 0 & 1 & 1 \\ 1 & 0 & 0 & 1 \\ 0 & 1 & 0 & 1 \end{array}\right].$$

Row i of S^0 is the share of the ith party, if the secret pixel is white(0). Row i of S^1 is the share of the ith party, if the secret pixel is black(1).

Lemma 1 *Let* $(\overline{S^0}, \overline{S^1})$ *be the basis matrices that realize a VCS for a general access structure* $(\mathcal{Q}, \mathcal{F})$ *on the set of participants* $\overline{\mathcal{P}} = \{r+1, \ldots, r+n\}$ *with pixel expansion* m^**. Also let* (V^0, V^1) *be the basis matrices realizing a* (k, r)*-threshold VCS on a set of parties* $\mathcal{P} = \{1, 2, \ldots, r\}$ *with pixel expansion* m*. If we write* $V^0 = [C_1^0, C_2^0, \ldots, C_m^0]$ *and* $V^1 = [C_1^1, C_2^1, \ldots, C_m^1]$*, where* C_i^b *denotes the* i*th column of the matrix* V^b*,* $b \in \{0, 1\}$ *then the following matrices* (S^0, S^1) *are given by*

$$S^0 = \left[\begin{array}{cccc|ccc} m^*C_1^0 & m^*C_2^0 & \ldots & m^*C_m^0 & m^*C_1^1 & \ldots & m^*C_m^1 \\ \overline{S^0} & \overline{S^0} & \ldots & \overline{S^0} & \overline{S^1} & \ldots & \overline{S^1} \end{array}\right] \quad \textit{and}$$

$$S^1 = \left[\begin{array}{cccc|ccc} m^*C_1^0 & m^*C_2^0 & \ldots & m^*C_m^0 & m^*C_1^1 & \ldots & m^*C_m^1 \\ \overline{S^1} & \overline{S^1} & \ldots & \overline{S^1} & \overline{S^0} & \ldots & \overline{S^0} \end{array}\right]$$

realize a VCS on the participant set $\overline{\mathcal{P}} \cup \mathcal{P} = \{1, 2, \ldots, r, r+1, \ldots, r+n\}$ *in which any minimal qualified set is the union of a minimal qualified set in* $\mathcal{Q}_{min}$ *with a minimal qualified set in the given threshold scheme.*

Proof First, we prove the security condition for the above construction. If we can prove the security condition for any maximal forbidden set in the resulting access structure then we are done. Now any maximal forbidden set F can arise in two different ways.

Case 1: Let $F = \overline{\mathcal{P}} \cup D$, where D is a $(k-1)$ subset of $\mathcal{P}$. Now since D is a forbidden set in the (k, r)-threshold VCS and therefore it follows that the restricted matrices $V^0[D]$ and $V^1[D]$ are equal up to a column permutation. Without loss of generality, let $V^0[D] = V^1[D]$. Then the matrices obtained from them by replicating each column exactly m^* times should be equal. That is in our notation, $[m^*C_1^0[D] \cdots m^*C_m^0[D]] = [m^*C_1^1[D] \cdots m^*C_m^1[D]]$. Thus, when restricted to F the left (resp. right) portion of $S^0[F]$ is equal to the right (resp. left) portion of $S^1[F]$. This proves the security condition for this case.

Case 2: Let $F = \overline{F} \cup \mathcal{P}$, where $\overline{F} \in \underline{F_M}$. Now since $\overline{F}$ is a forbidden set, we may assume that the restricted matrices $(\overline{S^0}[\overline{F}], \overline{S^1}[\overline{F}])$ are equal. Now it is easy to see that when restricted to F the left (resp. right) portion of $S^0[F]$ is equal to the right (resp. left) portion of $S^1[F]$. This completes the proof of security condition.

For proving the contrast condition, we first observe that the pixel expansion of the resulting VCS is $2mm^*$. Let $\overline{\alpha}$ and α denote the relative contrasts of the $(\mathcal{Q}, \mathcal{F})$-VCS and (k, r)-threshold VCS, respectively. Let X denote a minimal qualified set in the resulting access structure. Then $X = Q \cup B$, where $Q \in \mathcal{Q}_{min}$ and B is a k subset of $\mathcal{P}$.

Thus, we have, $w(S_X^1) - w(S_X^0) \geq w(S_Q^1) - w(S_Q^0) \geq \overline{\alpha} m^*$. This proves the theorem. □

Theorem 1 *The matrices* $(S_{12\ldots u}^0, S_{12\ldots u}^1)$ *from a basis matrices for* $\mathbf{t}_{12\ldots u}$*-*$(k_{12\ldots u}, \mathbf{n}_{12\ldots u})^*$*-VCS with pixel expansion* $2^{u-1} \cdot m_1 m_2 \cdots m_u$*, where* m_i *is the pixel expansion of the* (t_i, n_i)*-VCS on the participant set* C_i*.*

Proof Proof of the theorem follows from the construction method in Sect. 3 and applying Lemma 1, $(u-1)$ times.

Remark: There exists VCS for any compartmented access structure $\boldsymbol{t}$-$(k, \boldsymbol{n})^*$-VCS with pixel expansion $2^{u-1} \cdot m_1 m_2 \cdots m_u$, where m_i is the pixel expansion of the (t_i, n_i)-VCS on the participant set $\mathcal{C}_i$. Moreover, letting $u = 2$, $|\mathcal{C}_1| = 1$ and $|\mathcal{C}_2| = n-1$ with $t_1 = 1$ we have Theorem 2.3 in [3]. Also, if $u = 2$ and $|\mathcal{C}_1| = t$ and $|\mathcal{C}_2| = n - t$ with $t_1 = t$ we have Theorem 4 in [9].

3.1 Finding Closed Form for Relative Contrast

Theorem 1 ensures the existence of a compartmented VCS. However, finding the closed form of relative contrast is a combinatorially challenging problem. In this section, we come up with a closed form of relative contrast for compartmented access structure using canonical threshold VCS. First let us define what a canonical VCS [6] is.

Sometimes the basis matrices realizing a VCS can be described in a very convenient manner by means of column multiplicities. The basis matrices are said to be in *canonical form* if a column of Hamming weight i appears f times in a matrix S^0 or S^1, then every weight i column occurs in the matrix exactly f times. When the basis matrices are in canonical form then the VCS is called *canonical VCS*.

To calculate the closed form of the relative contrast, we need the following lemmas.

Lemma 2 *Let (S_i^0, S_i^1) denote canonical basis matrices for a (t_i, n_i)-VCS with pixel expansion m_i. Let $\mu^b_{l_i,i}$ denote the number of occurrences of each column of weight l_i in the basis matrix S_i^b, where $i = 1, 2, \ldots, u$ and $b \in \{0, 1\}$. For a minimal qualified set X, let us further assume $\mathbf{W_i}^b = (C^b_{X,i,1}, C^b_{X,i,2}, \ldots, C^b_{X,i,m_i})$, where $C^b_{X,i,j}$ is the "OR" of the jth column of the restricted matrix $S_i^b[X]$. Let Z_i^b denote the number of zeros in $\mathbf{W}_i^b$.*

Then $Z_i^b = \sum_{l_i=0}^{n_i-t_i} \binom{n_i - t_i}{l_i} \mu^b_{l_i,i}$, where $i = 1, 2, \ldots, u$, $b \in \{0, 1\}$.

Proof Let us understand the logic behind the counting principle. Let us suppose $\mu^0_{l_i,i}$ denote the frequency with which each column of weight l_i occurs in basis matrix S_i^0. Let X be a minimal qualified set having t_i many parties in it. Observe that for a column vector $\boldsymbol{C}$ if at least one of its entries is 1 then the "OR" of all the entries of the column becomes 1. Thus the "OR" value is 0 if and only if all the entries in the column are 0. We note that a column with weight l_i when restricted to X will have all zero entries if the l_i many 1's appear at positions which are not indexed by the indices in X. Therefore, those l_i many 1's can appear anywhere among the $(n_i - t_i)$ entries left in the column $\boldsymbol{C}$. Now, each column of weight l_i appears exactly $\mu^0_{l_i,i}$ times. Therefore, total number of columns with weight l_i and having zeros at the positions which are indexed by X is given by the expression $\binom{n_i - t_i}{l_i}\mu^0_{l_i,i}$. Also, observe that if a column has weight $n_i - t_i$ or greater, then it is not possible to get all zeros at the positions indexed by X. At least one 1 must appear somewhere. Varying l_i over $\{0, 1, \ldots, n_i - t_i\}$, we get the required expression for $b = 0$. The same is the argument for $b = 1$.

Lemma 3 *Let* $W^b_{12\ldots u} = (C^b_{X,12\ldots u,1}, C^b_{X,12\ldots u,2}, \ldots, C^b_{X,12\ldots u,u})$, *where* $C^b_{X,12\ldots u,j}$ *denotes "OR" of the jth column of the restricted matrix* $S^b_{12\ldots u}[X]$, *for the minimal qualified set X of the access structure* $\boldsymbol{t}_{12\ldots u}$-$(k_{12\ldots u}, \boldsymbol{n}_{12\ldots u})^*$-*VCS having pixel expansion* $m_{12\ldots u}$. *Let* $Z^b_{12\ldots u}$ *denote the number of zeros in* $\mathbf{W}^b_{12\ldots u}$, *where* $b \in \{0, 1\}$.*Then*

$$Z^0_{12\ldots u} = \sum_{x=(x_1,\ldots,x_u):wt(x)\ even} Z_1^{x_1} Z_2^{x_2} \ldots Z_u^{x_u}$$

and

$$Z^1_{12\ldots u} = \sum_{x=(x_1,\ldots,x_u):wt(x)\ odd} Z_1^{x_1} Z_2^{x_2} \ldots Z_u^{x_u}.$$

Proof Let us first write the minimal qualified set X as the disjoint union of parts $X_1, X_2, \ldots, X_u$, where each X_i is a subset of the ith-level participants. Applying Lemma 2 separately for each X_i and then patching them up keeping in mind the construction given in Sect. 3 we get the result.

The following theorem is now straightforward.

Theorem 2 *For a given compartmented access structure* $\boldsymbol{t}_{12\ldots u}$-$(k_{12\ldots u}, \boldsymbol{n}_{12\ldots u})^*$, *there exists a* $\boldsymbol{t}_{12\ldots u}$-$(k_{12\ldots u}, \boldsymbol{n}_{12\ldots u})^*$-*VCS with pixel expansion* $m = 2^{u-1}m_1m_2\cdots m_u$ *and relative contrast* $\frac{\prod_{i=1}^{u}(Z_i^0 - Z_i^1)}{m}$.

4 Non-expansible Construction

We follow the methodology of Yang [18] to construct non-expansible, probabilistic $\boldsymbol{t}$-$(k, \boldsymbol{n})^*$-VCS. We first give a generic construction method of probabilistic VCS realizing any general access structure and then we describe a method that realizes $\boldsymbol{t}$-$(k, \boldsymbol{n})^*$-VCS with better contrast. In the probabilistic scheme, the secret image will be correctly recovered only with a certain probability. Let Q be a qualified set. Let $P_{w/w}(Q)$ be the probability of correctly reconstructing a white pixel in white from the S^0 matrix and $P_{w/b}(Q)$ be the probability of incorrectly reconstructing a black pixel in white from the S^0. Similarly, we define the notations $P_{b/b}(Q)$ and $P_{b/w}(Q)$. To capture the idea of probability of correct reconstruction of the secret image, we take the differences of the following probabilities:

$$P_{b/b}(Q) - P_{b/w}(Q) \quad (1)$$
$$\text{and} \quad P_{w/w}(Q) - P_{w/b}(Q). \quad (2)$$

Now a scheme is said to be β -probabilistic if there exists a positive constant β such that for any qualified set Q

$$P_{b/b}(Q) - P_{b/w}(Q) \geq \beta \quad (3)$$
$$\text{and} \quad P_{w/w}(Q) - P_{w/b}(Q) \geq \beta. \quad (4)$$

We now give the definition of a β-probabilistic VCS for general access structure as follows.

Definition 2 Let $(\mathcal{Q}, \mathcal{F})$ denote a general access structure on a set of n participants. A β-probabilistic $(\mathcal{Q}, \mathcal{F})$ is defined by two collection of $n \times 1$ matrices C_W and C_B such that

1. There exists $\beta > 0$, such that for any minimal qualified set $Q \in \mathcal{Q}_{min}$ it must hold that $P_{b/b}(Q) - P_{b/w}(Q) \geq \beta$.
2. For any maximal forbidden set $F \in \mathcal{F}_{max}$, the two restricted collections $\{M^0[F] : M^0 \in C_W\}$ and $\{M^1[F] : M^1 \in C_B\}$ contain the same matrices with equal frequencies.

4.1 Construction of Probabilistic VCS

Let $(\mathcal{Q}, \mathcal{F})$ denote a general access structure on a set of n participants. Ateniese et al. [4] gave a generic construction of basis matrices based on the combinatorial design Cumulative Array. Due to the space limitation, we only state the results.

Theorem 3 (adopted from [4]) *Let $(\mathcal{Q}, \mathcal{F})$ denote a general access structure on a set of n participants. Let the number of maximal forbidden sets be t. Then there exists a visual cryptographic scheme realized by the basis matrices (S^0, S^1) with pixel expansion 2^{t-1}.*

If we define $C_W = \{\boldsymbol{c} : \boldsymbol{c}\ is\ a\ column\ of\ S^0\}$ and $C_B = \{\boldsymbol{d} : \boldsymbol{d}\ is\ a\ column\ of\ S^1\}$, then it is easy to see that these collections realize a probabilistic VCS on $(\mathcal{Q}, \mathcal{F})$ with relative contrast $\gamma = \frac{1}{2^{|\mathcal{F}_{max}|-1}} = \beta$.

Thus we have the following theorem.

Theorem 4 *Given $(\mathcal{Q}, \mathcal{F})$ on a set of n participants, there exists a β-probabilistic VCS with $\beta = \frac{1}{2^{|\mathcal{F}_{max}|-1}}$, where $\mathcal{F}_{max}$ denotes the set of all maximal forbidden sets for the access structure $(\mathcal{Q}, \mathcal{F})$.*

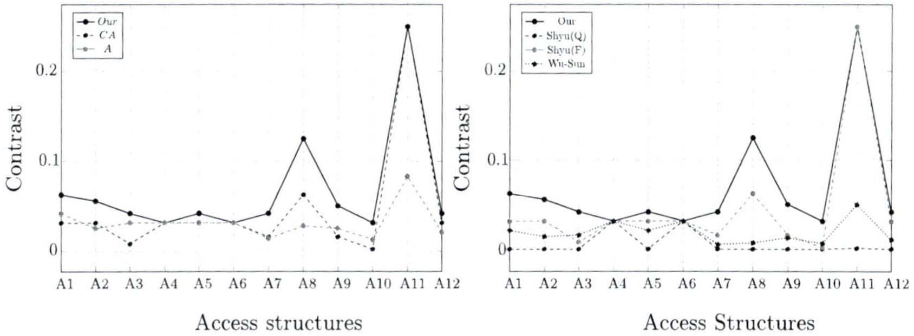

(a) Graph comparing contrasts of schemes in Table 1

(b) Graph comparing contrasts of schemes in Table 2

Fig. 1 Performance of our scheme

Corollary 1 *For any given compartmented access structure* $\boldsymbol{t}$-$(k, \boldsymbol{n})^*$, *there exists a* β-*probabilistic* $\boldsymbol{t}$-$(k, \boldsymbol{n})^*$-*VCS with pixel expansion* $m = 2^{sum} - 1$, *where* $sum = \binom{n_1}{t_1 - 1} + \binom{n_2}{t_2 - 1} + \cdots + \binom{n_u}{t_u - 1}$ *and* $\beta = \frac{1}{m}$.

Proof The proof follows from the fact that in case of compartmented access structure, $|\mathcal{F}_{max}| = \binom{n_1}{t_1 - 1} + \binom{n_2}{t_2 - 1} + \cdots + \binom{n_u}{t_u - 1}$. □

Based on the construction as described in Sect. 3 and the method proposed in [18], we are going to state the following Theorem (Fig. 1).

Theorem 5 *For a given compartmented access structure* $\boldsymbol{t}_{12\ldots u}$-$(k_{12\ldots u}, \boldsymbol{n}_{12\ldots u})^*$-*VCS, there exists a* β-$\boldsymbol{t}_{12\ldots u}$-$(k_{12\ldots u}, \boldsymbol{n}_{12\ldots u})^*$-*VCS with* $\beta = \frac{\prod_{i=1}^{u}(Z_i^0 - Z_i^1)}{m}$.

4.2 Experiments and Comparisons

In this section we provide few numerical evidence in Table 1 to justify that our direct construction method for compartmented access structure works better in terms of both pixel expansion as well as relative contrast (or β-probability) than the scheme obtained through Corollary 1 in Sect. 4.1, obtained as a particular case of general access structure. We also compare our results with the results of [1]. We compare our results with existing random grid constructions in Table 2.

Table 1 Comparison among the schemes: In the table m_{our} (our proposed scheme), m_{CA} (obtained from [4]) and m_A (obtained from [1]) stand for the pixel expansions for the corresponding compartmented access structures appeared in the first column of this table. Further β_{our} (Theorem 5), β_{CA} (Sect. 4.1) and β_A are the β probabilities for the corresponding constructions

Access structure	m_1	m_2	m_3	m_{our}	m_{CA}[4]	m_A[1]	β_{our}	β_{CA}	β_A
$A_1 : (2, 2) - (4, (2, 4))^*$	2	4		16	32	24	1/16	1/32	1/24
$A_2 : (2, 2) - (4, (3, 3))^*$	3	3		18	32	40	1/18	1/32	1/40
$A_3 : (2, 3) - (5, (2, 4))^*$	2	6		24	128	32	1/24	1/128	1/32
$A_4 : (2, 4) - (6, (2, 4))^*$	2	8		32	32	32	1/32	1/32	1/32
$A_5 : (2, 3) - (5, (3, 3))^*$	3	4		24	32	32	1/24	1/32	1/32
$A_6 : (2, 2, 2) - (6, (2, 2, 2))^*$	2	2	2	32	32	32	1/32	1/32	1/32
$A_7 : (2, 2) - (4, (3, 4))^*$	3	4		24	64	72	1/24	1/64	1/72
$A_8 : (1, 2) - (3, (3, 4))^*$	1	4		8	16	36	1/8	1/16	1/36
$A_9 : (2, 2) - (4, (5, 2))^*$	5	2		20	64	40	1/20	1/64	1/40
$A_{10} : (3, 2) - (5, (5, 2))^*$	16	2		32	512	80	1/32	1/512	1/80
$A_{11} : (1, 2) - (3, (5, 2))^*$	1	2		4	4	12	1/4	1/4	1/12
$A_{12} : (1, 2, 2) - (5, (2, 3, 2))^*$	1	3	2	24	32	48	1/24	1/32	1/48

5 Conclusion

We have considered an efficient direct construction of visual cryptographic scheme for compartmented access structure having a closed form of both pixel expansion as well as relative contrast. We also considered the probabilistic construction to gain efficiency.

Table 2 Comparisons of contrast among the proposed scheme and related non-expansible methods (for compartmented access structures). The values of $Shyu(Q)$ are obtained from Algorithm 2 and $Shyu(F)$ from Algo. 3 in [15]

Access structure	Our	$Shyu(Q)$ [15]	$Shyu(F)$ [15]	Wu-Sun [17]
$A_1 : (2, 2) - (4, (2, 4))^*$	1/16	$1/2^{17}$	1/32	1/48
$A_2 : (2, 2) - (4, (3, 3))^*$	1/18	$1/2^{23}$	1/32	1/72
$A_3 : (2, 3) - (5, (2, 4))^*$	1/24	$1/2^{16}$	1/128	1/64
$A_4 : (2, 4) - (6, (2, 4))^*$	1/32	1/32	1/32	1/32
$A_5 : (2, 3) - (5, (3, 3))^*$	1/24	$1/2^{12}$	1/32	1/48
$A_6 : (2, 2, 2) - (6, (2, 2, 2))^*$	1/32	1/32	1/32	1/32
$A_7 : (2, 2) - (4, (3, 4))^*$	1/24	$1/2^{40}$	1/64	1/192
$A_8 : (1, 2) - (3, (3, 4))^*$	1/8	$1/2^{14}$	1/16	1/144
$A_9 : (2, 2) - (4, (5, 2))^*$	1/20	$1/2^{27}$	1/64	1/80
$A_{10} : (3, 2) - (5, (5, 2))^*$	1/32	$1/2^{37}$	1/512	1/160
$A_{11} : (1, 2) - (3, (5, 2))^*$	1/4	$1/2^{10}$	1/4	1/20
$A_{12} : (1, 2, 2) - (5, (2, 3, 2))^*$	1/24	$1/2^{22}$	1/32	1/96

References

1. Adhikari, A.: Linear algebraic techniques to construct monochrome visual cryptographic schemes for general access structure and its applications to color images. Des. Codes Cryptogr. **73**(3), 865–895 (2014)
2. Adhikari, A., Dutta, T.K., Roy, B.: A new black and white visual cryptographic scheme for general access structures, In: INDOCRYPT 2004. LNCS, vol. 3348, pp. 399–413. Springer, Berlin (2004)
3. Arumugam, S., Lakshmanan, R., Nagar, A.K.: On $(k, n)^*$-visual cryptography scheme. Des. Codes Cryptogr. **71**(1), 153–162 (2014)
4. Ateniese, G., Blundo, C., De Santis, A., Stinson, D.R.: Visual cryptography for general access structures. Inf. Comput. **129**, 86–106 (1996)
5. Cimato, S., De Prisco, R., De Santis, A.: Probabilistic visual cryptography schemes. Comput. J. **49**(1), 97–107 (2006)
6. D'Arco, P., De Prisco, R., De Santis, A.: Measure-independent characterization of contrast optimal visual cryptography schemes. J. Syst. Softw. **95**, 89–99 (2014)
7. Dutta, S., Adhikari, A.: XOR based non-monotone t-$(k, n)^*$-visual cryptographic schemes using linear algebra. In: ICICS 2014. LNCS, vol. 8958, pp. 230–242. Springer, Berlin (2014)

8. Dutta, S., Singh, R.R., Adhikari, A.: Constructions and Analysis of Some Efficient t-$(k, n)^*$-Visual Cryptographic Schemes Using Linear Algebraic Techniques. Des. Codes Cryptogr. **80**(1), 165–196 (2016)
9. Guo, T., Liu, F., Wu, C.K., Ren, Y.W., Wang, W.: On (k, n) visual cryptography scheme with t essential parties, ICITS 2013. LNCS, vol. 8371, pp. 56–68. Springer, Berlin (2014)
10. Ito, R., Kuwakado, H., Tanaka, H.: Image size invariant visual cryptography. IEICE Trans. Fundam. Electron. Commun. Comput. Sci. **E82A**(10), 2172–2177 (1999)
11. Kafri, O., Keren, E.: Encryption of pictures and shapes by random grids. Opt. Lett. **12**(6), 377–379 (1987)
12. Naor, M., Shamir, A.: Visual Cryptography, EUROCRYPT 1994. LNCS, vol. 950, pp. 1–12. Springer, Berlin (1994)
13. Praveen, K., Rajeev, K., Sethumadhavan, M.: On the Extensions of $(k, n)^*$-Visual Cryptographic Schemes, SNDS 2014. CCIS **420**, 231–238 (2014)
14. Shyu, S.J.: Image encryption by random grids. Patt. Recogn. **40**(3), 1014–1031 (2007)
15. Shyu, S.J.: Visual cryptograms of random Grids for General Access Structures. IEEE Trans. Circuits Syst. Video Techn. **23**(3), 414–424 (2013)
16. Simmons, G.J.: How to (really) share a secret, CRYPTO 1988. LNCS **403**, 390–448 (1990)
17. Wu, X., Sun, W.: Visual secret sharing for general access structures by random grids. IET Inf. Secur. **6**(4), 299–309 (2012)
18. Yang, C.-N.: New visual secret sharing schemes using probabilistic method. Pattern Recognit. Lett. Elsevier **25**, 481–494 (2004)

Design of Hash Function Using Two Dimensional Cellular Automata

Tanushree Haldar and Dipanwita Roy Chowdhury

Abstract This paper proposes a new hash function using two-dimensional cellular automata. The cellular structure of 5-neighborhood 2D CA is exploited to use it as the basic building block of the hash function. It has been shown that the excellent randomness along with the inherent diffusion makes 2D CA an ideal candidate of Hash function. This work also focuses on the security analysis of the Hash function. We demonstrate Birthday attack and the analysis shows that the design is secure against this attack.

Keywords Two-dimensional cellular automata · Keccak · Hash function · Hamming distance · Avalanche effect · Birthday attack

1 Introduction

Cryptographic Hash function is a widely used concept in the field of information security and its significance is ever-growing. Digital signature, random sequence generator are few prominent application of this. Cryptographic hash functions can be developed from scratch or existing block ciphers can be utilized to design the building blocks. Most of the commonly used hash function utilizes Merkle-Damgard design principles. Over the year, this design principle is studied and the weaknesses of Merkle-Damgard design are exposed specially against the collision attack. When the weaknesses of standard hash functions like SHA-1 and MD5 were exposed, NIST, USA started a global competition to determine the most effective, coherent, and secure hash algorithm as Advanced Hash Standard.

T. Haldar (✉) · D. R. Chowdhury
Department of Computer Science and Engineering, IIT Kharagpur, Kharagpur 721302, West Bengal, India
e-mail: tanushree.besu.2012@gmail.com

D. R. Chowdhury
e-mail: drc@cse.iitkgp.ernet.in

D. Giri et al. (eds.), *Proceedings of the Fifth International Conference on Mathematics and Computing*, Advances in Intelligent Systems and Computing 1170, https://doi.org/10.1007/978-981-15-5411-7_3

John von Neumann was the first person to conceptualize Cellular Automata. It has been used in public-key cryptography, symmetric key, block cipher, and stream cipher. 1D CA has been used for designing hash function. However, there are very limited work on 2D CA. The architecture of this new hash function is inspired by the design of Keccak (SHA-3) [1]. In the internal transformation of Keccak, three linear and one nonlinear operations are performed on a three-dimensional structure over twenty-four rounds. The aim of this design is to simplify the complex design without compromising the security by using 2D CA as a building block. Here we have explored the suitability of 2D CA as the building block of the proposed hash function. The simple and consistent design of CA is very much suitable for VLSI technology. Also, this adaptive interlinked structure is appropriate for hardware and software implementation due to simple operation involved. Our objective is to build a coherently computable f-function and in turn, a new One Way Hash Function. Security analysis of the 2D CA-based hash function is done to confirm that the design is secure against modern-day attacks.

The content of this paper is arranged as the following. A brief outline of CA which is used as a basic tool to design the proposed hash function is discussed in Sect. 2. The design architecture of the new hash function is explained in the Sect. 3. Section 4 describes the design rationale and choice of parameters of 2D CA-based hash function. Security analysis is carried out in Sect. 5 followed by conclusion in Sect. 6.

2 Preliminaries

The simple structure of CA along with its randomness makes it very attractive to the researchers. CA are a discrete grid of cells with a specific geometry. Each cell comprises a memory unit (Flip-Flop) and a combinatoric logic. Cells can have values from a finite set S. The cells are updated with each clock pulse concurrently. A function f:$S^k \rightarrow S$ is defined as local transition rule based on which the cell is updated. The transition function f completely depends on local neighborhoods of cell. Here the number of neighbors is represented by k. For example, for 3-neighborhood CA ith cell at nth clock pulse evolves as follows:

$$S^{n+1} = f(S^n_{i-1}, S^n_i, S^n_{i+1}) \tag{1}$$

CA which is able to produce all non-null states, starting from any non-null state, is named as maximum length CA. The characteristic polynomial of an n-bit maximal length CA becomes a primitive polynomial of degree n. For example, the primitive polynomial Z^4+Z+1 over GF(2) represents the maximal length CA [1, 0, 1, 0] where rule-90 is denoted by '0' while rule-150 is denoted by '1'.

A 2D CA is an abstraction of 1D CA. Here the cells are organized in a two-dimensional lattice with interconnections among the adjacent cells. The neighborhood function defining the next state of a cell $\boldsymbol{C}_{ij}$, depends on four of its neighbors

Fig. 1 Five-Neighborhood 2D CA

	top	
left	self	right
	bottom	

and itself (five-neighborhood dependency) as shown in Fig. 1. Thus, the next state ***S*** of the ***(i,j)***th cell of a 2D CA is given by

$$S_{i,j}^{n+1} = f(S_{i,j}^{n}, S_{i-1,j}^{n}, S_{i,j-1}^{n}, S_{i+1,j}^{n}, S_{i,j+1}^{n}) \quad (2)$$

Since f is a Boolean function of five variables, there are $2^5 = 32$ distinct neighborhood configurations. The decimal equivalent of the 5 -bit tuple representing the neighborhood dependency in anticlockwise defines the 2D CA rule, e.g., $< self, top, right >$ is the rule 25 <11001>.

3 Construction of 2D CA-Based Hash Function

3.1 Why 2D CA

The basic requirement for a cryptographic hash function is to generate a unique digest and provide optimal security against the known attacks. One of the most logical ways to achieve this is to induce random sequences. Traditionally, CA has been studied in the context of pseudo-random sequence generation, cryptographic applications and its capability is proven in this context. However, it can be observed that the pseudo-random sequences generated using CA has a varied level of randomness and maximum length[1] CA is the one with the highest randomness [4]. Also, it is more likely to observe a CA attaining maximum length if it is designed combining multiple rules (i.e., hybrid CA). Both linearity and non-linearity are combined with the Maximum Length hybrid CA while designing the building blocks for the hash function in order to make it even more secure. While linear operations provide security against most of the side-channel attack, i.e., power attack, timing attack, nonlinear operations induce further complexity in order to provide greater security [8–10]. Here in [16], it is shown that the 2D CA produces better randomness than that of 1D CA. In literature, there is very few research in 2D CA. In this paper, we explore the use of 2D CA as a basic building block of cryptographic hash function.

In this design, we have chosen domain extension algorithm which is analogous to the sponge function over Merkle–Damgard(MD) [11, 12] structure. Most of the standard hash function follows the MD structure. However, MD was proven to be weak

[1]A CA having cycle length $2^n - 1$ is called maximum length CA.

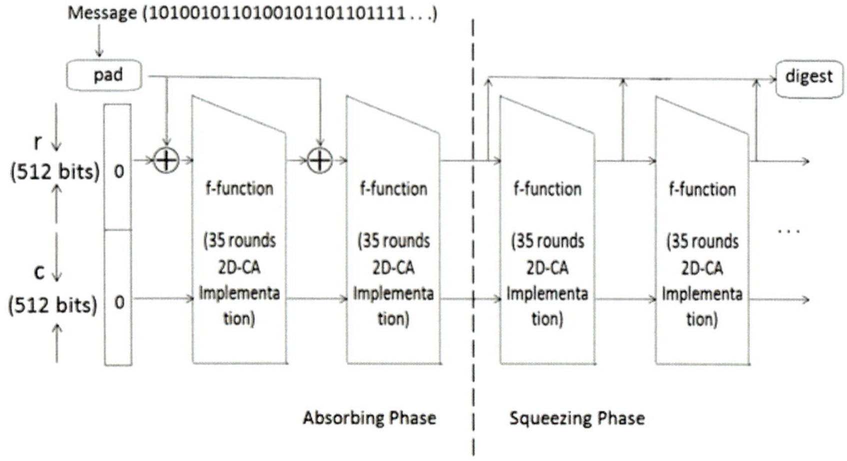

Fig. 2 Construction of 2D CA-based hash function

against different generic attacks such as multi-collision attack [14], herding attack [15] and its hardware implementation is also proven to be inefficient. On the other hand, Sponge, designed by Bertoni provides security depending on its configurable parameters. It has been observed that Sponge construction has several advantages over MD, ranging from efficient hardware implementation to a variable number of output bits. In our proposed hash algorithm, we have adopted Sponge function in order to achieve reasonable security while keeping the hardware implementation simpler and flexibility of configuration.

F-function

Here we briefly sketch the construction of 2D CA-based hash function. A constant internal transformation is iterated to generate variable length output digest shown in Fig. 2. The internal state matrix is divided into two parts, where c is the capacity and r is the bit-rate. The internal state transformation is considered as f function and it operates on a 32×32, i.e., total 1024 bits block. In the beginning, the state matrix is set to all zeros and first 512 bits of the block will be considered as rate or r and rest 512 bits will be considered as capacity or c. At first, the text message is taken as input and is converted to a binary bit string. This bit string is then padded with 10*1 format so that length of the whole message becomes a multiple of 512 with the only restriction that the final message block should not be all zero. In the next step, the padded message is divided into blocks of size 512. Each of these blocks will be first XORed with the internal state matrix. When the first block will come and XORed with the internal state matrix, only the top 16 rows of the state matrix will contain nonzero binary values. The rest 16 rows will contain only zero values. As the CA evolves the zero rows got diffused with nonzero values. The f-function needs only 16 rounds to diffuse all 16 zero rows as nonzero rows. However, in this design 2D CA has been implemented for 35 rounds. Many statistical analysis is done based on the output of different rounds of f-function, which is described in the subsequent sections. Here nonlinear hybrid 2D CA has been implemented. The nonlinear rules are applied only at three positions. Linear rules violate the security whereas the use of only nonlinear rules increases the complexity of the circuit and also prone to side-channel attack. This non-linearity injection point has been decided based on experiments. As 2D CA are used for random number generations and reversibility of 2D CA is a difficult problem [5] so it is used as a tool to design this new hash function.

4 Design Rationale

Through experiments, we concluded the column numbers <7 16 25> of the first row to inject nonlinear CA rule. The aim of this experiment was to estimate the maximum length Nonlinear Hybrid Two-Dimensional Cellular Automata. The nonlinear function is injected according to Algorithm 1 [4]. Non-linearity is injected at the top row so that it will affect the value of the bottom rows naturally.

First, we have tried with many linear 2D CA rule-sets. The behavior of the f function has been observed. The experiment started with linear maximum length 1D CA [16] and it is converted to five-neighborhood 2D CA rule set. That five-neighborhood 2D CA rule set is again modified to estimate the maximum length. The analysis of f-function has been done based on randomness, rate of diffusion, avalanche effect.

Finding Nonlinear 2D CA for Longer Period

The maximum length linear 1D CA rule set < 150, 150, 90, 150, 90, 150, 90 ,. . .> has been considered. It has three neighbors called left, self, and right.

$$Q_i^{t+1} = Q_{i-1}^t \oplus Q_i^t \oplus Q_{i+1}^t \qquad \text{Rule 150}$$

$$Q_i^{t+1} = Q_{i-1}^t \oplus Q_{i+1}^t \qquad \text{Rule 90}$$

The three-neighborhood 1D CA has been converted to five-neighborhood 2D CA. In Fig. 3, it is shown that rule 90 has only left dependency and right dependency. The top and bottom dependency has been introduced for rule 90 along with its left and right dependency to convert it to rule 15. For rule 150, only top dependency has been included and it is converted to rule 29. The 2D linear rule set matrix is shown below (Table 1).

The column numbers <7 16 25> of the first row have been considered to inject nonlinear CA rule according to Algorithm 1 [4]. Figure 3 is considered as input ruleset F_L. After implementation of the final 2D CA nonlinear hybrid rule set, a very good random output pattern Fig. 4 has been generated.

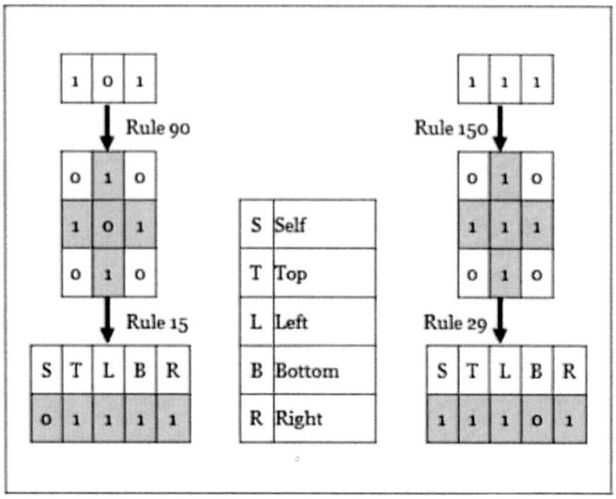

Fig. 3 Construction of 2D CA rule from 1D CA rule

Table 1 Final linear 2D CA rule set

29	29	15	29	15	29	15	29	15	29	.	.	.	.	.	.	.	.	.	29	15	29	15	29	15	29	15	29	15	29	15	29

4.1 Diffusion Analysis of F-Function

If any change of single character in the plaintext leads to the change of multiple characters in the ciphertext, and any change of single character in ciphertext leads to change of multiple characters in plaintext then it is called diffusion. Eventually, this leads to the diffusion of the frequency of letters [digraphs, etc.] in the plaintext over many characters in the ciphertext. This results in the requirement of a larger set of ciphertext for a reasonable statistical attack.

Algorithm 1 NHCA Synthesize Algorithm

INPUT: A maximum length LHCA with ruleset F_L. A position j to inject non-linearity and the set of cells of the LHCA S.

OUTPUT: A maximum length NHCA ruleset F_N.

$F_N \leftarrow F_L$

Let $F_N = \{f_{n-1}, \dots, f_0\}$

$\chi \subset S : \forall x \in \chi, x \notin N(j)$ ▷ select a subset from S

$P \leftarrow f_N(\chi)$ ▷ f_N is non-linear function

$f_j \leftarrow f_j \oplus P$

$(f_j \xrightarrow{P} f_{j+1})$ ▷ Apply shifting operation

$f_j \leftarrow f_j \oplus P$

return F_N

Figure 5 shows how the nonzero rows of the state matrix affect the zero rows. Here the 2D CA rules having top dependencies have been implemented. So with

Fig. 4 2D CA implementation with final NHCA ruleset

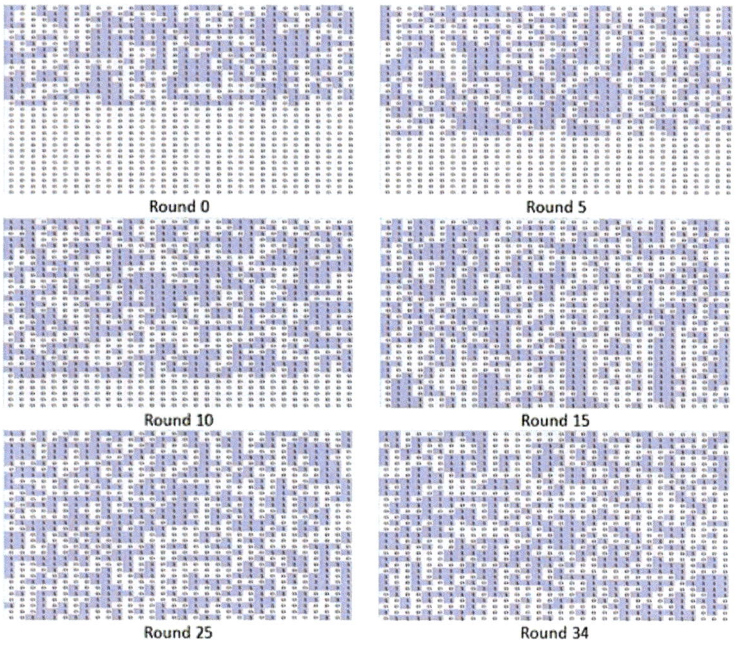

Fig. 5 Rate of diffusion from round 0 to round 34

number of rounds, the zero rows got diffused. After round 16, all 16 zero rows of the state matrix will be diffused. However, in the proposed design f-function with 35 rounds has been implemented. The reason behind these extra 19 rounds is to make sure that the output from the f-function satisfies avalanche criteria. In the following experiments, it is also checked that the amount of randomness does not decrease with the number of rounds.

4.2 Avalanche Effect

One of the most desired properties of Cryptographic Hash Functions is the avalanche effect. According to this phenomena, a minimal change in the input sequence (say one bit) grows until it affects half of the output hash. A cryptographic morphing is complete if all of the input bits have an impact on every single output bit. It has two criteria called Strict Avalanche Criterion[1] and Bit independence Criterion.[2]

Choice of Parameter: Number of Rounds
To fix the number of rounds, one experiment has been done on the f-function. We have taken only one random sample message block, implemented f-function on that and got the digest. Then each bit of the message block has been flipped at a time and sent to the f-function. All the one bit flipped messages and corresponding digests have been stored in a lookup table along with the initial message and the digest. In the next step, the Hamming distance between the initial digest and all other digests have been calculated. It generates a 1D array of size 512 and the value of the array will be the Hamming distance between the digests of the 1-bit flipped messages and the digest of the original message.

The same experiment has been carried out with a different round of the f-function. The results are shown in Fig. 6. Initially, the Hamming distance between the messages is very low. From round 1 to round 9, the Hamming distance lies approximately between 10 and 130. It is clear that the Hamming distance has increased with the number of rounds. From round 28 to round 32, the Hamming distance lies approximately between 230 and 260. From round 32 to round 35, it saturates and most of the digests have the Hamming distance between 245 and 256 which meets the avalanche criterion. This experiment has been performed over several thousand random sample messages. From the experimental results, we have fixed the number of rounds of the f-function as 35 in the proposed design.

What Properties Should a Cryptographic Hash Function Have?
Any cryptographic hash function should satisfy below properties.

- Deterministic: The f function designed for the hash function is deterministic. It produces the same digest for same input message.
- Computationally simple and fast: The primitive of the proposed hash function is nonlinear 2D CA, which is considered as a computationally simple and hardware friendly crypto primitive.
- Messages and digests should be uncorrelated: One-bit change in the message should result in half of the bits change in the digest. It has been explained in Sect. 4.2 in detail. To meet this avalanche criteria, the number of rounds of the f function has been fixed to 35.

[1]Strict Avalanche Criterion: If a change of a bit in the input sequence leads to the change of every output bits with a probability of 50%, then it is said to observe the Strict Avalanche Criteria.

[2]Bit independence Criterion: If a change in a bit i of the input sequence leads to change in the output bit j and k where j and k changes independently for all i, j and k then it is said to be Bit independence Criterion.

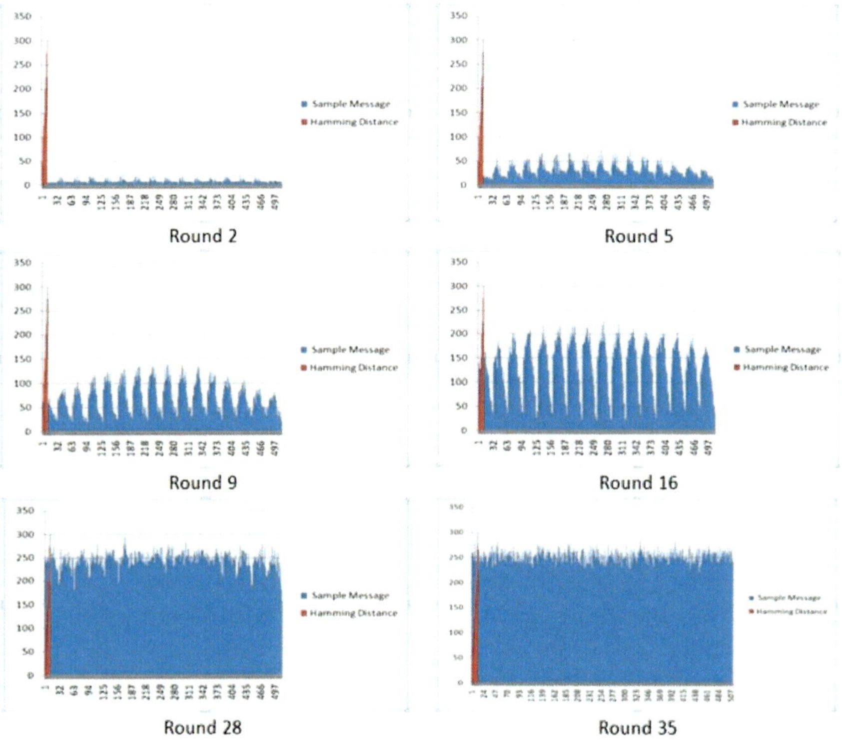

Fig. 6 Avalanche effect from round 2 to round 34

- Preimage resistance: For every digest, it should be infeasible to computationally find the corresponding input message. For example, if Z is the digest where the corresponding input is unknown and M^0 is the preimage where $hash(M^0) = Z$, it should be infeasible to computationally find M^0 only from Z. As reversibility of 2D CA is a difficult problem [5] it ensures that the hash function is non-invertible. The complexity of this attack is 2^N where N is the length of the digest. To make the hash function secure against this type of attack, length of the digest is made 512 which makes this attack computationally infeasible.
- Second-preimage resistance: If M^1 is the given message and the digest is $hash(M^1)$, it should require an infinite computation power to find another message, say M^2 for which the digest matches the first digest, i.e., $hash(M^1) = hash(M^2)$. The complexity of this attack for the proposed hash function is 2^{512}.
- Collision resistance: It should be infeasible to computationally find two input messages M^1 and M^2, where the digest is the same, i.e., $hash(M^1) = hash(M^2)$. Birthday Attack is one type of Collision resistance attack. It has a complexity of 2^{256} for the 2D CA-based hash function

5 Cryptanalysis of 2D CA-Based Hash Function

In this paper, we will discuss Birthday Attack and analyze how the proposed hash function is secure against the attack.

5.1 Birthday Attack

Birthday attacks are a class of cryptographic attack that exploits the mathematics behind the birthday phenomenon in probability theory.

Let $f : P \rightarrow Q$ be a hash function. According to birthday attack, a collision is expected within $r^{1/2}$ number of experiments, here the size of the range of f is denoted by r. Some random points $z^1, ..., z^k$ are picked from P and $y^i = f(z^i)$ is computed for $i = 1, ..., k$. The attack is considered successful if there exists a collision, i.e., $y^i = y^j$ for $z^i \neq z^j$ and k is the total number of experiments.

Experimental Results of Birthday Attack on 2D CA-based Hash Function
Let $\{0, 1\}^n \rightarrow \{0, 1\}^{512}$ defines the domain and range of 2D CA-based hash function, where n > 1024 and the required number of trials expected to meet a collision is approximately 2^{256}.

A birthday attack has been performed on the proposed hash function. For the ease of computation, first 64 bits of the hash digest have been considered as the output digest of the hash function for the experiment. No collision is found over 2^{32} number of random sample messages. The original length of the digest is 512 which is long enough for brute-force attack with complexity 2^{512}. The complexity of birthday attack for this hash function is 2^{256}, which is computationally infeasible.

We have analyzed the random sample messages, which are considered for birthday attack and the respective digests. The data are stored in a lookup table. The correlation is tried to be found between the messages to make sure that they are chosen randomly. The Hamming distances between two messages are considered as the correlation parameter. The Hamming distance is computed between every pair of sample messages. It is plotted in Fig. 7 for 10 random samples. The sample messages are plotted along X and Z axis. The corresponding Hamming distances are plotted along Y axis. It shows that the Hamming distances between the sample messages vary between 150 and 1000. So it can be said that the sample messages are random. The same experiment has been carried out for sample digests. Hamming distances between the sample digests are varying between 250 and 450. This randomness property is required for any cryptographic hash function. So it can be said that the proposed design is secured against Birthday attack.

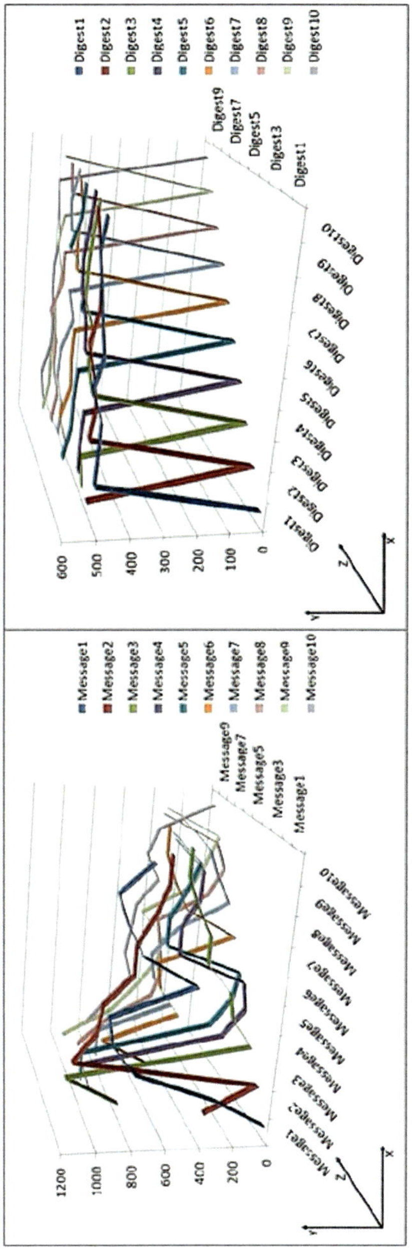

Fig. 7 Correlation results of birthday attack on 10 random sample messages

6 Conclusion

The full implementation of the proposed hash function is done using a five- neighborhood nonlinear hybrid 2D CA. The number of rounds of the primitive functions is decided based on the statistical analysis of the hash output. Security analysis of the proposed hash function has been done. Birthday Attack is carried out on the 2D CA-based hash function. The birthday attack is based on finding collisions, where collision resistance is the basic criterion for any hash function. The statistical analysis of the hash digests is done to show the strength of the hash function against this attack.

References

1. Bertoni, G. et al.: Keccak sponge function family main document. Submission to NIST (Round 2) 3: 30 (2009)
2. Ghosh, S., Sengupta, A., Saha, D., Chowdhury, D.R.: A scalable method for constructing non-linear cellular automata with period 2^n-1. In: Was, J., Sirakoulis, G.C., Bandini, S. (eds.) Cellular Automata. ACR. Lecture Notes in Computer Science, vol. 8751. Springer, Cham (2014)
3. Daemen, J., Rijmen, V.: The Design of Rijndael. Springer (2002)
4. Hell, M., Johansson, T., Meier, W.: Grain: a stream cipher for constrained environments. Int. J. Wire. Mob. Comput. **2**, 86–93 (2007)
5. Bertoni, G., Daemen, J., Peeters, M., Assche, G.V.: The keccak sha-3 submission. Submission to NIST (Round 3) (2011)
6. Tomassini, M., Sipper, M., Perrenoud, M.: On the generation of high-quality random numbers by two-dimensional cellular automata. IEEE Trans. Comput. **49**, 1146–1151 (2000). EPFL-ARTICLE-28657
7. Damgärd, I.B.: A Design Principle for Hash Functions. In: Brassard, G. (ed.) CRYPTO 1989. LNCS, vol. 435, pp. 416–427. Springer, Heidelberg (1990)
8. Merkle, R.C.: One way hash functions and DES. In: Brassard, G. (ed.) CRYPTO 1989. LNCS, vol. 435, pp. 428–446. Springer, Heidelberg (1990)
9. Chabaud, F., Joux, A.: Differential collisions in SHA-0. In: Krawczyk, H. (ed.) CRYPTO 1998. LNCS, vol. 1462, pp. 56–71. Springer, Heidelberg (1998)
10. Kelsey, J., Kohno, T.: Herding hash functions and the nostradamus attack. In: Vaudenay, S. (ed.) EUROCRYPT 2006. LNCS, vol. 4004, pp. 183–200. Springer, Heidelberg (2006)
11. Jarkko KARI Mathematics Department, University of Turku, Finland ,Physica D: Nonlinear Phenomena Volume 45, Issues 1-3, 2 September 1990, Pages 379-385, (1990) , North-Holland
12. Daemen, J., Govaerts, R., Vandewalle, J.: A framework for the design of one-way hash functions including cryptanalysis of damgärd's one-way function based on a cellular automaton. In: Matsumoto, T., Imai, H., Rivest, R.L. (eds.) ASIACRYPT 1991. LNCS, vol. 739, pp. 82–96. Springer, Heidelberg (1993)
13. Mihaljevi'c, M.J., Zheng, Y., Imai, H.: A cellular automaton based fast one-way hash function suitable for hardware implementation. In: Imai, H., Zheng, Y. (eds.) PKC 1998. LNCS, vol. 1431, pp. 217–233. Springer, Heidelberg (1998)
14. Cattell, K., Muzio, J.C.: Synthesis of one-dimensional linear hybrid cellular automata. IEEE Trans. Comput.-Aid. Des. Integr. Circuits Syst. **15**, 325–335 (1996)
15. Meier, W., Staffelbach, O.: Analysis of pseudo random sequences generated by cellular automata. In: Davies, D.W. (ed.) EUROCRYPT 1991. LNCS, vol. 547, pp. 186–199. Springer, Heidelberg (1991)

16. Coron, J.-S., Dodis, Y., Malinaud, C., Puniya, P.: Merkle-damgärd revisited: How to construct a hash function. In: Shoup, V. (ed.) CRYPTO 2005. LNCS, vol. 3621, pp. 430–448. Springer, Heidelberg (2005)

Anonymous Fine-Grained User Access Control Scheme for Internet of Things Architecture

Soumya Banerjee, Vanga Odelu, Ashok Kumar Das, Samiran Chattopadhyay, and Debasis Giri

Abstract With the rapid growth of wireless technology, Internet of Things (IoT) became very popular in both industrial as well as consumer product domains. While there is a lot of available platforms and technologies for IoT, the access control issue is often overlooked in the IoT security research. An effective access control depends on the proper user authentication mechanism. Thus, access control in this scenario is an emerging and challenging problem in the IoT environment. In this paper, we design an anonymous fine-grained user access control mechanism for IoT architecture. In the proposed scheme, the user authentication is performed by the smart device node based on the user attributes, which enables fine-grained access control over the authorized data. We utilize the widely accepted formal verification tool, called the Automated Validation of Internet Security Protocols and Applications (AVISPA), to formally prove the security of the proposed scheme. Additionally, we also provide a detailed informal security analysis of the scheme. Finally, we perform a

S. Banerjee (✉) · S. Chattopadhyay
Department of Information Technology, Jadavpur University, Salt Lake City
700 098, Kolkata, India
e-mail: soumyaBanerjee@outlook.in

S. Chattopadhyay
e-mail: samirancju@gmail.com

V. Odelu (✉)
Department of Computer Science and Information Systems, Birla Institute of Technology and Science, Pilani Hyderabad Campus, Hyderabad 500 078, India
e-mail: odelu.vanga@hyderabad.bits-pilani.ac.in

A. K. Das
Center for Security, Theory and Algorithmic Research, International Institute of Information Technology, Hyderabad 500 032, India
e-mail: iitkgp.akdas@gmail.com; ashok.das@iiit.ac.in

D. Giri
Department of Information Technology, Maulana Abul Kalam Azad University of Technology, Kolkata 700064, India
e-mail: debasis_giri@hotmail.com

D. Giri et al. (eds.), *Proceedings of the Fifth International Conference on Mathematics and Computing*, Advances in Intelligent Systems and Computing 1170, https://doi.org/10.1007/978-981-15-5411-7_4

simulation study using the broadly used NS3 network simulator to show the practical impact on the proposed scheme on various network parameters.

1 Introduction

Internet of Things (IoT) refers to the heterogeneous network comprising of the numerous devices that directly or indirectly enable exchange data over the Internet. These devices include a wide range of spectrum, such as Radio-Frequency Identification (RFID) tags to the ever common smartphones to the Internet-enabled consumer appliances. By the year 2020, it is estimated that the number of IoT devices will approach to fifty billion [14]. The promise of ubiquitous connectivity has powered an unprecedented growth in the adoption of IoT devices, which in turn has exposed the associated security and privacy challenges [3, 6, 12]. A generic IoT architecture shown in Fig. 1 dictates that the data is transmitted through heterogeneous networks. To ensure the integrity of the sensitive and private data transmission over potentially insecure networks, it is essential to adapt security protocols in the IoT environment. To deal with these, encryption and access control mechanisms need to be integrated into IoT protocols [25]. Furthermore, there are some situations where not all users should have access to all resources. For example, in a smart medicare system, a user being a doctor should not be able to access the database of financial information, medical histories or any diagnostic devices. Looking up access privileged every time, user access is cumbersome and inefficient. Thus, fine-grained access control (FGAC) mechanism needs to be integrated into the IoT environment to secure private and confidential data.

In 2008, Jeong et al. [17] proposed an One-Time Password (OTP) based user authentication scheme for smart homes. However, their scheme fails to guarantee mutual authentication between the involved participants. Their scheme also lacks anonymity and untractability properties as user identities are transmitted in plaintext. Hunumanathappa et al. [16] presented a three-way user authentication scheme for ubiquitous computing devices. Their scheme supports device attestation through encrypted pass-phrases. More recently, Zhou et al. [29] presented a lightweight anonymous user authentication protocol for smart home environment using only one-way hash and bitwise XOR operations. Challa et al. [7] also presented a user authentication scheme for IoT applications using Elliptic Curve Cryptography (ECC) signatures.

In 2001, utilizing the concept introduced in [13], Yu et al. [28] presented the first FGAC scheme for distributed Wireless Sensor networks (WSNs). More recently, Chatterjee et al. [9] presented an FGAC scheme in WSNs, where only the users possessing the requisite set of attributes can decrypt the encrypted information from the cluster heads. Lounis et al. [21] presented an architecture for secure dissemination of data generated by medical sensor networks. However, FGAC mechanism remains unexplored in the IoT environment. In this paper, we present a scheme that uses Key-Policy Attribute-Based Encryption (KP-ABE) with AND access structure to achieve

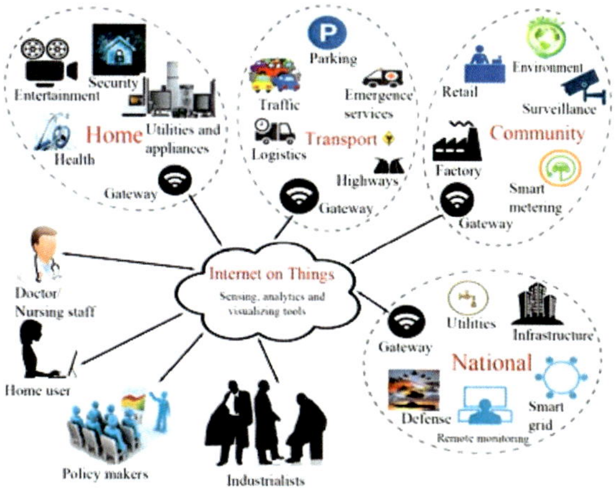

Fig. 1 Authentication model for IoT applications (reproduced from [7])

FGAC in IoT architecture. Under KP-ABE, the senders can encrypt messages under a set of attributes defined by access structures, only users that satisfy the access structures can decrypt the resultant ciphertext.

The rest of the paper is organized as follows. In Sect. 2, we discuss the system models as well as the contributions made in the paper. In Sect. 3, we discuss various phases related to the proposed scheme. In Sect. 4, we discuss the supported dynamic properties related to the proposed scheme. Next, we provide the security analysis in Sect. 5. Sections 5.2 and 6 comprise the results of AVISPA and NS3 simulations of the proposed scheme, respectively. Discussion of the performance with related schemes is provided in Sect. 7. Finally, Sect. 8 concludes the work with some future research scopes.

2 System Models and Research Contributions

In this section, we describe the systems models including the network and threat models, and the research contributions.

2.1 Network Model

In this paper, we utilize the generalized network model reported in [3] and described in Fig. 1. IoT architecture is composed of different subsystems. The subsystems, which are comprised of many sensors and actuators, are connected to the public Internet through a dedicated gateway node(s). Collectively, the sensors and actuators can be grouped together as smart devices. In order to gain access to a smart device, a user must register with the associated gateway node [14]. Then, only registered legal users can enjoy services from the smart devices provided that the mutual authentication between them is achieved.

2.2 Threat Model

We adopted the de facto standard Dolev-Yao (DY) threat model [11] for the proposed scheme. According to this model, an adversary, say $\mathcal{A}$, has absolute control over the communication channel, and consequently he/she can eavesdrop, modify, and delete any transmission over the public channels. Furthermore, it is assumed that $\mathcal{A}$ has the capability to capture and extract sensitive information from a user's lost or stolen smart card through the power analysis attacks [18, 22]. We assume that the gateway nodes are trusted and will not be compromised by $\mathcal{A}$; otherwise, the whole network will be compromised. As pointed out [4, 27], we can avert the scenarios of physical capture of the singular gateway node ($GWNs$) by physical securing. However, it is assumed that some of the smart devices can be compromised through physical capture by $\mathcal{A}$, leading to $\mathcal{A}$ learning all the credentials stored in those captured IoT devices.

2.3 Research Contributions

The research contributions are listed below:

- We propose an anonymous user authentication scheme for IoT architecture. The proposed scheme enables fine-grained access control over the authorized data.
- We provide formal security verification of the proposed scheme for using the broadly used AVISPA tool. The simulation results for the OFMC and CL-AtSe backends ensure that the scheme is safe against man-in-the-middle and replay attacks. In addition, through informal security analysis, we show that the proposed scheme has potential to resist several other known attacks such as impersonation, stolen smart card, offline guessing, privileged-insider, ephemeral secret leakage and stolen verifier attacks. Finally, we show that the proposed scheme preserves anonymity and untraceability properties.
- We also simulate the proposed scheme using the widely accepted NS3 simulator in order to measure the scheme's impact on network performance like throughput and end-to-end delay.

Table 1 Notations used in this paper

Symbol	Description
U, GWN, Dev	Mobile user, gateway node, and smart device, respectively
ID_X, EID_x	Identity or Encrypted identity of an entity X
q	A large prime number
G	Generator point for elliptic curve group over finite field Z_q
$k \cdot P$	Scalar multiplication of an elliptic curve point P, $k \cdot P = P + P + \cdots (k\, times)$
(k_x, Q_x)	Private and public key of an entity X , respectively, where $Q_x = k_x \cdot G$
LTK_X	Long-term private key of an entity X
$\hat{e}(\cdot, \cdot)$	A non-singular bilinear pairing such that $\hat{e}(aG, bG) = \hat{e}(G, G)^{ab}$ with $a, b \in Z_q^*$
Q_{sys}	System public key, $Q_{sys} = \hat{e}(G, G)^{LTK_{sys}}$
T_i, S_i, C_i	Public key, user component key, and device dynamic key for attribute i, respectively
$p(\cdot)$	A randomly chosen polynomial with $p(0) = LTK_{GWN}$
$Gen(\cdot), Rep(\cdot)$	Fuzzy biometric generation and reproduction functions, respectively
BIO_u	Biometrics input for a mobile user U
σ_u	Derived biometric secret from BIO_u
τ_u	Public reproduction parameter associated with BIO_u
et	Tolerance threshold for $Rep(\cdot)$
$h(\cdot)$	A collision-resistant cryptographic one-way hash function
$E[\cdot]_k/D[\cdot]_k$	Symmetric-key encryption/decryption under the pre-shared key k
$\|\|, \oplus$	String concatenation and bitwise exclusive (XOR) operations, respectively

3 The Proposed Scheme

The proposed scheme consists of four phases, namely, (1) setup, (2) device enrollment, (3) user registration, and (4) login and user authentication. The proposed scheme also efficiently supports the required dynamic properties which are discussed in Sect. 4. The notations used in this paper are tabulated in Table 1.

The different entities involved in the proposed scheme are the users (U), the gateway nodes (GWN), and the smart devices (Dev). It is worth noting that the GWN acts as the attribute authority, which a trusted entity in the IoT environment. Figure 2 gives a high-level overview of the proposed scheme. The GWN executes the system setup and then defines the attribute public keys. During smart device enrollment, the GWN saves the attribute public keys into Dev. When a user U registers in the system, the user component keys for each authorized attribute, are also generated by the GWN and then are stored into the user's smart card, say SC_u.

During authentication and session key agreement, U logins in the system and then sends authentication request message M_1 directly to the smart device Dev. On receiving a valid authentication request, Dev calculates a session key as well as

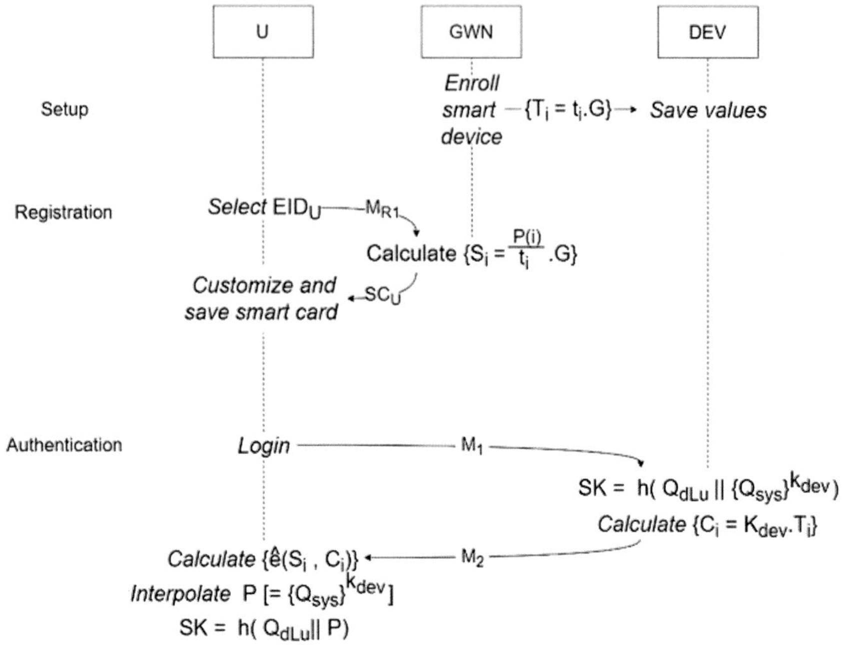

Fig. 2 High-level overview of the proposed scheme

the device dynamic keys for all the attributes necessary to access them. The device dynamic keys are then sent to the user U. If U is authorized for all the necessary attributes to access Dev, he/she can combine the device dynamic keys with the user component keys in order to calculate the same session key. As only a legitimate user with necessary attributes can recreate the session key, he/she is implicitly authenticated by the smart devices. In the following, the detailed mechanism of the proposed scheme is provided.

3.1 Setup Phase

In this phase, the GWN initializes its parameters and also generates the attribute public keys.

3.1.1 Initial Setup

The GWN begins the setup phase by selecting a bilinear group pairing defined by $\hat{e} : G_1 \times G_1 \rightarrow G_2$, where G_1 is an additive cyclic group on elliptic curve and G_2 is a multiplicative cyclic group over finite field with both of prime order q. Let

G be a generator for G_1 and $a, b \in \mathbb{Z}_q^*$. Then, there exists an efficient mapping $\hat{e}(\cdot, \cdot)$ such that $\hat{e}(aG, bG) = \hat{e}(G, G)^{ab}$ and $\hat{e}(G, G) \neq 1$, where 1 is identity in G_2. The GWN also defines a cryptographically secure random number generator $RNG_{seed}(\cdot)$. In addition, the GWN select a private key LTK_{GWN} and calculates its public key $Q_{GWN} = LTK_{GWN}.G$, and declares Q_{GWN} as the pubic key. The system public key is also computed as $Q_{sys} = \hat{e}(G, G)^{LTK_{GWN}}$.

3.1.2 Attribute Public Key Generation

The attribute authority (here, the GWN) executes the following steps. For each attribute $i \in [1, n]$ supported by it, the GWN selects $t_i \in \mathbb{Z}_q$ and then computes $T_i = t_i.G$. After that, the GWN saves $\langle \{t_i\}_{i \in [1,n]} \rangle$ in its database, where n is the number of attributes supported by the authority.

3.2 IoT Device Enrollment Phase

The IoT smart devices can be enrolled into the system dynamically at any time after the setup phase. The steps required to enroll a smart device, Dev under the proposed scheme are described below.

Step 1. The GWN selects an identity ID_{dev} for device Dev, generates a random number r_{dev}, and calculates $LTK_{dev} = h(Q_{sys} \oplus h(ID_{dev} \| r_{dev}))$. ID_{dev} and its long-term key LTK_{dev} are then loaded into the smart device Dev prior to its deployment.

Step 2. The GWN lists ID_{Dev} among its list of available devices and the attribute public keys as discussed in Sect. 3.1.2.

3.3 User Registration Phase

A user U needs to be registered with the GWN in order to access the services from a designated smart device, say Dev. The following steps are essential to register for a user U at the GWN:

Step 1. U chooses identity ID_u, generates a random number r_u, and calculates $EID_u = h(ID_u \parallel r_u)$, and then securely sends it to the GNW as the registration request message M_{R1}.

Step 2. The GWN (attribute authority) randomly selects a degree d polynomial $p(\cdot)$ such that $p(0) = LTK_{GWN}$ and calculates $S_i = (p(i)/t_i)G$ for each eligible attribute i for the user U. The GWN securely issues the smart card SC_u containing $\{EID_u, \{S_i\}_{i \in [1,n]}, DeviceList\}$, where $DeviceList$ lists the identities of the smart devices that U is authorized to access those devices.

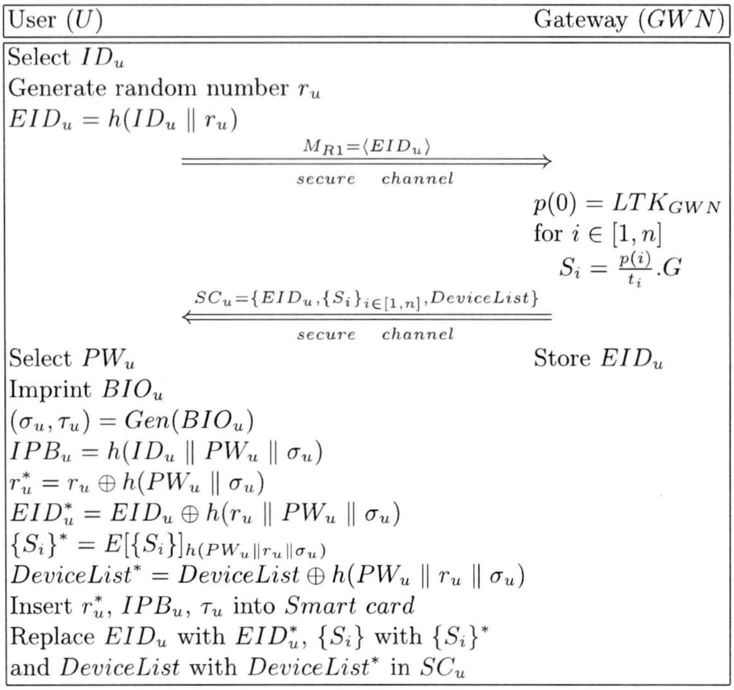

Fig. 3 Summary of user registration phase

Step 3. U on receiving SC_u chooses a password PW_u and imprints his/her biometric BIO_u at a particular terminal. SC_u then calculates σ_u and τ_u using the fuzzy generator function $Gen(\cdot)$ [10] and the identity verification token $IPB_u = h(ID_u \| PW_u \| \sigma_u)$. SC_u also calculates $r_u^* = r_u \oplus h(PW_u \parallel \sigma_u)$ and saves r_u^*, IPB_u, and τ_u. U also saves a counter $count$ with $count = 0$ into SC_u. Here, we use the fuzzy extractor method for user biometric verification [10]. A fuzzy extractor contains the following two procedures:

Gen: It is a probabilistic generation method which takes user personal biometrics and outputs a pair (σ_u, τ_u), where σ_u is known as biometric secret key and τ_u is the public reproduction parameters associated with biometrics BIO_u, that is, $(\sigma_u, \tau_u) = Gen(BIO_u)$.

Rep: It is a deterministic reproduction method which a noisy biometrics, say BIO_u' and τ_i and constructs the original biometric secret key σ_i, that is, $\sigma_i = Rep(BIO_u', \tau_u)$ as long as the Hamming distance between Bio_u and Bio_u' is less than or equal to an error tolerance threshold value et.

Step 4. SC_u finally replaces EID_u, $\{S_i\}$ and $DeviceList$ with the calculated $EID_u^* = EID_u \oplus h(r_u \parallel PW_u \parallel \sigma_u)$, $\{S_i\}^* = E[\{S_i\}]_{h(PW_u \| r_u \| \sigma_u)}$ and $DeviceList^* = DeviceList \oplus h(PW_u \parallel r_u \parallel \sigma_u)$ in its memory.

Figure 3 summarizes the overall procedure of our user registration phase.

User (U)	IoT smart device (Dev)
$\langle\{ EID_u^*, \{S_i\}^*, DeviceList^*, r_u^*, IPB_u, \tau_u\}_{SC_u}, count\rangle$	$\langle\{C_i\}, ID_{dev}, LTK_{dev}\rangle$
Enter ID_u, PW_u	
Imprint Bio_u	
Compute $\sigma_u = Rep(BIO_u, \tau_u)$	
$IPB'_u = h(ID_u \parallel PW_u \parallel \sigma_u)$	
If $(IPB_u \neq IPB'_u)$, terminate	
Compute $r_u = r^* \oplus h(PW_u \parallel \sigma_u)$,	
$EID_u = EID_u^* \oplus h(r_u \parallel PW_u \parallel \sigma_u)$,	
$\{S_i\} = D[\{S_i\}^*]_{h(PW_u \parallel r_u \parallel \sigma_u)}$,	
$DeviceList = DeviceList^* \oplus h(PW_u \parallel r_u \parallel \sigma_u)$	
Choose $k_u \in \mathbb{Z}_q$	Set $TS_{dev} = TS_{current}$
Compute $Q_u = k_u.G$,	If $TS_{dev} - TS_u > \Delta T$, terminate
$Q_{dLu} = k_u.Q_{LTK_{dev}}$,	Compute $Q_{dLu} = LTK_{dev}.Q_u$,
$DID_u = RNG_{EID_u}(count)$	$(DID_u \parallel IDTS) = id\_token \oplus h(TS_u \parallel Q_{dLu})$
Set $TS_u = TS_{current}$	If $(h(ID_{dev} \parallel TS_u) \neq IDTS)$, terminate
Compute $IDTS = h(ID_{dev} \parallel TS_u)$,	Choose $k_{dev} \in \mathbb{Z}_q$
$id\_token = (DID_u \parallel IDTS) \oplus h(TS_u \parallel Q_{dLu})$	Compute $SK = h(Q_{dLu} \parallel \{Q_{sys}\}^{k_{dev}})$,
$\xrightarrow{M_1 = \langle id\_token, Q_u, TS_u\rangle}$	$cert = h(SK \parallel TS_{dev})$
Check freshness of TS_{dev}	
For each attribute $i \in A^{dev} \bigcap A^u$, compute $\hat{e}(S_i, C_i)$	For each attribute $i \in A^{dev}$ compute $C_i = k_{dev}.T_i$
Interpolate $P\ (= \hat{e}(g, g)^{k_{dev}p(0)})$	Logs DID_u into $access\_table$
Compute $SK = h(Q_{dLu} \parallel P)$	$\xleftarrow{M_2 = \langle\{C_i\}, cert, TS_{dev}\rangle}$
If $h(SK \parallel TS_{dev}) \neq cert)$ $terminate$	
Set $count = count + 1$	
Store session key SK shared with Dev.	Store session key SK shared with U.

Fig. 4 Summary of login and user authentication phase

3.4 Login and User Authentication Phase

A registered user U can login and authenticate himself/herself and then securely negotiate a session key with an accessed IoT smart device Dev provided that he/she posses the correct attributes. The following steps need to be executed for this purpose. This phase is also briefed in Fig. 4.

Step 1. U first provides his/her identity ID_u and password PW_u, and then imprints biometric BIO_u. The smart card SC_u calculates $\sigma_u = Rep(BIO_u, \tau_u)$ and $IPB'_u = h(ID_u \parallel PW_u \parallel \sigma_u)$. Only if $IPB'_u = IPB_u$, the login is successful and entered credentials are authenticated. SC_u then calculates $r_u = r_u^* \oplus h(PW_u \parallel \sigma_u)$ and recovers EID_u, $\{S_i\}$ and $DeviceList$ as $EID_u = EID_u^* \oplus h(r_u \parallel PW_u \parallel \sigma_u)$, $\{S_i\} = D[\{S_i\}^*]_{h(PW_u \parallel r_u \parallel \sigma_u)}$ and $DeviceList = DeviceList^* \oplus h(PW_u \parallel r_u \parallel \sigma_u)$. Next, U selects the public key $Q_{LTK_{dev}}$ of accessed smart device Dev from $DeviceList$ and a short term random key k_u, and calculates $Q_u = k_u.G$, $Q_{dLu} = k_u.(Q_{LTK_{dev}}$, the dynamic identity $DID_u = RNG_{EID_u}(count)$ and $id\_token = (DID_u \parallel IDTS) \oplus h(TS_u \parallel Q_{dLu})$, where $IDTS = h(ID_{dev} \parallel TS_u)$ and TS_u is the current time stamp. Finally, the message $M_1 = \langle id\_token, Q_u, TS_u \rangle$ is sent to the smart device Dev by SC_u via open channel.

Step 2. The smart device Dev on receiving the message M_1, verifies its freshness by the condition $TS_{dev} - TS_u \leq \Delta T$, where ΔT is the maximum transmission delay and TS_{dev} the current timestamp of Dev. If it satisfies, Dev calculates $Q_{dLu} = LTK_{dev}.Q_u$, and retrieves DID_u and TSQ_u from $id\_token \oplus h(TS_u \parallel Q_{dLu})$. Only if $h(ID_{dev} \parallel TS_u) = IDTS$, Dev selects a short term random key k_{dev} and calculates the session key $SK = h(Q_{dLu} \parallel Q_{sys}^{k_{dev}})$ shared with the user U, $cert = h(SK \parallel TS_{dev})$ and $\{C_i = k_{dev}.T_i\}_{i \in A^{dev}}$, where A^{dev} denotes the attribute set associated with Dev. Dev then logs DID_u into its $access\_table$ and finally, it sends the message $M_2 = \langle \{C_i\}, cert, TS_{dev} \rangle$ to the user U through open channel.

Step 3. U on receiving the message M_2, SC_u checks the freshness of the message by verifying the timeliness of TS_{dev}. After that for each attribute $i \in A_i^C \bigcap A_i^u$, SC_u calculates $\hat{e}(S_i, C_i)$ $\{= \hat{e}(g, g)^{k_{dev} p(i)}\}$ and then interpolates all the intermediate values to calculate P $\{= \hat{e}(g, g)^{k_{dev} p(0)} = \hat{e}(g, g)^{k_{dev}.LTK_{GWN}} = Q_{sys}^{k_{dev}}\}$. SC_u also computes the session key $SK = h(Q_{dLu} \parallel P)$ shared with Dev and checks if $h(SK \parallel TS_{dev}) = cert$. If it is so, U considers the session key SK as valid and updates its $count$ as $count = count + 1$.

4 Dynamic Properties

In this section, we discuss the following properties that the proposed scheme offers.

4.1 Password and Biometric Update Phase

For security reasons, it is a good practice to update password/biometrics by a registered user U. This phase works independent of the GWN and it requires the following steps:

Step 1. The user U first supplies his/her identity ID_u and password PW_u, and imprints biometrics BIO_u. The smart card SC_u then calculates $\sigma_u = Rep(BIO_u, \tau_u)$ and $IPB_u' = h(ID_u \parallel PW_u \parallel \sigma_u)$. If $IPB_u' = IPB_u$, the login is successful. SC_u then proceeds to calculate $r_u = r_u^* \oplus h(PW_u \parallel \sigma_u)$ and recovers the values EID_u, $\{S_{k,i}\}$ and $DeviceList$. SC_u asks U to supply new credentials.

Step 2. U selects a new password PW_u^{new} and also imprints new biometric Bio_u^{new}, if needed. In case, U does not want to update biometrics, he/she can keep the current biometrics BIO_u, which will be treated as Bio_u^{new}. SC_u then calculates σ_u^{new} and τ_u^{new} using the fuzzy generator function $Gen(\cdot)$, and the identity verification token IPB_u^{new} as $h(ID_u \parallel PW_u^{new} \parallel \sigma_u^{new})$. SC_u also calculates $r_u^{*new} = r_u \oplus h(PW_u \parallel \sigma_u)$,

$EID_u^{*new} = EID_u \oplus h(r_u \parallel PW_u^{new} \parallel \sigma_u^{new})$, $\{S_{k,i}\}^{*new} = E[\{S_{k,i}\}]_{h(PW_u^{new} \parallel r_u \parallel \sigma_u^{new})}$ and $DeviceList^{*new} = DeviceList \oplus h(PW_u^{new} \parallel r_u \parallel \sigma_u^{new})$.

Step 3. SC_u replaces EID_u^*, $\{S_{k,i}\}^*$, $DeviceList$, R_u^*, IBP_u and τ_u with EID_u^{*new}, $\{S_{k,i}\}^{*new}$, $DeviceList^{*new}$, r_u^{*new}, IPB_u^{new} and τ_u^{new}, respectively.

4.2 Maintenance Phase

Both user U and gateway node GWN need to periodically perform some operations in order to maintain security and accountability in the system. In this section, we provide the following relevant mechanisms for the proposed scheme.

4.2.1 Gateway Node

The GWN being the attribute authority needs to periodically update the attribute public keys as described in Sect. 3.1.2 controlled by it. This will ensure that the users must perform their maintenance tasks to continue the services in the IoT environment from the desired IoT smart devices.

4.2.2 Smart Device

As the smart devices are resource constrained, and also these lack the ability to interpret $access\_table$, the contents of $access\_table$ should be transferred to the gateway node GWN. The attribute set associated with the devices needs to be updated during this operation.

4.2.3 Mobile User

In order to continue the services, the users must periodically update their corresponding sets of attribute public keys. The following requisite steps are described below;

Step 1. A user U logs into the system as described in Sect. 3.4, and then sends the message $M_1 = \langle id\_token, Q_u, TS_u \rangle$ to the gateway node GWN. In this case, $id\_token = (EID_u \parallel count \parallel TS_u) \oplus k_u.Q_{GWN}$.

Step 2. The GWN on receiving M_1 verifies the freshness of timestamp TS_u, recovers EID_u from id_token, and runs the steps described in Sect. 3.3 in order to generate $\{S_i^{new}\}_{i \in [1,n]}$. It then saves the value of $count$ in the $user\_information$ table. The GWN also calculates $attribute\_token = E[\{S_i^{new}\}]_{LTK_{GWN} \cdot Q_u}$ and sends the message $M_2 = \langle attribute\_token \rangle$ to U through public channel.

Step 3. U on receiving M_2 decrypts it with the computed key $k_u.Q_{GWN}$, encrypts $\{S_i^{new}\}$ as D_u^{*new} and $\{S_i^{new}\}^*$, and saves the updated values into the smart card SC_u. SC_u also resets $count$ to zero in this case.

4.3 Smart Card Revocation Phase

In the proposed scheme, the user U is authenticated implicitly and he/she then reconstructs $Q_{sys}^{k_{dev}}$, which is encrypted under the set of attributes $\{A^{dev}\}$. From Sect. 4.2.1, it is known that the public keys of the attributes are changed periodically. In order to revoke U, the GWN can simply drop the relevant entry from $user\_information$ table. This will stop U from updating his/her smart card SC_u through the steps described in Sect. 4.2.3. Thus, U will be unable to reconstruct $Q_{sys}^{k_{dev}}$ in the future.

4.4 Dynamic IoT Device Addition Phase

New smart devices can be dynamically registered into the system at any time after the setup phase through the steps described in Sect. 3.2. After their deployment, a user can access the services from those smart devices provided that the GWN allows that user to access those devices.

5 Security Analysis

In this section, we perform security analysis of the proposed scheme. In Sect. 5.1, through informal security analysis, we show the proposed scheme's resistance against various known attacks. In addition, in Sect. 5.2, we simulate the proposed scheme under the widely accepted formal security verification tool AVISPA (Automated Validation of Internet Security Protocols and Applications)[26]. The results from AVISPA backends, namely OFMC and CL-AtSe, ensure that the proposed scheme is secure against replay and man-in-the-middle attacks.

5.1 Informal Security Analysis

This section demonstrates that the proposed scheme is secure against a passive/active adversary $\mathcal{A}$ under following well-known attacks.

5.1.1 Replay Attack

Assume that an adversary $\mathcal{A}$ captures and replays the transmitted message M_1 to the IoT smart device Dev. But, Dev checks the time stamp and rejects the message. In the unlikely scenario that the message is replayed within the time window ΔT, $\mathcal{A}$ will receive $M_2 = \{\{C_i\}, cert, TS_{dev}\}$, but the response to M_2 depends on the session-related random secret k_u and also the proper set of attributes. Thus, without the knowledge of user component keys, it is computationally hard to compute the session key SK by $\mathcal{A}$. Therefore, the proposed scheme successfully prevents the replay attack, and thus, it also prevents unauthorized access.

5.1.2 Impersonation Attack

Suppose $\mathcal{A}$ attempts to impersonate a user, a smart device or the GWN. From the above discussion, user impersonation is infeasible by the adversary $\mathcal{A}$ as he/she cannot construct the valid login request with the knowledge of the user component keys. In case of smart device, say Dev, computing Q_{dLu} is again computational hard by $\mathcal{A}$ without the knowledge of the device secret key LTK_{dev}. Therefore, $\mathcal{A}$ cannot compute the valid response message M_2 too. Since the GWN does not play any role in the session key establishment step, it cannot be impersonated. Hence, the proposed scheme is secure against impersonation attack.

5.1.3 Man-in-the-Middle Attack

Assume that $\mathcal{A}$ intercepts the message M_1 and tries to modify it. We have $M_1 = \{id\_token, Q_u, TS_i\}$, and Q_u and TS_u are encapsulated with id_token, which itself is secured using the secret K_u. $\mathcal{A}$ cannot modify M_1 even if he/she uses current timestamp. A similar attempt by $\mathcal{A}$ to intercept and modify the message M_2 will also fail without the proper knowledge of the device secret keys. Thus, the proposed scheme is secure against man-in-the-middle attack.

5.1.4 Smart Card Lost/Stolen and Offline Guessing Attacks

Assume that $\mathcal{A}$ recovers the information EID_u^*, $\{S_u\}^*$, $DeviceList^*$, r_u^*, IPB_u and τ_u from the lost or stolen smart card SC_u of a registered legal user U through power analysis attack as discussed in the threat model. Except IPu and τ_u, none of these credentials are stored in plaintext forms, and it requires user identity, password and biometric to decrypt and recover user component keys as well as the device list. IBP_u is the one-way hash of ID_u, PW_u and σ_u. For $\mathcal{A}$ to gain some insight through offline guessing attack, he/she needs to simultaneously guess three secret credentials ID_u, PW_u and σ_u (indirectly BIO_u), which is a computationally infeasible task. Thus, the proposed scheme is secure against these attacks.

5.1.5 Privileged-Insider Attack

In this attack, the adversary $\mathcal{A}$ becomes a privileged-insider user of the GWN. Assume $\mathcal{A}$ knows the information during registration phase, such as $EID_u = h(ID_u \parallel r_u)$ related to a registered legal user U. Now, we further assume that $\mathcal{A}$ has lost/stolen the smart card SC_u of U, and recoveres the stored values including EID_u^*, r_u^* and IPB_u using power analysis attacks, where $EID_u^* = EID_u \oplus h(r_u \parallel PW_u \parallel \sigma_u)$, $r_u^* = r_u \oplus h(PW_u \parallel \sigma_u)$ and $IPB_u = h(ID_u \parallel PW_u \parallel \sigma_u)$. However, it is computationally infeasible to guess any one of the secret values from the retrieved information as the proposed scheme protects lost smart card attack. Therefore, even the adversary is privileged insider of the GWN, he/she cannot gain any advantage with registration information to recover the user secret credentials, such as identity, password, and biometrics.

5.1.6 Ephemeral Secret Leakage (ESL) Attack and Perfect Forward Secrecy

In this attack, even if a smart device long-term secret key is revealed to the adversary $\mathcal{A}$, all the past sessions are secure. We assume that $\mathcal{A}$ somehow learns session key SK, where $SK = h(k_u.LTK_{dev}.G \parallel k_{dev}.Q_{sys})$. Due to hardness of inverting one-way cryptographic hash function and solving elliptic curve discrete logarithm problem (ECDLP), neither of k_u, k_{dev} or LTK_{dev} can be learned from the compromised session key SK. Furthermore, as the session key SK is composed of both long-term as well as session temporary secrets, all other past and future session keys still remain secure even with the exposure of the smart devices' long-term secret keys LTK_{dev}. In this way, the proposed scheme provides perfect forward secrecy and also protects ESL attack.

5.1.7 Stolen Verifier Attack

The gateway node GWN does not maintain any verification-specific information as it is not involved in the session key establishment procedure. Moreover, the smart device Dev does not maintain any verifier table as the user is granted access based on his/her attributes with the ability to reconstruct $Q_{sys}^{k_{dev}}$. Thus, the proposed scheme is secure against stolen verifier attack.

5.1.8 Resilience Against Smart Device Capture Attack

As discussed in the threat model, some IoT smart devices can be physically captured by an adversary $\mathcal{A}$. Suppose $\mathcal{A}$ physically captures a smart device Dev. Then, $\mathcal{A}$ can extract all the credentials, such as identity ID_{dev} and long-term key LTK_{dev} from Dev's memory, where $LTK_{dev} = h(Q_{sys} \oplus h(ID_{dev} \| r_{dev}))$. Due to the usage of random secret r_{dev}, the long-term secrets of all devices throughout the network are distinct. Thus, $\mathcal{A}$ can only compromise the session key $SK = h(Q_{dLu} \parallel \{Q_{sys}\}^{k_{dev}})$ between Dev and a user U where $Q_{dLu} = LTK_{dev}.Q_u$. However, all other session keys established between the same user U and other non-compromised smart devices are secure and distinct as the long-term secrets are distinct. This means that compromise of a smart device does not affect to compromise the secure communications among the user and other non-compromised smart devices. In this way, the proposed scheme is unconditionally secure against smart device capture attack.

5.1.9 Anonymity and Untractability

In the proposed scheme, we assume that the adversary $\mathcal{A}$ eavesdrops and monitors the messages $M_1 = \{id\_token, Q_u, TS_i\}$ and $M_2 = \{\{C_i\}, cert, TS_{dev}\}$. It is clear that none of the eavesdropped values contain any identity-related information of the user U or the smart device Dev except id_token. However, id_token is itself composed of dynamic identity DID_u which is also encapsulated with random secrets and current timestamps. Hence, the proposed scheme provides both user and smart device anonymity. In addition, all of the eavesdropped messages are either composed of random secrets or current timestamp, and consequently, these messages are always different across different sessions. As a result, $\mathcal{A}$ cannot also trace a user U.

5.2 *Formal Security Verification Using AVISPA*

AVISPA [26] is push-button tool for automated validation of security protocols. AVISPA verifies the resistance of a security protocol against replay and man-in-the-middle attacks. To verify a security protocol with AVISPA, it is realized using in HLPSL (High-Level Protocol Specification Language) [23], which is a role-oriented language. In AVISPA, there are two kinds of roles: (1) basic roles, which define the roles for the participants related to the protocol and (2) mandatory roles, which define the session and goal and environment. The HLPSL code is converted to the Intermediate Format (IF) using the HLPSL2IF translator [26]. The IF is then fed to one of the four available backends of AVISPA, which are OFMC, CL-AtSe, SATMC, and TA4SP. The details of AVISPA, HLPSL, and backends can be found in [26].

In this section, we have implemented the proposed scheme in HLPSL for the three basic roles $mobileuser$, $smartdevice$, and $gatewaynode$, which represent the roles for U, Dev, and GWN, respectively, during the registration and login and authentication phases. We have also implemented the compulsory roles for $session$, $goal$, and $enviroment$.

We have evaluated the proposed scheme under OFMC (On-the-fly Model checker) and CL-AtSe (Constraint-Logic-based Attack Searcher) backends [26] using the SPAN, the Security Protocol ANimator for AVISPA [1]. AVISPA implements the Dolev-Yao (DY) threat model [11] in its evaluation. The mentioned backends check

SUMMARY	SUMMARY
SAFE	SAFE
DETAILS	DETAILS
BOUNDED_NUMBER_OF_SESSIONS	BOUNDED_NUMBER_OF_SESSIONS
TYPED_MODEL	PROTOCOL
PROTOCOL	/home/soumya/span//testsuite/results/IOT.if
/home/soumya/span//testsuite/results/IOT.if	GOAL
GOAL	as specified
As specified	BACKEND
BACKEND	OFMC
CL-AtSe	STATISTICS
STATISTICS	TIME 828 ms
Analysed : 63 states	parseTime 0 ms
Reachable : 63 states	visitedNodes: 800 nodes
Translation: 0.06 seconds	depth: 6 plies
Computation: 0.00 seconds	

Fig. 5 Analysis of simulation results under OFMC and CL-AtSe backends

for replay and man-in-the middle attacks. The simulation results reported in Fig. 5 formally verifies that the proposed scheme is no vulnerable to replay or man-in-the middle attacks.

6 Practical Perspective: Simulation Study Through NS3

In this section, we evaluate the impact on network performances for the proposed scheme using the widely accepted network simulator tool, NS3 (3.28) on Ubuntu 16.04 LTS platform. The network performance of the proposed scheme has been quantized in terms of its network throughput (in bytes per second) and end-to-end delay (in seconds). In our simulation, each user attempts to establish the session keys with each smart device.

6.1 Simulation Parameters

We simulated several scenarios(described in Table 2). The smart devices are assumed to be randomly located between 20 and 100 m away from the access point which acts as the gateway node. The mobile users were permitted to move randomly within 150 m square centered over the access point. The users and the smart devices communicate over 2.4 Ghz Wi-Fi media. We have also simulated the network with different number of users and smart devices. For the simulation, it is assumed that each user has five attributes set. In Table 2, we listed all the simulation parameters needed for the

Table 2 NS3 simulation parameters

Parameter	Description	
Platform	NS3(3.28)/Ubuntu 16.04 LTS	
Network scenarios	No. of users	No. of devices
1	2	10
2	5	10
3	5	15
4	10	10
5	8	15
6	10	15
Mobility	random (0–3 m/s)	
Simulation time	1200 s	

proposed scheme. The remaining simulation parameters are set as per as guidelines of NS3.

6.2 *Simulation Results*

In this section, we present the simulation results and discuss their implications.

6.2.1 Impact on Network Throughput

Fig. 6a plots the graph of network throughput for all the scenarios. The different scenarios are plotted along the horizontal axis. The network throughput is calculated as $\frac{N_{pkt} * size}{T_{total}}$, where N_{pkt} denotes the total number of received packets, $size$ is the size of the received packets in bytes and T_{total} the total time taken (in seconds). From the figure, it is observed that the network throughput increases with the number of transmitted messages.

6.2.2 Impact on End-to-End Delay

Figure 6b plots the graph of end-to-end delay for all the scenarios. The different scenarios are plotted along the horizontal axis. The end-to-end delay is calculated as $\sum_{i=0}^{N_{pkt}} (T_{recv_i} - T_{send_i})/N_{pkt}$, where N_{pkt} is the total number of received packets, T_{send_i} and T_{recv_i} are the time when the i_{th} packet was transmitted and received, respectively. From this figure, it is also observed that the end-to-end delay increases with the increasing number of transmitted message. This is mainly due to the increased number of messages which further results in congestion.

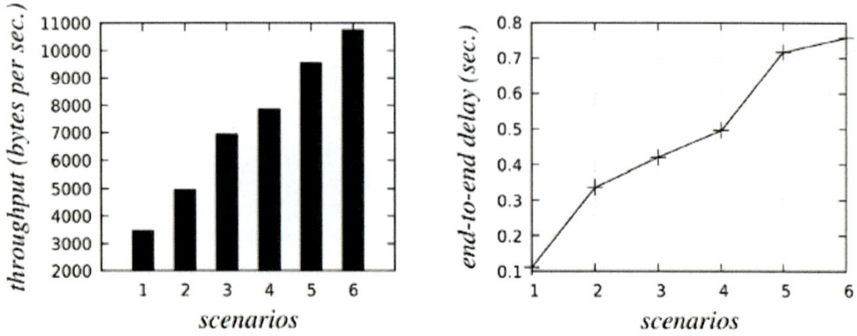

Fig. 6 **a** Throughput (bytes per second) **b** End-to-end delay (seconds)

7 Discussions and Future Work

In the proposed scheme, the storage space in the smart card as well as authentication message sizes are both linearly proportional to the number of attributes possessed by the user. Guo et al. [15] and Attrapudung et al. [2] presented ABE variants with constant size key size and ciphertext size, respectively. The proposed scheme incurs more computation overhead due to ABE. The concept of outsourced encryption/decryption [19, 20] might be employed to decrease the computation overhead. Additionally, the proposed scheme utilizes KP-ABE with AND access structure. Bethencourt et al. [5] presented Ciphertext-Policy Attribute-Based Encryption (CP-ABE). Ostrovsky et al. [24] introduced non-monotonic access structures for attribute-based encryption. For IoT architectures, a policy hidden CP-ABE with support for non-monotonic access structures could be more desirable. Chase et al. [8] presented an ABE scheme with multi-authority system. In a distributed architecture like IoT, a multi-authority ABE scheme may be advantageous. We intend to consolidate these features to present a comprehensive ABE solution for IoT as a future research scope.

8 Conclusion

In this paper, we proposed a novel attribute-based fine-grained device access control scheme suitable for the IoT environment. The proposed scheme efficiently supports dynamic properties, such as password and biometric update, smart card revocation when it is unexpectedly lost or stolen, and dynamic IoT device addition. Through the security analysis, we showed that our scheme is secure against various well-known attacks. In addition, we simulated the proposed scheme for formal security verification using AVISPA tool, and the simulation results assure that the proposed scheme is also secure. Moreover, we simulated the proposed scheme for practical impact study using NS3, and the simulation results show that the scheme is practical and is also suitable for IoT environment.

References

1. SPAN, the Security Protocol ANimator for AVISPA. http://www.avispa-project.org. Accessed on September 2018
2. Attrapadung, N., Herranz, J., Laguillaumie, F., Libert, B., De Panafieu, E., Ràfols, C.: Attribute-based encryption schemes with constant-size ciphertexts. Theor. Comput. Sci. **422**, 15–38 (2012)
3. Atzori, L., Iera, A., Morabito, G.: The internet of things: a survey. Comput. Netw. **54**(15), 2787–2805 (2010)
4. Bertino, E., Shang, N., Jr., S.S.W.: An efficient time-bound hierarchical key management scheme for secure broadcasting. IEEE Trans. Dependable Sec. Comput. **5**(2), 65–70 (2008)
5. Bethencourt, J., Sahai, A., Waters, B.: Ciphertext-policy attribute-based encryption. In: IEEE Symposium on Security and Privacy (S&P'07), pp. 321–334 (2007)
6. Carlin, A., Hammoudeh, M., Aldabbas, O.: Intrusion detection and countermeasure of virtual cloud systems-state of the art and current challenges. Int. J. Adv. Comput. Sci. Appl. **6**(6) (2015)
7. Challa, S., Wazid, M., Das, A.K., Kumar, N., Reddy, A.G., Yoon, E.J., Yoo, K.Y.: Secure signature-based authenticated key establishment scheme for future IoT applications. IEEE Access **5**, 3028–3043 (2017)
8. Chase, M., Chow, S.S.: Improving privacy and security in multi-authority attribute-based encryption. In: 16th ACM Conference on Computer and Communications Security (CCS'09), pp. 121–130. Chicago, Illinois, USA (2009)
9. Chatterjee, S., Das, A.K.: An effective ecc-based user access control scheme with attribute-based encryption for wireless sensor networks. Sec. Commun. Netw. **8**(9), 1752–1771 (2015)
10. Dodis, Y., Reyzin, L., Smith, A.: Fuzzy extractors: how to generate strong keys from biometrics and other noisy data. In: Annual International Conference on the Theory and Applications of Cryptographic Techniques- Advances in Cryptology (Eurocrypt'04). LNCS, vol. 3027, pp. 523–540. Springer, Interlaken, Switzerland (2004)
11. Dolev, D., Yao, A.: On the security of public key protocols. IEEE Trans. Inf. Theory **29**(2), 198–208 (1983)
12. Ghafir, I., Prenosil, V., Alhejailan, A., Hammoudeh, M.: Social engineering attack strategies and defence approaches. In: IEEE 4th International Conference on Future Internet of Things and Cloud (FiCloud'16), pp. 145–149 (2016)
13. Goyal, V., Pandey, O., Sahai, A., Waters, B.: Attribute-based encryption for fine-grained access control of encrypted data. In: Proceedings of the 13th ACM conference on Computer and Communications Security (CCS'06), pp. 89–98. Alexandria, VA, USA (2006)
14. Gubbi, J., Buyya, R., Marusic, S., Palaniswami, M.: Internet of Things (IoT): a vision, architectural elements, and future directions. Future Gen. Comput. Syst. **29**(7), 1645–1660 (2013)
15. Guo, F., Mu, Y., Susilo, W., Wong, D.S., Varadharajan, V.: CP-ABE with constant-size keys for lightweight devices. IEEE Trans. Inf. Forensics Sec. **9**(5), 763–771 (2014)
16. Hanumanthappa, P., Singh, S.: Privacy preserving and ownership authentication in ubiquitous computing devices using secure three way authentication. In: International Conference on Innovations in Information Technology (IIT'12), pp. 107–112. IEEE (2012)
17. Jeong, J., Chung, M.Y., Choo, H.: Integrated otp-based user authentication scheme using smart cards in home networks. In: 41st Annual Hawaii International Conference on System Sciences, pp. 294–294. IEEE (2008)
18. Kocher, P., Jaffe, J., Jun, B.: Differential power analysis. In: Annual International Cryptology Conference, pp. 388–397. Springer (1999)
19. Lai, J., Deng, R.H., Guan, C., Weng, J.: Attribute-based encryption with verifiable outsourced decryption. IEEE Trans. Inf. Forensics Sec. **8**(8), 1343–1354 (2013)
20. Li, J., Huang, X., Li, J., Chen, X., Xiang, Y.: Securely outsourcing attribute-based encryption with checkability. IEEE Trans. Parallel Distrib. Syst. **25**(8), 2201–2210 (2014)

21. Lounis, A., Hadjidj, A., Bouabdallah, A., Challal, Y.: Healing on the cloud: secure cloud architecture for medical wireless sensor networks. Future Gen. Comput. Syst. **55**, 266–277 (2016)
22. Messerges, T.S., Dabbish, E.A., Sloan, R.H.: Examining smart-card security under the threat of power analysis attacks. IEEE Trans. Comput. **51**(5), 541–552 (2002)
23. von Oheimb, D.: The high-level protocol specification language hlpsl developed in the eu project avispa. In: Proceedings of 3rd APPSEM II (Applied Semantics II) Workshop (APPSEM'05), pp. 1–17. Frauenchiemsee, Germany (2005)
24. Ostrovsky, R., Sahai, A., Waters, B.: Attribute-based encryption with non-monotonic access structures. In: Proceedings of the 14th ACM conference on Computer and Communications Security (CCS'07), pp. 195–203. Alexandria, VA, USA (2007)
25. Roman, R., Najera, P., Lopez, J.: Securing the Internet of Things. Computer **44**(9), 51–58 (2011)
26. Vigano, L.: Automated security protocol analysis with the AVISPA tool. Electron. Notes Theor. Comput. Sci. **155**, 61–86 (2006)
27. Wazid, M., Das, A.K., Odelu, V., Kumar, N., Susilo, W.: Secure Remote User Authenticated Key Establishment Protocol for Smart Home Environment. IEEE Trans. Dependable Sec. Comput. (2018)
28. Yu, S., Ren, K., Lou, W.: Fdac: toward fine-grained distributed data access control in wireless sensor networks. IEEE Trans. Parallel Distrib. Syst. **22**(4), 673–686 (2011)
29. Zhou, L., Li, X., Yeh, K.H., Su, C., Chiu, W.: Lightweight iot-based authentication scheme in cloud computing circumstance. Future Gen. Comput. Syst. **91**, 244–251 (2019)

Preventing Fault Attack on Stream Ciphers by Fault Detection and Correction

Swapan Maiti and Dipanwita Roy Chowdhury

Abstract Stream ciphers have been popular targets of fault-based cryptanalysis. Almost all of the winners of the eSTREAM project have been cryptanalyzed successfully using fault attack techniques. There exist a very little research work in the contemporary literature to prevent fault attacks on stream ciphers and most of them are cipher specific. Most of the existing related work concentrates only on fault detection. In this work, we propose a generic strategy to prevent fault attacks on a stream cipher by fault detection and correction. The scheme guarantees 100% coverage of detection and correction in both single-bit and double-bit fault. The scheme is validated taking Grain-128 as a hardware efficient stream cipher in the eSTREAM portfolio along with FPGA implementation.

Keywords Stream cipher · Fault analysis · Grain-128 · Error detection and correction · Parity bit checking

1 Introduction

Stream ciphers have gained popularity in recent years in resource-constrained environments. The eSTREAM project started in 2004, introduced a number of stream ciphers in hardware and software efficient environments in order to standardize stream ciphers. The winners of the eSTREAM portfolio ciphers are 7 candidates, 4 in software category HC-128, Rabbit, Salsa20/12, SOSEMANUK, and 3 in hardware category Grain v1, MICKEY 2.0, and Trivium. A number of cryptanalysis of the ciphers are studied by the research community. Algebraic or statistical analysis is one kind of cryptanalysis of strength of the ciphers. Side-Channel Attack (SCA) on stream ciphers is another powerful cryptanalysis that is used to cryptanalyze all kinds

S. Maiti (✉) · D. R. Chowdhury
Indian Institute of Technology Kharagpur, Kharagpur, India
e-mail: swapankumar_maiti@yahoo.co.in

D. R. Chowdhury
e-mail: drc@cse.iitkgp.ac.in

D. Giri et al. (eds.), *Proceedings of the Fifth International Conference on Mathematics and Computing*, Advances in Intelligent Systems and Computing 1170, https://doi.org/10.1007/978-981-15-5411-7_5

of symmetric ciphers, which includes power analysis, fault analysis, and timing analysis. Fault attacks constitute one of the most interesting Side-Channel Attacks. In this kind of attack, faults are injected during cipher operations. After injecting fault, the attacker analyzes the fault-free and faulty ciphertext or key-streams to deduce full or partial information about the secret key. Most of the ciphers in the eStream portfolio are susceptible to fault attacks [2, 5]. There exist some fault attacks on authenticated encryption stream ciphers, and the countermeasures in literature [1, 8, 11].

In general for symmetric ciphers, the prevention schemes follow two different ideas either with dual modular redundant approach or coding-theoretical countermeasures for error detection [9]. Error detection exploits different forms of redundancy, hardware, temporal, or memory. With redundant memory, faulty bits are computed from fault-free and faulty cipher states and then by comparing faults are detected [6]. A number of countermeasures have been presented against fault attacks on AES [3, 7, 9, 10]. Although a number of research has been carried out to prevent fault attacks on AES, a very little attempt has been made for stream ciphers. Most of the existing related work on stream ciphers concentrates only on fault detection. However, here we consider fault detection as well as correction which indeed results in the fault-resilient cipher.

The main contribution of this work can be summarized below.

- A generic strategy for preventing fault attack on a stream cipher following the method of error detection and correction.
- Design architecture with redundant memory that exploits the idea of double parity and gives a countermeasure against single-bit fault as well as a double-bit fault.
- Verified the proposed scheme on a case study of Grain-128 [4].
- An FPGA implementation of Grain-128 with single/double-bit fault detection and correction.

The organization of the rest of the paper is as follows. A general model of a stream cipher and the attack environment for preventing fault attack is shown in Sect. 2. Section 3 describes the countermeasure against fault attacks on a stream cipher using fault detection and correction. The case study of the proposed scheme on Grain-128 is furnished in Sect. 4. Finally, the paper is concluded in Sect. 5.

2 General Model of a Stream Cipher and the Attack Environment

A stream cipher needs two functions, namely a state update function (U) and an output function. In general, the function U constructed with a linear and a nonlinear function updates the cipher state at each clock cycle. Based on a few bits (T) of the cipher state (S), the output function produces z as cipher output as shown in Fig. 1(b). After loading a key and an IV (initialization vector), the cipher is allowed to

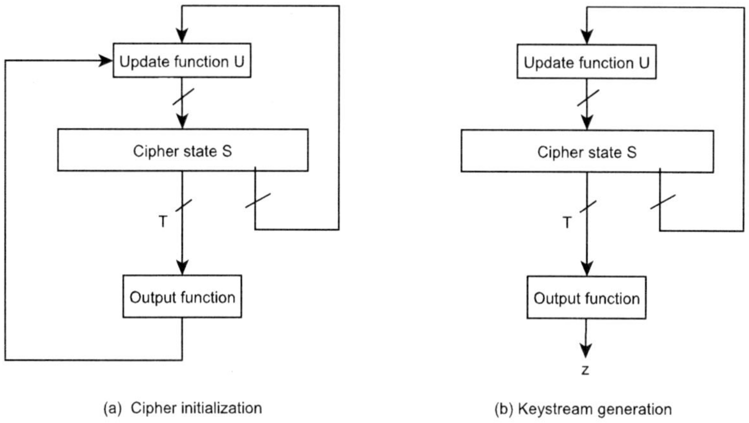

Fig. 1 An overview of a stream cipher

run for a number of clock cycles. During this time, the output z is fed back as input to the update function (U) without producing any keystream. This is called cipher initialization as shown in Fig. 1a.

After initialization, the cipher produces a keystream bit as output at each clock cycle.

2.1 *Fault Model*

We propose a strategy to prevent fault attack on such a stream cipher by fault detection and correction using some redundant storage. The underlying idea behind the fault attacks is to inject faults into the state of the cipher and form some equations by observing the output difference of faulty and fault-free output. By carefully choosing the fault locations, the attackers are able to reduce the algebraic degree of the equations and hence, recover the internal state of the cipher, and eventually retrieve the secret key.

In the fault model, the fault assumed is a single-bit or double-bit fault. We assume that the attacker injects fault in the cipher state or in the redundant storage, and the fault injection is localized which justifies single/double-bit faults. Our strategy to countermeasures against fault attacks is as follows. While the cipher sets up its state, some check bits are simultaneously generated and stored in some redundant storage. Before transferring the cipher state bits to the update function and the output function, the check bits are again generated and are compared with the previously generated check bits. If a fault occurs in the cipher state, then the faulty bit(s) is/are detected by this comparison. After fault detection, the corrected bits are transferred to the output function and to the state update function. Therefore, the output function produces the correct cipher output and the state update function sets up the correct next state.

Even if the fault occurs in the redundant storage, this fault will not propagate to the cipher state by our strategy.

3 Countermeasure Using Fault Detection and Correction

In this section, we propose a formal method for single-bit and double-bit fault detection and correction using some redundant storage with matrix manipulation.

3.1 *Single-Bit Fault*

Let the cipher be constructed with 256 memory elements and the cipher state be of 256-bit length. Let the present state of the cipher be S denoted by $\langle x_0, x_1, x_2, \ldots, x_{255} \rangle$ at time step t. The state values of the cipher are arranged in a 16×16 matrix as shown in Fig. 2a, such that 1st row of the matrix represents 1st two bytes of the cipher state S, 2nd row represents 2nd two bytes of S, and so on.

For generating check bits, we perform row-sum and column-sum operations in GF(2). Let the generated check bits by row-sum and column-sum be assigned to R and C, respectively. For all j, $0 \le j \le 15$, the elements of jth row are added and the computed value is assigned to R_j. For all k, $0 \le k \le 15$, the elements of kth column are added and the computed value is assigned to C_k. Suppose, a single-bit fault is induced into the ith cell of the cipher blocks for some i, $0 \le i \le 255$. The cipher changes its present state. Let the faulty cipher state be S'. The state S' becomes

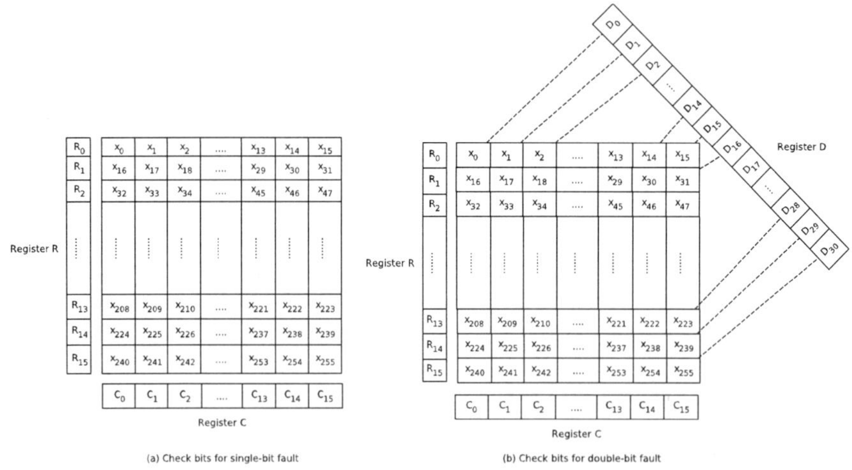

Fig. 2 Computing check bits by 16×16 Matrix representation of the cipher state

$\langle x_0, x_1, \ldots, x_{i-1}, x_i \oplus 1, x_{i+1}, \ldots, x_{255}\rangle$, where x_i has changed to its complement (i.e., $x_i \oplus 1$). As earlier discussion, the values of S' are arranged in a 16×16 matrix, and row-sum and column-sum operations are performed in GF(2) for all rows and all columns of the matrix. First row-sum is assigned to R'_0, second row-sum is assigned to R'_1, and so on. First column-sum is assigned to C'_0, second column-sum is assigned to C'_1, and so on. We perform $r_j = R_j \oplus R'_j$ for all $j, 0 \leq j \leq 15$ and $c_k = C_k \oplus C'_k$ for all $k, 0 \leq k \leq 15$.

For all j, $0 \leq j \leq 15$, r_j, and for all k, $0 \leq k \leq 15$, c_k are all 0's if no fault occurs in the cipher state. Since the position of the value of ith cell in the matrix representation is in jth row and kth column such that $i = 16 * j + k$, if a fault is induced into the ith cell of the cipher for some i, $0 \leq i \leq 255$, then for all j, and for all k, $0 \leq j, k \leq 15$, r_j and c_k are all 0's except r_u and c_v where $u = i \ quot \ 16$, $v = i \ rem \ 16$. Both $r_u = 1$ and $c_v = 1$. For example, if 36th cell contains a faulty bit (i.e., $x_{36} \oplus 1$), then $r_2 = 1$ and $c_4 = 1$.

An element of a matrix has a unique row number and a unique column number with respect to its position in the matrix. Therefore, a pair (j, k), $0 \leq j, k \leq 15$, corresponds to the ith cell of the cipher such that $i = 16 * j + k$. Clearly, there exists a bijection (one-to-one and onto mapping) from (j, k) for all j and all k, $0 \leq j, k \leq 15$ to i for all $i, 0 \leq i \leq 255$ with $i = 16 * j + k$. Observing all the pairs (r_j, c_k), $0 \leq j, k \leq 15$, we can easily identify that the bit value of the ith cell is a faulty bit iff both $r_j = 1$ and $c_k = 1$ and hence, the correct value of the ith cell is its complement value.

3.2 Design Architecture

The architectural view of the proposed scheme for single-bit fault is shown in Fig. 3. Let the state of the cipher be S denoted by $\langle x_0, x_1, x_2, \ldots, x_{255}\rangle$ at time step t. Each of the two registers R and C used in this architecture is of length 16-bits. In a clock pulse, the values $\langle x_0, x_1, x_2, \ldots, x_{255}\rangle$ available as inputs of the cipher cells by a combinational circuit, are stored in S as shown in Fig. 3. The combinational circuit to set up the contents of the registers R and C computes the values which can be obtained after arranging the values $\langle x_0, x_1, x_2, \ldots, x_{255}\rangle$ available as inputs of the cipher cells in a 16×16 matrix as shown in Fig. 2a and computing row-sums and column-sums by addition modulo 2 operations (or by XOR operations). The outputs of the combinational circuit, which are available as inputs to the registers R and C are stored in the registers by the same clock pulse (i.e., at time step t).

Now, if a single-bit fault is induced in the state S, the faulty cipher state values including the faulty bit propagate to the combinational circuit for computing R' and C' as shown in Fig. 3. After arranging 256 cipher state values in a 16×16 matrix representation, 16-bit value R' and 16-bit value C' are computed by the row-sums and column-sums of the matrix. The combinational circuit (i.e., Comparator CMP) in the fault detection unit takes 16-bit value R' and 16-bit value C' and the contents of the two 16-bit registers R and C as inputs. The circuit produces 32 output bits

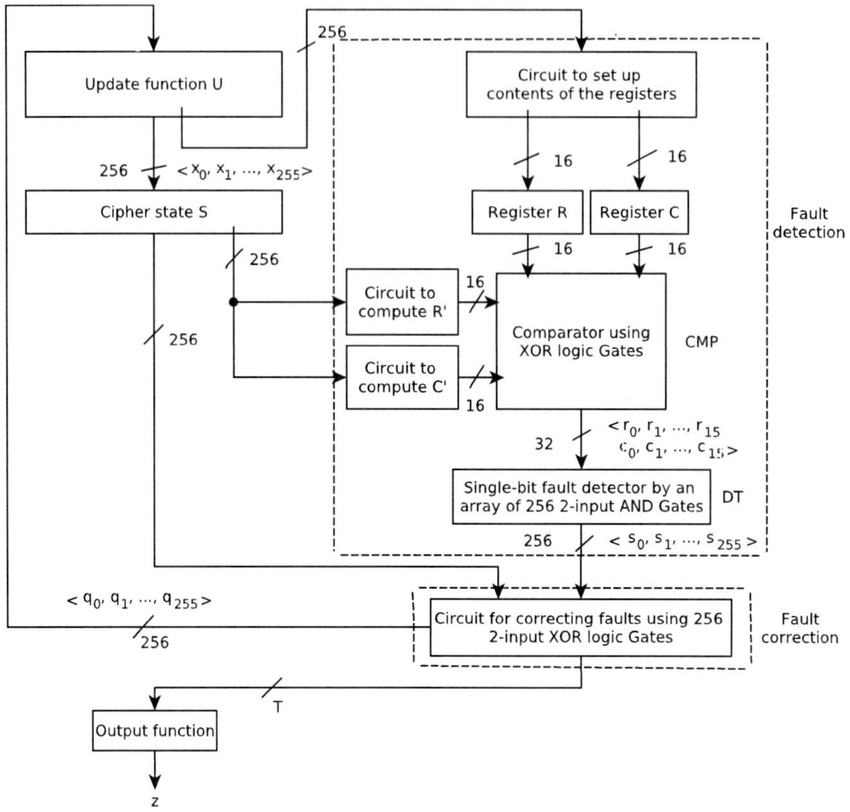

Fig. 3 Overview of a stream cipher with fault-resiliency

denoted by $\langle r_0, r_1, \ldots, r_{15}, c_0, c_1, \ldots, c_{15}\rangle$ by performing XOR operations of the values denoted by R' and C' with the corresponding bits of the registers R and C. The combinational circuit (i.e., Detector DT) which consists of 256 2-input AND Gates, each AND Gate takes one input from r_j, $0 \leq j \leq 15$ and other input from c_k, $0 \leq k \leq 15$, produces 256 outputs denoted by $\langle s_0, s_1, \ldots, s_{255}\rangle$. If s_i is 1 for some i, $0 \leq i \leq 255$, then the circuit detects that the ith cell of the cipher contains a faulty bit.

The combinational circuit in the fault correction unit takes 256 values $\langle s_0, s_1, \ldots, s_{255}\rangle$ from the fault detection unit and 256 values from the cipher cells as shown in Fig. 3. If the line s_i $(0 \leq i \leq 255)$, is 1, then the corresponding XOR Gate in the circuit provides the complement of the value coming from the ith cell of the faulty cipher state as output, and if the selection line s_i is 0 then the XOR Gate provides the value of the ith cell. Finally, the circuit produces 256 output values denoted by $\langle q_0, q_1, \ldots, q_{255}\rangle$ which are the correct state of the cipher at time step t. These values are used to set up the correct next state of the cipher and also to the output function.

Therefore, even if the cipher state contains a faulty bit, the output function produces the correct cipher output and the next state of the cipher will be the correct next state.

In our approach, since both row-sums (R) and column-sums (C) are stored, the appearance of a single-bit fault changes both R and C values and so it guarantees the detection of fault. Once detected, the simple correction circuit identifies the location of fault and the output bit as well as the next state bit are corrected.

In this case, even if the single-bit fault appears either in register R or C, then the detection logic circuit (i.e., the circuit DT with 256 2-input AND gates) will not propagate this fault because all AND gates will produce all 0's, and so it will not affect the correction logic. Hence, the cipher output will be the correct output and the next state of the cipher will be the correct next state, and therefore the adversary will not get any advantage from this scenario. Hence, it achieves 100% coverage of fault detection and correction.

3.3 *Double-Bit Fault*

In this section, we show that our prevention scheme for single-bit fault can easily be extended for double-bit fault with some more extra logic and more redundant storage. Let the present state of the cipher be S at time step t, denoted by $\langle x_0, x_1, x_2, \ldots, x_{255}\rangle$. The state values of the cipher are arranged in a 16×16 matrix as shown in Fig. 2(b), such that first row of the matrix represents first two bytes of the cipher state S, second row represents second two bytes of S, and so on. We perform row-sum, column-sum, and diagonal-sum operations in GF(2). Row-sums and column-sums are stored in registers R and C as in Sect. 3.1. For all l, $0 \le l \le 30$, the elements of lth diagonal are added and the computed value is assigned to D_l as shown in Fig. 2(b). Suppose, double-bit fault is induced into the cipher blocks. Therefore, the cipher state S has changed to S', say, where two bits of S' are faulty bits. As earlier discussion, the values of S' are arranged in a 16×16 matrix, and row-sum, column-sum, and diagonal-sum operations are performed in GF(2) for all rows, all columns, and all diagonals of the matrix. First row-sum is assigned to R'_0, second row-sum is assigned to R'_1, and so on. First column-sum is assigned to C'_0, second column-sum is assigned to C'_1, and so on. First diagonal-sum is assigned to D'_0, second diagonal-sum is assigned to D'_1, and so on. We perform $r_j = R_j \oplus R'_j$ for all j, $0 \le j \le 15$ and $c_k = C_k \oplus C'_k$ for all k, $0 \le k \le 15$ and $d_l = D_l \oplus D'_l$ for all l, $0 \le l \le 30$.

For all j, $0 \le j \le 15$, r_j and for all k, $0 \le k \le 15$, c_k and for all l, $0 \le l \le 30$, d_l are all 0's if no fault occurs. An element of a matrix has a unique row number, a unique column number, and a unique diagonal number with respect to its position in the matrix. Therefore, if single-bit fault occurs, then the faulty bit affects r_j for some j, and c_k for some k, $0 \le j, k \le 15$ and d_l for some l, $0 \le l \le 30$. If double-bit fault occurs and if they are not in the same row , same column and same diagonal of the matrix, then also they individually affect two different r_j, and two different c_k, $0 \le j, k \le 15$, and two different d_l, $0 \le l \le 30$. If double-bit fault occurs and if they are in the same row, then they individually affect two different c_k and two different

d_l, they do not affect in r_j, and if they are in same column, then they individually affect two different r_j and two different d_l, they do not affect in c_k, and if they are in same diagonal, then they individually affect two different r_j and two different c_k, they do not affect in d_l.

Any element $x_i, 0 \leq i \leq 255$ in the matrix has a unique (j, k, l), where j is a row number, k is a column number, and l is a diagonal number with $i = 16 * j + k$ for some j, and for some k, $0 \leq j, k \leq 15$, and $l = j + k$. There exists a bijection (one-to-one and onto mapping) from $(j, k, j + k)$ for all j and all $k, 0 \leq j, k \leq 15$ to i for all $i, 0 \leq i \leq 255$ with $i = 16 * j + k$, since there exists a bijection from (j, k) for all j and all $k, 0 \leq j, k \leq 15$ to i for all $i, 0 \leq i \leq 255$ with $i = 16 * j + k$. For any faulty bit, at least two elements in its corresponding (r_j, c_k, d_l) configuration are 1's. For example, if 33th cell contains a faulty bit (i.e. $x_{33} \oplus 1$), then $r_2 = 1$ and $c_1 = 1$ and $d_3 = 1$. If 33th cell contains a faulty bit (i.e., $x_{33} \oplus 1$) and 45th cell contains a faulty bit (i.e., $x_{45} \oplus 1$), then $r_2 = 0$ and $c_1 = 1$ and $d_3 = 1$ (by 1st faulty bit) and $r_2 = 0$ and $c_{13} = 1$ and $d_{15} = 1$ (by 2nd faulty bit), they simultaneously do not affect r_2. Observing all the configurations (r_j, c_k, d_l) or (r_j, c_k, d_{j+k}), $0 \leq j, k \leq 15$, we can easily identify which bits are faulty bits and hence, the correct values of them are their complement values.

The design architecture for this case will be the same as that for the single-bit fault as shown in Fig. 3, with addition of one more 31-bit register D and one more combinational circuit to compute a 31-bit value D'. The comparator CMP needs 63 XOR logic gates to produce 63 output bits denoted by $\langle r_0, r_1, \ldots, r_{15}, c_0, c_1, \ldots, c_{15}, d_0, d_1, \ldots, d_{30}\rangle$. The detector circuit DT needs 768 (i.e., 3×256) 2-input AND gates and 256 3-input OR gates to produce 256 output bits denoted by $\langle s_0, s_1, \ldots, s_{255}\rangle$. The ith cell corresponds to the configuration (r_j, c_k, d_{j+k}) already discussed above. To produce the output bit s_i, three 2-input AND gates, and one 3-input OR gate is needed, where first AND gate takes (r_j, c_k) as inputs, second AND gate takes (r_j, d_{j+k}) as inputs and third AND gate takes (c_k, d_{j+k}) as inputs. The OR gate takes outputs of the three AND gates as inputs and produces s_i as output.

In this case, even if the double-bit fault appears either in register R or C or D then the detection logic circuit (i.e., the circuit DT) will not propagate this fault because all AND gates will produce all 0's and so it will not affect the correction logic. Hence, the cipher output will be the correct output and the next state of the cipher will be the correct next state, therefore the adversary will not get any advantage from this scenario. Hence, it achieves 100% coverage of fault detection and correction.

4 Case Study on Grain-128

In this section, Grain-128 [4] is chosen as a primitive to provide a case study for the proposed scheme. The block diagram of Grain-128 is shown in Fig. 4. In Grain-128 [4], the contents of the NFSR are denoted by $b_i, b_{i+1}, \ldots, b_{i+127}$, the contents of the LFSR are denoted by $s_i, s_{i+1}, \ldots, s_{i+127}$. The linear and nonlinear update functions,

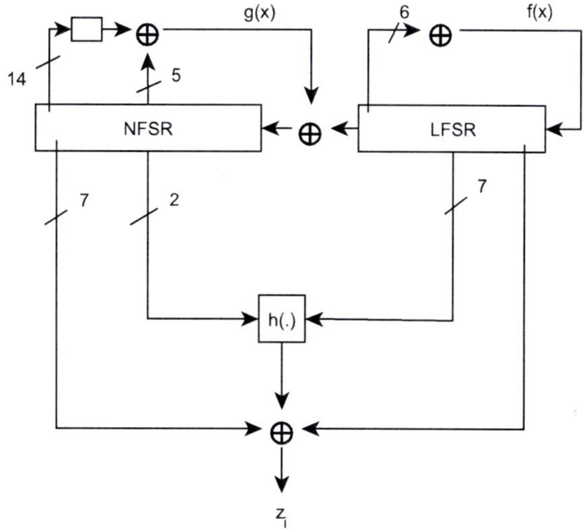

Fig. 4 Structure of Grain-128

nonlinear filter function $h(.)$, which takes two input bits from the NFSR and seven input bits from the LFSR, and the output z_i are defined as

$$\begin{aligned}
s_{i+128} &= s_i + s_{i+7} + s_{i+38} + s_{i+70} + s_{i+81} + s_{i+96}\\
b_{i+128} &= s_i + b_i + b_{i+26} + b_{i+56} + b_{i+91} + b_{i+96} + b_{i+3}b_{i+67} + b_{i+11}b_{i+13}\\
&\quad + b_{i+17}b_{i+18} + b_{i+27}b_{i+59} + b_{i+40}b_{i+48} + b_{i+61}b_{i+65} + b_{i+68}b_{i+84}\\
h(.) &= b_{i+12}s_{i+8} + s_{i+13}s_{i+20} + b_{i+95}s_{i+42} + s_{i+60}s_{i+79} + b_{i+12}b_{i+95}s_{i+95}\\
z_i &= b_{i+2} + b_{i+15} + b_{i+36} + b_{i+45} + b_{i+64} + b_{i+73} + b_{i+89} + h + s_{i+93}
\end{aligned}$$

4.1 Prevention of Single-Bit Fault

Here, we apply the proposed scheme to prevent fault attack on Grain-128. Let the state of the cipher be S denoted by $\langle x_0, x_1, x_2, \ldots, x_{255}\rangle$ at time step t, such that $\langle x_0, \ldots, x_{127}\rangle$ represents the state of 128-bit NFSR and $\langle x_{128}, \ldots, x_{255}\rangle$ represents the state of 128-bit LFSR.

Two registers R and C store the row-sums and column-sums of 256-bit cipher state arranged in 16×16 matrix form at time step t as discussed in the Sect. 3. If the LFSR or NFSR of the cipher contains a faulty bit, then the faulty bit is detected instantly by comparing the row-sums ($R' : \{R'_0, R'_1, \ldots, R'_{15}\}$) and column-sums ($C' : \{C'_0, C'_1, \ldots, C'_{15}\}$) of faulty cipher state with that of fault-free sums stored in R and C registers as shown in Fig. 5. The comparator (CMP) produces 32-bit values denoted by $\langle r_0, r_1, \ldots, r_{15}, c_0, c_1, \ldots, c_{15}\rangle$ as shown in Fig. 6a. The circuit (DT)

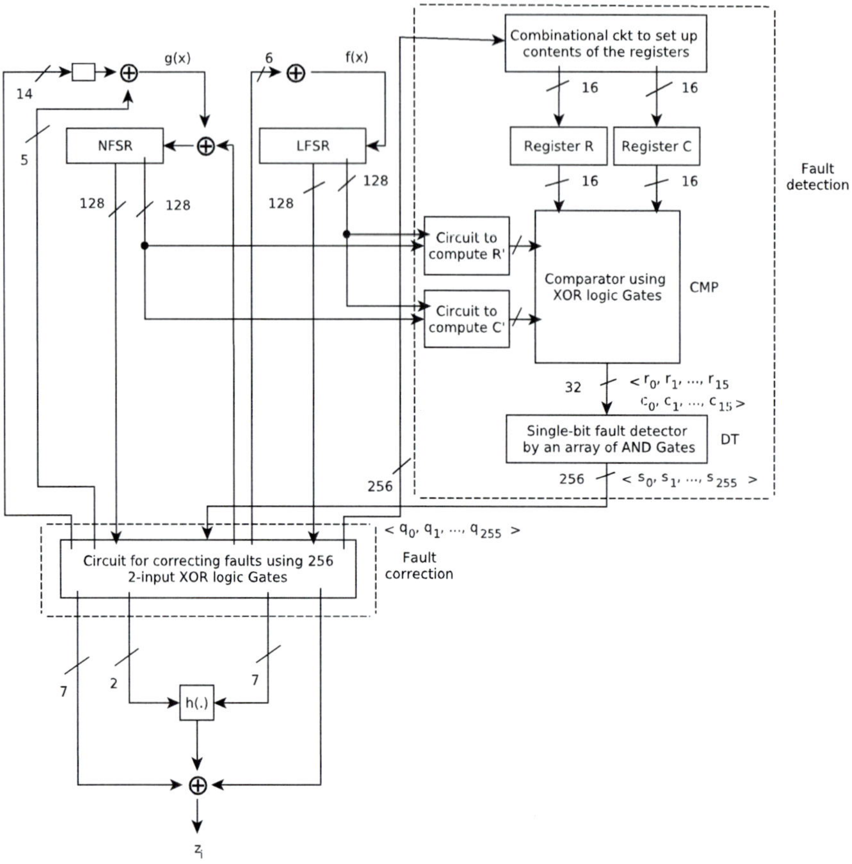

Fig. 5 Grain-128 with a single-bit fault detection and correction

for detection of fault produces 256-bit values denoted by $\langle s_0, s_1, \ldots, s_{255}\rangle$ using an array of 256 2-input AND gates as shown in Fig. 6b. The circuit for correcting faults using XOR logic Gates takes 256 values $\langle s_0, s_1, \ldots, s_{255}\rangle$ from the output of the detector (DT) and 128 Q values of NFSR and 128 Q values of LFSR and provides Q or $!Q$ as output of the corresponding 2-input XOR logic Gate based on the value of s_i, $0 \le i \le 255$, and hence provides the correct state values $\langle q_0, q_1, \ldots, q_{255}\rangle$ of the cipher as shown in Fig. 6c. Therefore, $h(.)$ function and z_i do not take the values directly from NFSR and LFSR block, rather they take values from $\langle q_0, q_1, \ldots, q_{255}\rangle$ and the feedback functions $g(x)$ and $f(x)$ take values from $\langle q_0, q_1, \ldots, q_{255}\rangle$ as shown in Fig. 5. Even the shifted values from one cell to another cell are also taken from $\langle q_0, q_1, \ldots, q_{255}\rangle$, which is not shown in the figure. Therefore, even if the cipher state contains a faulty bit, $h(.)$ produces the correct output and correct cipher output z_i, and thus the next states of NFSR and LFSR will give the correct cipher state.

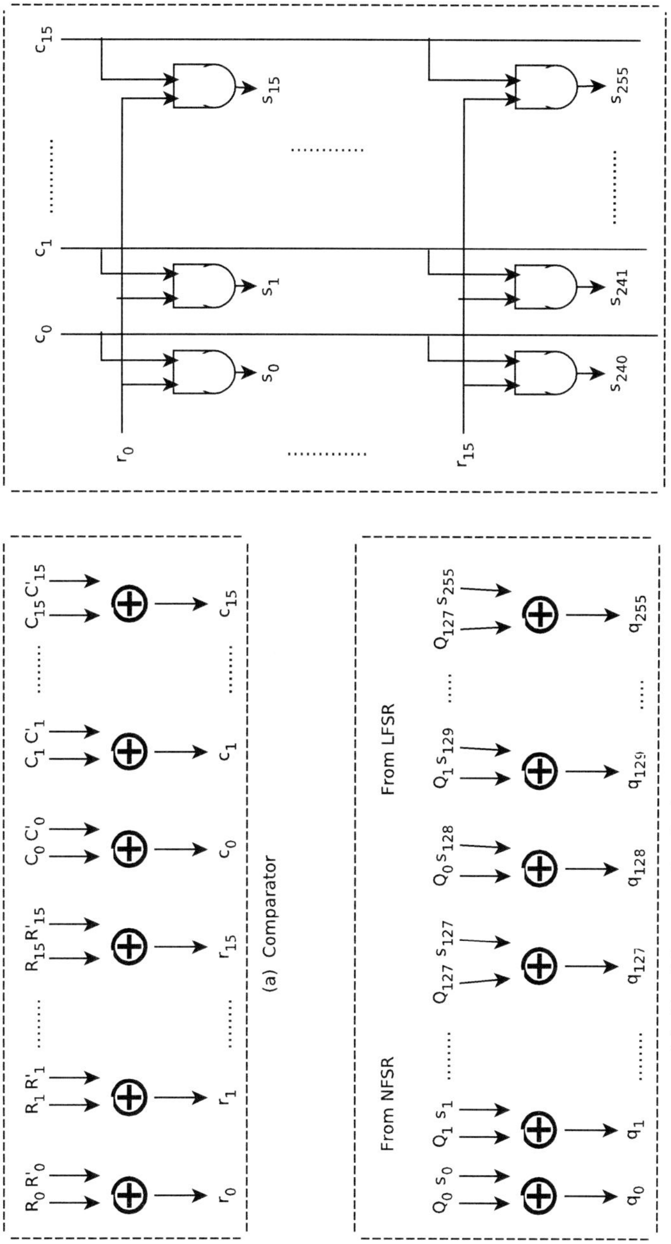

Fig. 6 Design architecture for fault prevention

4.2 Prevention of Double-Bit Fault

Here, three registers R, C, and D store the row-sums, column-sums, and diagonal-sums of 256-bit cipher state arranged in 16×16 matrix form at time step t as discussed in the Sect. 3.3. For double-bit fault in the cipher state, the comparator (CMP) produces 63-bit values denoted by $\langle r_0, r_1, \ldots, r_{15}, c_0, c_1, \ldots, c_{15}, d_0, d_1, \ldots, d_{30} \rangle$. The circuit (DT) for detection of fault produces 256-bit values denoted by $\langle s_0, s_1, \ldots, s_{255} \rangle$ using an array of 256 combinational units each containing three 2-input AND gates and one 3-input OR gate. The circuit for correcting faults (i.e., the fault correction logic circuit) is the same as shown in Fig. 5.

In our approach, the appearance of single-bit fault changes both R, C, and D values, and double-bit fault changes at least two of the values of R, C, and D.Â Therefore, the coverage for detection and location of fault(s) are 100% for Grain-128.

4.3 Hardware Implementation

This section discusses the hardware requirements of Grain-128[4] without fault-resilient module and also with fault-resilient module for single-bit/double-bit fault. We have simulated Grain-128 for single and double-bit fault on Xilinx Spartan 3 XC3S200-4FT256 FPGA device using Xilinx 14.2 synthesis tool, and our result is shown in Table 1. For single-bit fault correction, the number of slices required in

Table 1 Hardware comparison on FPGA

Logic utilization	Grain-128	Grain-128 with single-bit fault-resilient module	Grain-128 with double-bit fault-resilient module
Number of slices (synthesis/PPR)	144/227	343/481	445/591
Number of slice Flip-flops (synthesis/PPR)	247 /247	309/309	339/339
Number of 4-input LUTs (synthesis/PPR)	27/27	646/645	853/852
Number of bonded IOBs	10	10	10
Number of GCLKs	1	2	2
Average Fanout of Non-clock Nets	2.13	3.21	3.26
Maximum frequency estimated	178.094 MHz	77.513 MHz	72.648 MHz

post-place and route (PPR) is 481, and the speed estimated is 77.513 MHz and thus, throughput is 77.513 Mbps. The efficiency that is throughput to area ratio is $\frac{77.513}{481} = 0.161$. For double-bit fault correction, the speed estimated is 72.648 MHz and thus, throughput is 72.648 Mbps. The efficiency is $\frac{72.648}{591} = 0.123$. The implementation results show that the prevention of single/double-bit fault has been achieved at the cost of hardware and memory redundancy.

5 Conclusion

A new scheme has been proposed to prevent fault attack on a stream cipher by fault detection and correction. The proposed scheme has been verified on a case study of Grain-128. The FPGA implementation shows that the proposed design is effective to make a fault-resilient stream cipher at the cost of hardware and memory redundancy.

References

1. Adomnicai, A., Fournier, J.J.A., Masson, L.: Masking the lightweight authenticated ciphers ACORN and Ascon in software. IACR Cryptol. **2018**, 708 (2018), https://eprint.iacr.org/2018/708
2. Banik, S., Maitra, S., Sarkar, S.: A differential fault attack on the grain family of stream ciphers. In: Cryptographic Hardware and Embedded Systems - CHES 2012 - 14th International Workshop, Leuven, Belgium, September 9-12, 2012. Proceedings, pp. 122–139 (2012). https://doi.org/10.1007/978-3-642-33027-8_8
3. Bousselam, K., Natale, G.D., Flottes, M., Rouzeyre, B.: On countermeasures against fault attacks on the Advanced Encryption Standard. In: Fault Analysis in Cryptography, pp. 89–108 (2012). https://doi.org/10.1007/978-3-642-29656-7_6
4. Hell, M., Johansson, T., Maximov, A., Meier, W.: A stream cipher proposal: Grain-128. In: Proceedings 2006 IEEE International Symposium on Information Theory, ISIT 2006, The Westin Seattle, Seattle, Washington, USA, July 9-14, 2006, pp. 1614–1618 (2006). https://doi.org/10.1109/ISIT.2006.261549
5. Hojsík, M., Rudolf, B.: Floating fault analysis of trivium. In: Progress in Cryptology - INDOCRYPT 2008, 9th International Conference on Cryptology in India, Kharagpur, India, December 14–17, 2008. Proceedings, pp. 239–250 (2008). https://doi.org/10.1007/978-3-540-89754-5_19
6. Kermani, M.M., Reyhani-Masoleh, A.: Parity-based fault detection architecture of s-box for advanced encryption standard. In: 21th IEEE International Symposium on Defect and Fault-Tolerance in VLSI Systems (DFT 2006), 4–6 October 2006, Arlington, Virginia, USA, pp. 572–580 (2006). https://doi.org/10.1109/DFT.2006.50
7. Pahlevanzadeh, H., Dofe, J., Yu, Q.: Assessing CPA resistance of AES with different fault tolerance mechanisms. In: 21st Asia and South Pacific Design Automation Conference, ASP-DAC 2016, Macao, Macao, January 25–28, 2016, pp. 661–666 (2016). https://doi.org/10.1109/ASPDAC.2016.7428087
8. Salam, M.I., Simpson, L., Bartlett, H., Dawson, E., Wong, K.K.: Fault attacks on the authenticated encryption stream cipher MORUS. Cryptography **2**(1), 4 (2018). https://doi.org/10.3390/cryptography2010004

9. Schmidt, J., Medwed, M.: Countermeasures for symmetric key ciphers. In: Fault Analysis in Cryptography, pp. 73–87 (2012). https://doi.org/10.1007/978-3-642-29656-7_5
10. Schneider, T., Moradi, A., Güneysu, T.: Parti - Towards Combined Hardware Countermeasures Against Side-Channel and Fault-Injection Attacks. In: Advances in Cryptology - CRYPTO 2016 - 36th Annual International Cryptology Conference, Santa Barbara, CA, USA, August 14–18, 2016, Proceedings, Part II, pp. 302–332 (2016). https://doi.org/10.1007/978-3-662-53008-5_11
11. Zhang, X., Feng, X., Lin, D.: Fault attack on ACORN v3. IACR Cryptology **2017**, 855 (2017). http://eprint.iacr.org/2017/855

A Guests Managing System with Lattice-Based Verifier-Local Revocation Group Signature Scheme with Time-Bound Keys

Maharage Nisansala Sevwandi Perera and Takeshi Koshiba

Abstract In this paper, we propose a visitors managing system using group signature schemes. The paper focuses on situations that guests who are residing in a hotel are like to keep their activities such as accessing hotel amenities anonymously. Moreover, guests are generally staying in a hotel for a short period. Thus, they should be managed with a proper revocation mechanism. This paper proposes a system using a group signature scheme with Verifier-local revocation and time-bound keys.

Keywords Lattice-based group signatures · Almost-full anonymity · Dynamical-almost-full anonymity · Member registration · Verifier-local revocation · Time-bound keys

1 Introduction

The group signature schemes, first submitted by Chaum et al. [6] in 1991, allow any member of the group to represent the group without revealing his identity. In this manner, group members are privileged to sign messages for the sake of the group anonymously. Accordingly, signature verifiers can only validate the signatures, but cannot identify the signers. Besides, if necessary, an authority can open the signatures, and recognize the signers. These two features, *anonymity* and *traceability* open doors for practical employment of the group signature schemes. For instance, applications like key-card access systems, digital-right management systems, and e-commerce systems interest using group signature schemes. Most of the first group signature schemes used bilinear map settings. Security of those schemes will be broken once the quantum computers become a reality. Nearly a decade before, in 2010,

M. N. S. Perera (✉)
Graduate School of Science and Engineering, Saitama University, Saitama, Japan
e-mail: mnisansalasperera@gmail.com

T. Koshiba
Faculty of Education and Integrated Arts and Sciences, Waseda University, Tokyo, Japan
e-mail: tkoshiba@waseda.jp

D. Giri et al. (eds.), *Proceedings of the Fifth International Conference on Mathematics and Computing*, Advances in Intelligent Systems and Computing 1170, https://doi.org/10.1007/978-981-15-5411-7_6

Gorden et al. [10] put forward the first quantum resist group signature scheme using lattice assumptions. Lattice cryptography is one of the finest auspicious primitives against quantum computers. Lattice cryptography owns provable strong security in reliance on the worst-case hardness of the lattice problems. Moreover lattice cryptography claims efficient implementation. Thus, Gorden's scheme [10] showed strong security against quantum computers. However, the scheme in [10] has a noticeable disadvantage. The size of the group signature in the scheme in [10] increases with the number of group members N (linear-barrier problem). More specifically, signatures in Gorden's scheme have size $\mathcal{O}(N)$. Later, in 2012, Camenisch et al. [5] offered a more secure and efficient scheme with an anonymous attribute token system. However, this scheme still failed to resolve the linear-barrier problem. Finally, in 2013, Languillaumie et al. [12] suggested a scheme that overcomes the linear-barrier problem. Thus the sizes of the signatures and the group public key in their proposal, given in [12], are proportional to $\log N$. However, none of those above-mentioned group signature schemes from lattices were able to support member registration or revocation. All the three schemes are suitable for static groups, where both the number of group members and their keys are fixed at the time of setup. As a result, no new member can join later, or no existing member can be removed.

Later, in 2014, Langlois et al. [13] suggested the first group signature scheme from lattices that supports member revocation using a revocation method called *Verifier-local Revocation (VLR)*. Thus Langlois's VLR group signature scheme [13] is the first revocation scheme that is accepted to be quantum-resistant at that time. Even though the scheme presented in [13] has several remarkable advantages over the previous works, the security of the scheme is weaker since the scheme depends on a relaxed security notion titled *selfless-anonymity*. The scheme given in [20] offered a security notion known as *almost-full anonymity* for VLR group signature schemes. Even though the almost-full anonymity is stronger than the selfless-anonymity, it shows weaker security than the full-anonymity, suggested by Bellare et al. [3] for static groups. However, the almost-full anonymity can be seen as a reasonable solution for the security of VLR group signature schemes because realizing full-anonymity for VLR group signature schemes is a technically difficult chore. Later, in 2016, Libert et al. [14] formed a group signature scheme from lattice assumptions to facilitate member registration. Their scheme [14] allows new users to register to the group via a group joining protocol. However, even they have considered member registration, they have not focused on facilitating member revocation. Ling et al. [15] offered the first fully dynamic group signature scheme from lattices by using accumulators. Using accumulators seem to be less efficient than using VLR in large groups. Recently, a fully dynamic group signature scheme based on lattices was proposed in [19]. The scheme in [19] serves both user registration and member revocation. The scheme in [19] allows members to join via a joining protocol, and member revocation is managed by using VLR. Moreover, they suggested a new security notion, namely, *dynamical-almost-full anonymity*. The dynamical-almost-full anonymity is for fully dynamic (with user registration and member revocation) group signature schemes with VLR and it is an extended version of the almost-full anonymity [20].

Our Contribution
This paper proposes a group signature scheme for managing guest information in a hotel. For handling of guests' check-out, this paper employs Verifier-local Revocation (VLR) technique. The revocation method, VLR utilizes a token system to identify revoked members from active members in a group. Each member in a group has a token, and when a particular member is retiring or removing from the group, the group manager adds the token of that member to a list known as *Revocation List* (*RL*). Then he sends the latest *RL* to the signature verifiers. At the time of verifying, the verifiers scan and check whether signing member's token is not in the list RL. When we implement group signature schemes with VLR to a dynamic system like guest managing system where guests are joining the group temporarily, the length of the revocation list grows in a short time. As a result, the verifiers have to spend a long time to check RL to validate the signer. As a solution to this problem, we can use time-bound keys, which are proposed in [8] by Chu et al. Thus, this paper applies the time-bound keys to the scheme given in [19] and presents a secured and efficient scheme to manage guest information in a hotel.

2 Preliminaries

2.1 *Notations*

Throughout this paper we express the set of integers $\{1, \ldots, i\}$ by $[i]$ for any integer $i \geq 1$. Matrices are represented in bold uppercase letters such as $\mathbf{X}$, and vectors are in bold lowercase letters, such as $\mathbf{v}$. We consider only column vectors in this paper. While the concatenation of matrices $\mathbf{X} \in \mathbb{R}^{n\times m}$ and $\mathbf{Y} \in \mathbb{R}^{n\times p}$ is presented as $[\mathbf{X}|\mathbf{Y}] \in \mathbb{R}^{n\times(m+p)}$, the concatenation of vectors $\mathbf{v} \in \mathbb{R}^m$ and $\mathbf{z} \in \mathbb{R}^k$ is indicated as $(\mathbf{v}\|\mathbf{z}) \in \mathbb{R}^{m+k}$. If S is a finite set, we mean b is chosen uniformly at random from S by $b \xleftarrow{\$} S$ and if S is a probability distribution we mean b is drawn according to S by $b \xleftarrow{\$} S$. By $\|\mathbf{z}\|$ we denote the Euclidean norm of $\mathbf{z}$. By $\|\mathbf{z}\|_\infty$ we denote the infinity norm of $\mathbf{z}$. χ is a b-bounded distribution over $\mathbb{Z}$. In other words, samples output by χ are with norm at most b with overwhelming probability. Here $b = \sqrt{n}\omega(\log n)$.

2.2 *Lattices*

Let $\mathbf{V} = [\mathbf{v}_1|\cdots|\mathbf{v}_m] \in \mathbb{Z}_q^{r\times m}$ be linearly independent vectors in $\mathbb{Z}_q^r$ and q be a prime. We define the r-dimensional lattice $\Lambda(\mathbf{V})$ for $\mathbf{V}$ as

$$\Lambda(\mathbf{V}) = \{\mathbf{z} \in \mathbb{Z}^r \mid \mathbf{z} \equiv \mathbf{V}\mathbf{x} \bmod q \text{ for some } \mathbf{x} \in \mathbb{Z}_q^m\}.$$

$\Lambda(\mathbf{V})$ is the set of all linear combinations of columns of $\mathbf{V}$ and m indicates the rank of $\mathbf{V}$.

This paper considers a discrete Gaussian distribution in respect of a lattice. We define the Gaussian function centered in a vector $\mathbf{x}$ with parameter $s > 0$ as $\rho_{s,\mathbf{x}}(\mathbf{z}) = e^{-\pi\|(\mathbf{z}-\mathbf{x})/s\|^2}$ and the corresponding probability density function proportional to $\rho_{s,\mathbf{x}}$ as $D_{s,\mathbf{x}}(\mathbf{z}) = \rho_{s,\mathbf{x}}(\mathbf{z})/s^n$ for all $\mathbf{z} \in \mathbb{R}^n$. For a lattice Λ we specify the discrete Gaussian distribution as $D_{\Lambda,s,\mathbf{x}}(\mathbf{z}) = D_{s,\mathbf{x}}(\mathbf{z})/D_{s,\mathbf{x}}(\Lambda) = \rho_{s,\mathbf{x}}(\mathbf{z})/\rho_{s,\mathbf{x}}(\Lambda)$ for all $\mathbf{z} \in \Lambda$. As $\mathbb{Z}^m$ is also a lattice, we can declare a discrete Gaussian distribution for $\mathbb{Z}^m$. Thus, by $D_{\mathbb{Z}^m,\sigma}$, we express the discrete Gaussian distribution for $\mathbb{Z}^m$ around the origin with the standard deviation σ.

2.3 Lattice-Related Hardness Problems

2.3.1 Learning with Errors (LWE)

Definition 1 Learning With Errors (LWE) [18]: Let $n, m \geq 1$, and $q \geq 2$ be integers. For $\mathbf{s} \in \mathbb{Z}_q^n$ and χ, let the distribution $A_{\mathbf{s},\chi}$ gained by sampling uniformly random $\mathbf{a} \in \mathbb{Z}_q^n$ and selecting $e \leftarrow \chi$, and resulting the pair $(\mathbf{a}, \mathbf{a}^T \cdot \mathbf{s} + e)$.

Search-LWE and Decision-LWE are the two versions of LWE problems. For a given LWE samples Search-LWE is determined to find the secret s. Decision-LWE is for distinguishing LWE samples and samples selected according to the uniformly distribution. This paper uses the hardness of the problem Decision-LWE.

We say, for a prime power q, $b \geq \sqrt{n}\omega(\log n)$, and distribution χ, solving $LWE_{n,q,\chi}$ problem is at least as hard as solving $SIVP_\gamma$, where $SIVP$ means *Shortest Independent Vector Problem* and $\gamma = \tilde{O}(nq/b)$ [9, 22].

2.3.2 Short Integer Solution ($SIS_{n,m,q,\beta}$)

In 1996, SIS was first explained in seminal work of Ajtai [2]. SIS problem is for finding a sufficiently short nontrivial integer combination of given uniformly random elements of a certain large finite additive group, which sums to zero [18].

Definition 2 Short Integer Solution ($SIS_{n,m,q,\beta}$ [18, 22]): For given m uniformly random vectors $\mathbf{a}_i \in \mathbb{Z}_q^n$, which forms the columns of a matrix $\mathbf{A} \in \mathbb{Z}_q^{n\times m}$, SIS requires to find the vector $\mathbf{z} \in \mathbb{Z}^m$, which is a nonzero vector and satisfies $\|\mathbf{z}\| \leq \beta$ and $\mathbf{Az} = 0 \mod q$.

We say, for any m, β, and for any $q \leq \sqrt{n}\beta$, solving $SIS_{n,m,q,\beta}$ problem with non-negligible probability is at least as hard as solving $SIVP_\gamma$ problem, for some $\gamma = \beta \cdot O(\sqrt{n})$ [9].

2.4 Lattice Algorithms

To construct the propose scheme, this paper uses algorithms SampleD [9, 16] and GenTrap [1, 9, 16]. The algorithm SampleD is a randomized nearest-plane algorithm and it samples from a discrete Gaussian $D_{\Lambda,s,\mathbf{c}}$ over any lattice Λ. GenTrap is a preimage sampleable trapdoor function (PSTF).

- SampleD($\mathbf{T_A}, \mathbf{A}, \mathbf{u}, \sigma$): On inputs a vector $\mathbf{u}$ in the image of $\mathbf{A}$, a trapdoor $\mathbf{T_A}$, and $\sigma = \omega(\sqrt{n \log q \log n})$, SampleD results $\mathbf{z} \in \mathbb{Z}^m$ sampled from the distribution $D_{\mathbb{Z}^m,\sigma}$. The output $\mathbf{z}$ should assure $\mathbf{A} \cdot \mathbf{z} = \mathbf{u} \bmod q$.
- GenTrap(n, m, q): GenTrap(n, m, q) results a matrix $\mathbf{A} \in \mathbb{Z}_q^{n \times m}$ and a trapdoor $\mathbf{T_A}$ for any given integers $n \geq 1, q \geq 2$, and sufficiently large $m = O(n \log q)$. The distribution of the resulted $\mathbf{A}$ is negl(n)-far from the uniform distribution.

2.5 One-Time Signature Scheme

Our new scheme uses one-time signature scheme $\mathcal{OTS}$ = (OGen, OSign, OVer) [17]. $\mathcal{OTS}$ schemes are digital signature schemes based on one-way functions. While OGen is an algorithm for generating keys, OSign and OVer are algorithms for producing signatures and verifying signatures. OGen takes (1^n), and creates a signing, verification key pair (**osk**, **ovk**). For **osk** and a message M, OSign produces a signature Σ. OVer takes as inputs **ovk**, a message M, and a signature Σ and outputs $\top$ or $\bot$ [7].

3 Applying VLR Group Signature Schemes with Time-Bound Keys to the Guest-Managing System

This section first discusses the primitives used to construct the new VLR group signature scheme with time-bound keys and then explains the application of the guest-managing system. Thus, it defines the general VLR group signature scheme and the dynamical-almost full anonymity with the scheme proposed in [19]. Next, it describes time-bound keys. Finally, it discusses the application of time-bound keys and VLR scheme on guest managing system.

3.1 VLR Group Signature Schemes

Even though general group signature schemes have four algorithms, KeyGen, Sign, Verify, and Open, group signature schemes with VLR have of only the first three algorithms. Since VLR group signature schemes have an *implicit tracing algorithm* which requires to execute Verify for each member until the signer is revealed, those schemes do not need the algorithm Open for tracing signers.

- KeyGen(n, N): KeyGen is a randomized PPT algorithm, and on input of n and N it produces a group public key **gpk**, a set of group members secret keys $\mathbf{gsk} = (\mathbf{gsk}[0], \ldots, \mathbf{gsk}[N-1])$, and a set of group members revocation tokens $\mathbf{grt} = (\mathbf{grt}[0], \ldots, \mathbf{grt}[N-1])$. While $n \in \mathbb{N}$ is the security parameter, N is the number of group users.
- Sign(**gpk**, **gsk**[d], M): Sign is a randomized algorithm that creates a group signature Σ on a given message M. Sign takes secret signing key **gsk**[d], the group public key **gpk**, and a message $M \in \{0, 1\}^*$ as inputs.
- Verify(**gpk**, RL, Σ, M): Verify is a deterministic algorithms. With the given group public key **gpk** Verify checks if the input signature Σ is valid for the input message M and validates the signer not being revoked using the latest RL.

3.2 Dynamical-Almost-Full Anonymity

In the full-anonymity game [3] between an adversary and a challenger, in the beginning, the challenger gives all the secret signing keys of the members to the adversary. However, for VLR group signature schemes we cannot achieve the full-anonymity because VLR group signatures have tokens and tokens cannot be given to the adversary. If any token is gained by the adversary, he can execute the algorithm Verify with the token he has and can identify whether the owner of the token generated the signature or not. In previous VLR group signature schemes [13], tokens for the members are formed using a part of the secret signing keys of the members. Thus, we cannot allow the adversary to get any secret signing keys also because he can obtain the tokens from the secret signing keys. Hence, VLR group signatures rely on the selfless-anonymity which restricts revealing any information related to the challenging signature of the adversary.

The scheme for the VLR group signatures with both member registration and revocation, given in [19], provides stronger security than the selfless-anonymity called *dynamical-almost-full anonymity*. At the beginning of the anonymity game of the dynamical-almost-full anonymity, the challenger gives all the present members' secret signing keys to the adversary. Thus, the dynamic-almost-full anonymity is applicable only for the schemes where the tokens are generated separately to the secret signing keys. Then the anonymity game of the dynamical-almost-full anonymity enables the adversary insert new members. However, at the time of adversary registering as a new user, we do not provide his revocation token to him. The adversary can request revocation token of any member later. At the challenging phase, we generate the challenging signature only for the noncorrupted members added by the adversary. Thus we do not create a challenging signature for members whose token is revealed to the adversary and who is not added by the adversary as new members.

3.3 Time-Bound Keys

When members are joining a group temporarily, the size of the revocation list (RL) increases quickly. Thus the verifiers have to check whether the signer of a given signature is valid by scanning a long list. Checking a long list decreases the efficiency of Verify. Chu et al. [8] suggested a solution called time-bound keys for reducing the cost of the revocation check. In Chu's scheme, the group members' keys and the signatures have expiration dates. At the time of adding a new user to the group, the group manager should determine an expiration date for the new member keys. Thus every registered member's secret key has an expiration date, and only the members having non-expired keys are allowed to generate signatures. This cuts down the cost of checking the expired members' (naturally revoked members) revocation statue at the signature verification. Since expired members are not allowed to generate valid signatures, the group manager does not need to add their revoked details to the revocation list RL.

To efficiently compare two dates, Chu et al. [8] have used the date format as "YYMMDD" in integer form. For instance, the date December 24, 2019 is indicated as "191224". Moreover, $t_1 > t_2$ represents that the date t_1 is later than the date t_2. In Chu's scheme, the key-expiration date, which is selected by the group manager, is represented by t_r. Moreover, at the signature generation, the signer is allowed to choose a signature expiration date t_s which should satisfy $t_r > t_s$ and $t_s \geq t_v$, where t_v is the verification date. A valid signature should satisfy $t_r > t_s \geq t_v$.

3.4 VLR Group Signature Scheme with Time-Bound Keys for a Guest-Managing System

When a guest G_i proceeds the hotel check-in, G_i provides check-in information including information regarding his stay. The receptionist or the hotel manager or the system issues G_i a key-card with some details that he is allowed to access. The key-card only can be used until the leaving date. This restriction can be applied using time-bound keys. Thus the key-card have an expiration date (t_{r_i}). The guest G_i can utilize the hotel amenities (bar, pool, spa, and gym) using the key-card until he leaves. However, the key-card will not reveal any personal information like name, room number of the guest. The system can only verify the key-card is not expired and is valid to use the facility. All the amenities have a limited period of utilizing them. For instance, the pool can be used only for two hours at a time. To control the period of using amenities, again the time-bound keys can be used. Each time the guest is accessing an amenity can be regarded as issuing a signature. Thus, the expiration time of using the pool is the expiration time of the signature (t_s).

We use the date format as "YYYYMMDDhhmm". If G_i's check-out date and time is August 09, 2018 morning 10, then his key-card expiration time (t_r) is 201808091000. If he accesses the gym in August 07, 2018 evening 4:00 which is allowed only for two hours, then his signature expiration time will be 201808071800. After finishing the workout in the gym, he has to enter his card to go out of the gym. At that time system will validate his signature expiration time with the leaving time. If he has used the gym for more than two hours, then he has to pay the penalty. If he used an expired key-card, then he will not be able even to enter the gym. When accessing the gym, the key-card generates a signature with signature expire time (t_s), and the system validates that the signature expiration time is greater than or equal to the current time (t_v). When he leaves, the system again checks that the signature expiration time is greater than or equal to the current time (t_v). For instance, as discussed before, the guest enters the gym at 4 pm on August 07, 2018 and the gym allows only two hours of work-out. Then t_s is 201808071800 and entering time (t_v) is 201808071600. Since $t_s > t_v$ and $t_r > t_s$ he is allowed to enter. If he finishes the workout at 5:50 pm then the system allows him to exit without any issue since the leaving time (201808071750) $t_v < t_s$. But if he tries to leave at 6:30 pm (201808071830), then system asks him to pay additional charge because $t_v > t_s$ (he used the gym extra 30 min).

Using the time-bound keys, we can manage the use of amenities and the room in a hotel guest-managing system as discussed.

4 New Scheme

This section first, defines the underlying interactive protocol that the proposed scheme uses to confirm the validity of the signers. Then this section provides the description of the new scheme.

4.1 Supporting Zero-Knowledge Protocol

In this section, we briefly explain the zero-knowledge proof system that our scheme uses as underlying interactive protocol. Let COM be the statistically hiding and computationally binding commitment scheme [11]. In our scheme, matrices **F**, **A**, **B**, **V**, **G**, **H** and vectors **u**, **v**, $\mathbf{c}_1$, $\mathbf{c}_2$ are the public parameters. The witness of the prover consists of vectors **x**, $\mathsf{bin}(\mathbf{z})$, **r**, **s**, $\mathbf{e}_1$, and $\mathbf{e}_2$. The goal of the prover is to prove $\mathbf{F} \cdot \mathbf{x} = \mathbf{H}_{4n \times 2m} \cdot \mathsf{bin}(\mathbf{z})$ (as given in [14]) and $\mathbf{V} \cdot (\mathbf{A} \cdot \mathbf{r}) + \mathbf{e}_1 = \mathbf{v} \mod q$ and ($\mathbf{c}_1 = \mathbf{B}^T \mathbf{s} + \mathbf{e}_1, \mathbf{c}_2 = \mathbf{G}^T \mathbf{s} + \mathbf{e}_2 + \lfloor q/2 \rfloor \mathsf{bin}(\mathbf{z}_i)$) (as discussed in [21]) to the verifier. Here $\mathbf{H}_{n \times n\lceil \log q \rceil} \in \mathbb{Z}^{n \times n\lceil \log q \rceil}$ is a "power-of-2" matrix and $\mathbf{z} = \mathbf{H}_{n \times n\lceil \log q \rceil} \cdot \mathsf{bin}(\mathbf{z})$ for any $\mathbf{z} \in \mathbb{Z}_q^n$.

4.2 Description of the New Scheme

The new scheme has algorithms, KeyGen, Join, Sign, Verify, Open, and Revoke. The proposed scheme is constructed based on the scheme given in [19]. Using the time-bound keys the methods in the scheme given in [19] are modified and the new scheme is presented. Thus, the algorithms KeyGen (setup), Open, and Revoke are same as the algorithms given in [19]. The algorithms Join, Sign, and Verify are modified by adding the generation and validation of the time-bound keys.

The security parameter is denoted by λ. We denote the maximum number of predicted members in a group by $N = 2^{\ell}$. Then we choose lattice parameter $n = \mathcal{O}(\lambda)$, prime modulus $q = \tilde{\mathcal{O}}(\ell n^3)$, Gaussian parameter $\sigma = \Omega(\sqrt{n \log q} \log n)$, dimension $m = 2n\lceil \log q \rceil$, infinity norm bounds $\beta = \sigma\omega(\log m)$, and $b = \sqrt{n}\omega(\log n)$.

The algorithms of the new scheme are as follows.

Setup: The randomized algorithm $\mathsf{KeyGen}(1^n, 1^N)$ works as follows.

1. Obtain $\mathbf{A} \in \mathbb{Z}_q^{n \times m}$ and a trapdoor $\mathbf{T_A}$ by executing the PPT algorithm $\mathsf{GenTrap}(n, m, q)$.
2. Sample vector $\mathbf{u} \xleftarrow{\$} \mathbb{Z}_q^n$.
3. Execute $\mathsf{GenTrap}(n, m, q)$ to obtain the encryption and decryption keys $\mathbf{B} \in \mathbb{Z}_q^{n \times m}$ and a trapdoor $\mathbf{T_B}$.
4. Select a matrix $\mathbf{F} \xleftarrow{\$} \mathbb{Z}_q^{4n \times 4m}$.
5. Finally, output ($\mathbf{A}$, $\mathbf{B}$, $\mathbf{F}$, $\mathbf{u}$) as the group public key **gpk**, the group manager's (issuer/issuing manager) secret key $\mathbf{ik} := \mathbf{T_A}$, and the tracing manager's (opener) secret key $\mathbf{ok} := \mathbf{T_B}$.

Join: A guest (new user) i having a personal public key and private key pair ($\mathbf{upk}[i]$, $\mathbf{usk}[i]$) communicate with the group manager to join the group by following the steps below.

1. User i, the new user (guest), samples a discrete Gaussian vector $\mathbf{x}_i \leftarrow D_{\mathbb{Z}^{4m},\sigma}$, and computes $\mathbf{z}_i \leftarrow \mathbf{F} \cdot \mathbf{x}_i \in \mathbb{Z}_q^{4n}$. Then he makes a signature $\Sigma_{join} \leftarrow \mathsf{Sig}(\mathbf{usk}[i], \mathbf{z}_i)$ and passes $\mathbf{z}_i$, and Σ_{join} to the group manager. Here, since the user (guest) has a leaving date (check-out date and time) he sends that date and time also along with $\mathbf{z}_i$ and Σ_{join}. Thus the user (guest) i sends the group manager $\mathbf{z}_i$, Σ_{join}, and t_{r_i}.
2. The group manager confirms that $\mathbf{z}_i$ was not owned by any previous member and t_{r_i} is correct and later than the current date. Then he validates Σ_{join} is a valid signature created on $\mathbf{z}_i$, using $\mathsf{Vf}(\mathbf{upk}[i], \mathbf{z}_i, \Sigma_{join})$. He terminates if $\mathbf{z}_i$, and Σ_{join} is invalid. Otherwise he signs the user's index $d = \mathsf{bin}(\mathbf{z}_i)$ and t_{r_i}, using his secret key $\mathsf{bin}(\mathbf{z}_i)$ is the binary representation of $\mathbf{z}_i$. Then he generates the certificate for the index and the expiration date (check-out date) $\textit{cert-index}_i = \mathsf{Sign}(\mathbf{ik}, (\mathsf{bin}(\mathbf{z}_i), t_{r_i}))$. Thus, later the user (guest) i cannot fake his check-out date and time.

   The group manager selects $\mathbf{R}_i \xleftarrow{\$} \mathbb{Z}_q^{n \times 4n}$ and computes $\mathbf{w}_i = \mathbf{R}_i \cdot \mathbf{z}_i$. Then he samples a vector $\mathbf{r}_i \in \mathbb{Z}^m \leftarrow \mathsf{SampleD}(\mathbf{T_A}, \mathbf{A}, \mathbf{u} - \mathbf{w}_i, \sigma)$, and creates a certificate for the token $\textit{cert-token}_i = \mathsf{Sign}(\mathbf{ik}, (\mathbf{A} \cdot \mathbf{r}_i))$ $(\mathbf{ik} = \mathbf{T_A})$.

Next in the registration table, he records the information of the new user i as $reg[i] \leftarrow (i, d, \mathbf{upk}[i], \mathbf{z}_i, \Sigma_{join}, \mathbf{R}_i, \mathbf{w}_i, t_{r_i}, \mathbf{r}_i, 1)$ and makes the record active (1). Finally, the group manager transfers the new member's member-certificate $cert_i = (cert\text{-}index_i, cert\text{-}token_i, \mathbf{R}_i, (\mathbf{A} \cdot \mathbf{r}_i), t_{r_i})$.

Sign: $\mathsf{Sign}(\mathbf{gpk}, \mathbf{gsk}[i], cert_i, M, t_s)$ is a randomized algorithm and on the given message M it produces a signature Σ employing the signer (guest) secret key $\mathbf{gsk}[i] = \mathbf{x}_i$ as follows.

1. Let $\mathcal{H}_1$: $\{0, 1\}^* \to \mathbb{Z}_q^{n \times \ell}$, $\mathcal{H}_2$: $\{0, 1\}^* \to \{1, 2, 3\}^t$ and $\mathcal{G}$: $\{0, 1\}^* \to \mathbb{Z}_q^{n \times m}$. $\mathcal{H}_1$, $\mathcal{H}_2$, and $\mathcal{G}$ are hash functions modeled as a random oracle.
2. Parse **gpk** as $(\mathbf{A}, \mathbf{B}, \mathbf{F}, \mathbf{u})$.
3. Parse $cert_i$ as $(cert\text{-}index_i, cert\text{-}token_i, \mathbf{R}_i, (\mathbf{A} \cdot \mathbf{r}_i), t_{r_i})$.
4. If $t_s > t_{r_i}$ then return ε.
5. Run $\mathsf{OGen}(1^n) \to (\mathbf{ovk}, \mathbf{osk})$.
6. Encrypt the index $d = \mathsf{bin}(\mathbf{z}_i)$, where $\mathbf{z}_i = \mathbf{F} \cdot \mathbf{x}_i$.
   (a) Let $\mathbf{G} = \mathcal{H}_1(\mathbf{ovk}) \in \mathbb{Z}_q^{n \times 2m}$.
   (b) Sample $\mathbf{s} \leftarrow \chi^n$, $\mathbf{e}_1 \leftarrow \chi^m$ and $\mathbf{e}_2 \leftarrow \chi^\ell$.
   (c) Compute the ciphertext $(\mathbf{c}_1, \mathbf{c}_2)$ pair
   $(\mathbf{c}_1 = \mathbf{B}^T\mathbf{s} + \mathbf{e}_1, \mathbf{c}_2 = \mathbf{G}^T\mathbf{s} + \mathbf{e}_2 + \lfloor q/2 \rfloor \mathsf{bin}(\mathbf{z}_i))$.
7. Select $\rho \xleftarrow{\$} \{0, 1\}^n$, let $\mathbf{V} = \mathcal{G}(\mathbf{A}, \mathbf{u}, M, \rho) \in \mathbb{Z}_q^{n \times m}$.
8. Get $\mathbf{v} = \mathbf{V} \cdot (\mathbf{A} \cdot \mathbf{r}_i) + \mathbf{e}_1 \mod q$ ($||\mathbf{e}_1||_\infty \leq \beta$ with overwhelming probability).
9. Execute $\mathsf{Verify}(\mathbf{A}, (\mathsf{bin}(\mathbf{z}_i), t_{r_i}), cert\text{-}index_i)$ to prove that $cert\text{-}index_i$ is generated on $(\mathsf{bin}(\mathbf{z}_i), t_{r_i})$ and $\mathsf{Verify}(\mathbf{A}, (\mathbf{A} \cdot \mathbf{r}_i), cert\text{-}token_i)$ to prove that $cert\text{-}token_i$ is created on $(\mathbf{A} \cdot \mathbf{r}_i)$. Next make a proof as in Sect. 4.1, to show the signer is valid, above $\mathbf{v}$ is sincerely computed, and index is correctly encrypted. Then repeat the protocol given in Sect. 4.1 $t = \omega(\log n)$ times to make the soundness error of the interactive protocol negligible. Finally, make it non-interactive using the Fiat–Shamir heuristic as a triple, $\Pi = (\{CMT^{(k)}\}_{k=1}^t, CH, \{RSP^{(k)}\}_{k=1}^t)$, where $\text{CH} = (\{Ch^{(k)}\}_{k=1}^t) = \mathcal{H}_2(M, \{CMT^{(k)}\}_{k=1}^t, \mathbf{c}_1, \mathbf{c}_2)$.
10. Compute $\mathcal{OTS}$; $sig = \mathsf{OSig}(\mathbf{osk}, (\mathbf{c}_1, \mathbf{c}_2, \Pi))$.
11. Produce the signature $\Sigma = (\mathbf{ovk}, (\mathbf{c}_1, \mathbf{c}_2), \rho, \Pi, sig, \mathbf{v}, t_{r_i}, t_s)$.

Verify: $\mathsf{Verify}(\mathbf{gpk}, M, \Sigma, RL, t_v)$ is a deterministic algorithm and it proceeds as follows, where $RL = \{\{\mathbf{u}_i\}_i\}$ and t_v is the current time of validation.

1. Parse Σ as $(\mathbf{ovk}, (\mathbf{c}_1, \mathbf{c}_2), \rho, \Pi, sig, \mathbf{v}, t_{r_i}, t_s)$.
2. If $t_v > t_s$ or $t_s > t_{r_i}$ then return 0.
3. Get $\mathbf{V} = \mathcal{G}(\mathbf{A}, \mathbf{u}, M, \rho) \in \mathbb{Z}_q^{n \times m}$.
4. If $\mathsf{OVer}(\mathbf{ovk}, ((\mathbf{c}_1, \mathbf{c}_2), \Pi), sig) = 0$ then output 0.
5. Parse Π as $(\{CMT^{(k)}\}_{k=1}^t, \{Ch^{(k)}\}_{k=1}^t, \{RSP^{(k)}\}_{k=1}^t)$.
6. If $(Ch^{(1)}, \ldots, Ch^{(t)}) \neq \mathcal{H}_2(M, \{CMT^{(k)}\}_{k=1}^t, \mathbf{c}_1, \mathbf{c}_2)$ return 0 else continue.

7. Using the verification steps of the commitment scheme in Sect. 4.1 for $k = 1$ to t validate $RSP^{(k)}$ with respect to $CMT^{(k)}$ and $Ch^{(k)}$. Return invalid if any of the conditions fails.
8. Compute $\mathbf{e}_i^{'} = \mathbf{v} - \mathbf{V} \cdot \mathbf{u}_i \mod q$ for each $\mathbf{u}_i \in RL$ to certify whether there exists an index i such that $\|\mathbf{e}_i^{'}\|_\infty \leq \beta$. If so output invalid and abort.
9. Output valid.

Open: Open(**gpk**, **ok**, *reg*, *M*, Σ) acts as below. Here $\mathbf{ok} = \mathbf{T_B}$.

1. Let $\mathbf{G} = \mathcal{H}_1(\mathbf{ovk})$.
2. Use $\mathbf{T_B}$ to obtain $\mathbf{Y} \in \mathbb{Z}^{m \times 2m}$. Here $\mathbf{Y}$ is a small norm matrix and $\mathbf{B} \cdot \mathbf{Y} = \mathbf{G} \mod q$.
3. Compute $\mathsf{bin}(\mathbf{z}_i) = \lfloor(\mathbf{c}_2 - \mathbf{Y}^T \cdot \mathbf{c}_1)/(q/2)\rceil$, identify the signer querying the registration table *reg* with $\mathsf{bin}(\mathbf{z}_i)$, and return the signer's index.

Revoke: Revoke(**gpk**, **ik**, *i*, *reg*, RL) works as below.

1. Query *reg* for *i* to get the revocation token $(\mathbf{A} \cdot \mathbf{r}_i)$ of the revoking member.
2. Insert $(\mathbf{A} \cdot \mathbf{r}_i)$ to RL and modify $reg[i]$ to inactive (0).
3. Output RL.

5 Correctness and Security Analysis of the Scheme

This section provides an analysis of the correctness and the security of the scheme.

5.1 Correctness

Theorem 1 *For all* $t_r, t_s, t_v, RL, M \in \{0, 1\}^*$, $(\mathbf{gpk}, \mathbf{ok}, \mathbf{ik}) \leftarrow \mathsf{GKg}(1^\lambda)$ *and* $(\mathbf{gsk}[i], \mathbf{grt}[i], cert_i) \leftarrow \mathsf{Join}$,

$\mathsf{Verify}(\mathbf{gpk}, M, \mathsf{Sign}(\mathbf{gpk}, \mathbf{gsk}[i], \mathbf{grt}[i], M, t_s), RL, t_v) = \mathsf{valid} \iff \mathbf{grt}[i] \notin RL$ *and* $t_r > t_s \geq t_v$ *and*

$\mathsf{Open}(\mathbf{gpk}, \mathbf{ok}, reg, M, \mathsf{Sign}(\mathbf{gpk}, \mathbf{gsk}[i], \mathbf{grt}[i], M, t_s)) = i$.

The new scheme allows only new users whose public keys are not used before to join the group. The group manager validates the new user and issues a certificate with the key-expiration time (guest's check-out). Thus any user cannot pass the signature generation with a fake expiration date (t_r). Moreover, at the time of signature generation, first, the key-expiration is checked. If the signer cannot provide a valid signature expiration date, then he cannot proceed with generating signatures. This confirms that the no expired member can generate a signature. Next the signatures which are only produced by active and honest users with a valid signature expiration date are accepted by Verify. If the provided signature is expired, then the signature verification fails. Further, signatures created by users whose token is in *RL* are not accepted by Verify. Both Sign and Verify ensures this condition. Completeness of the underlying proof system ensures to accept legitimate signatures always. The soundness of the underlying proof system assure that revoked signers fail get acceptance for their signatures. Open returns the index of the signature owner with overwhelming probability. Open computes $\mathsf{bin}(\mathbf{z}_i)$, and get the information of the signer from *reg*.

Above discussion proves the correctness of the new scheme.

5.2 Anonymity (Dynamical-Almost-Full Anonymity)

Theorem 2 *Under the hardness of* $LWE_{n,q,\chi}$ *problem the new scheme is dynamical-almost-full anonymous in the random oracle mode.*

With the following sequence of games we prove the anonymity of the proposed scheme.

Game 0: First, the challenger C obtains a group public key **gpk** and other keys from $\mathsf{KeyGen}(1^n, 1^N)$. Next, C delivers **gpk** and all the group members' secret keys **gsk** to the adversary$A$. In the query phase, A can access opening for any desired signature and he can ask for any member revocation tokens. Furthermore, A can add new users to the group. When A requests to add a new user, C validates and adds the new user details to the registration table *reg* and a list called **HU**. But C will not give the revocation token of the new user. Thus, even the member certificate including the token and the key-expiration date and time is generated and saved, the member certification will not be given in the registration query. C gives a success message only for member registration of the adversary. Moreover, when A requests to reveal any user token, C traces the requested user's index by adding the request to a list called **TU**, and returns the member certificate *cert* which was saved in the registration table at the time of registering. In the challenge phase, A sends two indices (i_0, i_1) with a message M^*. Only if (i_0, i_1) are newly added (in **HU**) and are not used for querying tokens (not in **TU**), then C makes and responds back a signature $\Sigma^* = (\mathbf{ovk}^*, (\mathbf{c}_1^*, \mathbf{c}_2^*), \rho^*, \Pi^*, sig^*, \mathbf{v}^*, t_{r_i}^*, t_s^*) \leftarrow \mathsf{Sign}(\mathbf{gpk}, \mathbf{gsk}[i_b]^*, cert_{i_b}, M^*, t_s^*)$ for a random $b \leftarrow \{0, 1\}$. Finally, A provides $b' \in \{0, 1\}$ as the guess of b. If $b' = b$ then outputs 1 or 0 otherwise.

The adversary may try to win the game by sending the challenging indices with two different expiration dates. However, he cannot use this trick to win the game as the challenger will generate the challenging signature only if both indices having valid expiration dates. Both indices should satisfy $t_r > t_s$. Moreover, it should satisfy $t_r > t_s \geq t_v$. Thus the adversary cannot provide any indices with key-expiration dates that fail only at the time of validating and that passes signature generation. Thus, the adversary cannot attack the anonymity of the scheme using the time-bound keys.

Game 1: In this game, C makes a trivial amendment comparing to the above game **Game** 0. The challenger C creates $\mathcal{OTS}$ key pair $(\mathbf{ovk}^*, \mathbf{osk}^*)$ at the start of the game which is generated at the time of signature generation in general. If A request opening of a valid signature $\Sigma = (\mathbf{ovk}, (\mathbf{c}_1, \mathbf{c}_2), \rho, \Pi, sig, \mathbf{v}, t_r, t_s)$, where

ovk=**ovk***, then C aborts the game by providing a random bit. Besides, **ovk**=**ovk*** contradicts the strong unforgeability of $\mathcal{OTS}$. On the other hand, as adversary cannot know **ovk***, **ovk**=**ovk*** has negligible probability. Moreover, if A provides a valid signature, which satisfies **ovk**=**ovk***, then sig is a forged signature. We believe that A does not ask for opening of a valid signature with **ovk***. As a result, C halting the game is negligible.

Game 2: In this game, C creates the random oracle $\mathcal{H}_1$ at the start, which is generally obtained at the signature generation. In this game, at the beginning, C substitutes **B** and **G**, which are the encrypting matrices. C picks $\mathbf{B}^* \in \mathbb{Z}_q^{n\times m}$ and $\mathbf{G}^* \in \mathbb{Z}_q^{n\times \ell}$ in uniformly random. Then he sets $\mathcal{H}_1(\mathbf{ovk}^*)$=$\mathbf{G}^*$. To response the opening oracle queries of A with $\Sigma = (\mathbf{ovk}, (\mathbf{c}_1, \mathbf{c}_2), \rho, \Pi, sig, \mathbf{v}, t_r, t_s)$, C chooses $\mathbf{Y} \leftarrow (D_{\mathbb{Z}^m,\sigma})^\ell$, and computes $\mathbf{G} = \mathbf{B}^*\mathbf{Y} \in \mathbb{Z}_q^{n\times\ell}$. This **G** is used to response the opening and records (**ovk**, **Y**, **G**) to be reused if the adversary A repeats the same queries for $\mathcal{H}_1(\mathbf{ovk})$. Since the distributions of **G** is statistically close to the uniform over $\mathbb{Z}_q^{n\times\ell}$ [9], this game is indistinguishable from Game 1.

Game 3: In this game, without genuinely producing the non-interactive proof Π, without using the witness C simulates the proof. For this he executes the simulator for each $k \in [t]$ and then programs the random oracle $\mathcal{H}_1$, respectively. The challenging signature Σ^* is statistically close to the signature in the previous games as the proof system is statistically zero-knowledge. Thus, Game 3 is indistinguishable from previous game.

Game 4: C substitutes the naive revocation token with a vector sampled uniformly. In general, $\mathbf{v} = \mathbf{V} \cdot \mathbf{grt}[i_b] + \mathbf{e}_1 \mod q$. But in this game, C uniformly samples $\mathbf{t} \xleftarrow{\$} \mathbb{Z}_q^n$ and get $\mathbf{v} = \mathbf{V} \cdot \mathbf{t} + \mathbf{e}_1 \mod q$. **V** is uniformly random over $\mathbb{Z}_q^{m\times n}$ and $\mathbf{e}_1$ is sampled from the error distribution χ. Since C substitutes only $\mathbf{grt}[i_b]$ with **t** and the rest of the game is same as previous game, the two games are statistically indistinguishable.

Game 5: In this game, **v** is taken uniformly. In this game, C makes the revocation token details without depending on the challenging bit b. Accordingly, C gets $\mathbf{y} \xleftarrow{\$} \mathbb{Z}_q^m$. Then he sets $\mathbf{v} = \mathbf{y}$. In **Game** 4, the pair $(\mathbf{V}, \mathbf{v})$ is a proper $LWE_{n,q,\chi}$ instance. In this game C substitutes **v** with $\mathbf{y} \xleftarrow{\$} \mathbb{Z}_q^m$ which is sampled uniformly. The adversary can solve Decision-LWE problem, if he can distinguish **v** and **y**. Since the lattice problem $LWE_{n,q,\chi}$ is hard, Game 5 is indistinguishable from Game 4.

Game 6: In this game C amends the making of ciphertext $(\mathbf{c}_1^*, \mathbf{c}_2^*)$ uniformly. He sets $\mathbf{c}_1^* = \mathbf{x}_1$ and $\mathbf{c}_2^* = \mathbf{x}_2 + \lfloor q/2 \rfloor d_b$. The vectors $\mathbf{x}_1 \in \mathbb{Z}^m$ and $\mathbf{x}_2 \in \mathbb{Z}^\ell$ are uniformly random. d_b is the index of the challenging bit. The rest of the game is same as Game 5. Since $LWE_{n,q,\chi}$ is hard, Game 5 and Game 6 are indistinguishable.

Game 7: C creates Σ^* totally independent of the bit b. He picks uniformly random $\mathbf{x}_1' \in \mathbb{Z}_q^m$ and $\mathbf{x}_2' \in \mathbb{Z}_q^\ell$, and sets $\mathbf{c}_1^* = \mathbf{x}_1'$ and $\mathbf{c}_2^* = \mathbf{x}_2'$. Game 7 is statistically indistinguishable from Game 6. The advantage of the adversary in this game is 0 because the signature is totally independent from the challenger's bit b.

Hence, these games prove that the new scheme with the time-bound keys is secure with the dynamical-almost-full anonymity.

5.3 Traceability

Theorem 3 *Under the hardness of* ***SIS*** *problem, the new scheme is traceable in the random oracle model.*

We take an algorithm B that is capable of solving SIS problem with non-negligible probability. An adversary A having **gpk** and **ok** returns (M, Σ) in the traceability game. Moreover the adversary A has facility to add new users and replace members' personal public keys. Also, A is allowed to query any member secret signing key and revocation token. By using oracles, B answers the queries done by A.

Finally, the adversary A outputs a fake signature Σ^*=($\mathbf{ovk}^*$, $(\mathbf{c}_1^*, \mathbf{c}_2^*)$, ρ^*, Π^*, sig^*, $\mathbf{v}^*$, t_r^*, t_s^*) on message M^*. B opens the signature Σ^* and identify the index. The improved Forking Lemma [4] ensures that, with probability at least $1/2$, B can get 3-fork involving tuple $(M, \{CMT^{(k)}\}_{k=1}^t, \mathbf{c}_1, \mathbf{c}_2)$ running A up to $32 \cdot Q_H/(\varepsilon - 3^{-t})$ times with the same records. Namely, the first $\kappa^* - 1$ random oracle queries A returns the answer given for $Ch_1, \ldots, Ch_{\kappa^*-1}$ as the first run. From the κ^*th query onwards A obtains a fresh random oracle values $Ch'_{\kappa^*}, \ldots, Ch'_{Q\mathcal{H}_2}$ at each new run. A simple computation shows that: $Pr[\exists j \in \{1, \ldots, t\}] : \{Ch_i^{(1)}, Ch_i^{(2)}, Ch_i^{(3)}\} = \{1, 2, 3\}$ with probability $1 - (7/9)^t$. Under the condition of the existence of such index i, one parses the 3 forgeries corresponding to the fork branches to obtain $(RSP_i^{(1)}, RSP_i^{(2)}, RSP_i^{(3)})$.

We can use the knowledge extractor of the underlying argument system to get vectors $(\mathbf{x}, \mathsf{bin}(\mathbf{z}), \mathbf{r}, \mathbf{s}, \mathbf{e}_1, \mathbf{e}_2)$ which satisfy the conditions of underlying interactive protocol. Then B can get a vector which is the secret signing key of the forgery. Furthermore this vector is an answer for the SIS problem.

This confirms the traceability of the new scheme.

5.4 Non-frameability

Theorem 4 *If* ***SIS*** *problem is hard, then the new scheme is non-frameable in the random oracle model.*

We take an adversary A who can create a forgery signature that opens to a honest user i who did not generate it. Then we make a **PPT** algorithm B that figure out a solution for *SIS* problem. B knows authority keys and B creates the group public key **gpk** truly. Then B replies to the queries made by A. A can work as a corrupted group manager and registers a new user i to the group. With user i if A asks to make a signature on a message M, then B creates and returns the signature Σ=($\mathbf{ovk}$, $(\mathbf{c}_1, \mathbf{c}_2)$, ρ, Π, sig, $\mathbf{v}$, t_r, t_s). Finally, A outputs (M^*, Σ^*=($\mathbf{ovk}^*$, $(\mathbf{c}_1^*, \mathbf{c}_2^*)$, ρ^*, Π^*, sig^*, $\mathbf{v}^*$, t_r^*, t_s^*)), which opens to i^* who did not create Σ^*. B has a short vector $\mathbf{z}_{i^*} = \mathbf{F} \cdot \mathbf{x}_{i^*} \mod q$. B should have another short vector $\mathbf{z}_{i'} = \mathbf{F} \cdot \mathbf{x}_{i'} \mod q$ to solve SIS instance. To obtain such a vector, B replays A sufficient times and applies Improved Forking Lemma [4]. Then, B can get a short vector $\bar{\mathbf{x}}$, where $\mathbf{z}_{i^*} = \mathbf{F} \cdot \bar{\mathbf{x}} \mod q$. Moreover, $\bar{\mathbf{x}} \neq \mathbf{x}_{i^*}$ with overwhelming probability according to Stern-like proof of knowledge. A nonzero vector $\mathbf{h} = \mathbf{x}_{i^*} - \bar{\mathbf{x}}$ is an answer for SIS problem.

This confirms the non-frameability of the new scheme.

6 Conclusion

In this paper, we presented a VLR group signature scheme with time-bound keys that can be applied in the hotel guest-management system. The proposed scheme used an existing scheme and discussed the advantages and security when applying the time-bound keys. As a result, this paper provided a lattice-based VLR group signature scheme with time-bound keys. The proposed scheme can be applied for systems like staff management, subscribers of a telephone company, and visitors accessing facilities. However, this paper did not present an implementation of the proposed system.

Acknowledgements This work is supported in part by JSPS Grant-in-Aids for Scientic Research (A) JP16H01705 and for Scientic Research (B) JP17H01695.

References

1. Agrawal, S., Boyen, X., Vaikuntanathan, V., Voulgaris, P., Wee, H.: Functional encryption for threshold functions (or fuzzy IBE) from lattices. In: PKC 2012. LNCS, vol. 7293, pp. 280–297. Springer (2012)
2. Ajtai, M.: Generating hard instances of lattice problems. In: Proceedings of the Twenty-Eighth Annual ACM Symposium on Theory of Computing, pp. 99–108. ACM (1996)
3. Bellare, M., Micciancio, D., Warinschi, B.: Foundations of group signatures: formal definitions, simplified requirements, and a construction based on general assumptions. In: EUROCRYPT 2003. LNCS, vol. 2656, pp. 614–629. Springer (2003)
4. Brickell, E., Pointcheval, D., Vaudenay, S., Yung, M.: Design validations for discrete logarithm based signature schemes. In: PKC 2000. LNCS, vol. 1751, pp. 276–292. Springer (2000)
5. Camenisch, J., Neven, G., Rückert, M.: Fully anonymous attribute tokens from lattices. In: SCN 2012. LNCS, vol. 12, pp. 57–75. Springer (2012)
6. Chaum, D., Van Heyst, E.: Group signatures. In: EUROCRYPT 1991. LNCS, vol. 547, pp. 257–265. Springer (1991)
7. Chow, S.S., Wong, D.S.: Anonymous identification and designated-verifiers signatures from insecure batch verification. In: EuroPKI 2007. LNCS, vol. 4582, pp. 203–219. Springer (2007)
8. Chu, C.K., Liu, J.K., Huang, X., Zhou, J.: Verifier-local revocation group signatures with time-bound keys. In: Proceedings of the 7th ACM Symposium on Information, Computer and Communications Security, pp. 26–27. ACM (2012)
9. Gentry, C., Peikert, C., Vaikuntanathan, V.: Trapdoors for hard lattices and new cryptographic constructions. In: ACM 2008, pp. 197–206. ACM (2008)
10. Gordon, S.D., Katz, J., Vaikuntanathan, V.: A group signature scheme from lattice assumptions. In: ASIACRYPT 2010. LNCS, vol. 6477, pp. 395–412. Springer (2010)

11. Kawachi, A., Tanaka, K., Xagawa, K.: Concurrently secure identification schemes based on the worst-case hardness of lattice problems. In: ASIACRYPT 2008. LNCS, vol. 5350, pp. 372–389. Springer (2008)
12. Laguillaumie, F., Langlois, A., Libert, B., Stehlé, D.: Lattice-based group signatures with logarithmic signature size. In: ASIACRYPT 2013. LNCS, vol. 8270, pp. 41–61. Springer (2013)
13. Langlois, A., Ling, S., Nguyen, K., Wang, H.: Lattice-based group signature scheme with verifier-local revocation. In: PKC 2014. LNCS, vol. 8383, pp. 345–361. Springer (2014)
14. Libert, B., Ling, S., Mouhartem, F., Nguyen, K., Wang, H.: Signature schemes with efficient protocols and dynamic group signatures from lattice assumptions. In: ASIACRYPT 2016. LNCS, vol. 10032, pp. 373–403. Springer (2016)
15. Ling, S., Nguyen, K., Wang, H., Xu, Y.: Lattice-based group signatures: achieving full dynamicity with ease. In: ACNS 2017. LNCS, vol. 10355, pp. 293–312. Springer International Publishing, Cham (2017)
16. Micciancio, D., Peikert, C.: Trapdoors for lattices: simpler, tighter, faster, smaller. In: EUROCRYPT 2012, vol. 7237, pp. 700–718. Springer (2012)
17. Naor, D., Shenhav, A., Wool, A.: One-time signatures revisited: have they become practical? IACR Cryptology ePrint Archive 2005, p. 442 (2005)
18. Peikert, C.: A decade of lattice cryptography. Found. Trends Theor. Comput. Sci. **10**(4), 283–424 (2016). https://doi.org/10.1561/0400000074
19. Perera, M.N.S., Koshiba, T.: Achieving almost-full security for lattice-based fully dynamic group signatures with verifier-local revocation. In: ISPEC 2018. LNCS, vol. 11125, pp. 229–247. Springer (2018)
20. Perera, M.N.S., Koshiba, T.: Fully dynamic group signature scheme with member registration and verifier-local revocation. In: ICMC 2018. PROMS, vol. 253, pp. 399–415. Springer (2018)
21. Perera, M.N.S., Koshiba, T.: Zero-knowledge proof for lattice-based group signature schemes with verifier-local revocation. In: NBiS 2018. LNDECT, vol. 22, pp. 772–782 (2019)
22. Regev, O.: On lattices, learning with errors, random linear codes, and cryptography. In: STOC 2005, pp. 84–93. ACM Press (2005)

Theory and Application of Computationally-Independent One-Way Functions: Interactive Proof of Ability—Revisited

Sabyasachi Dutta and Kouichi Sakurai

Abstract We introduce the concept of computationally-independent pair of one-way functions (CI-OWF). We also provide two rich classes of examples of such functions based on standard assumptions. We revisit two-party interactive protocols for proving possession of computational power and existing two-flow challenge-response protocols. We analyze existing protocols for proof of computation power and propose a new two-flow protocol using CI-OWF based on *square* Diffie–Hellman problem. We observe that our work is related to Komargodski's work [TCC 2016] on leakage resilient one-way functions.

Keywords One-way functions · Zero-knowledge proof · Computational independence · Proof of computation ability · Leakage resilient

1 Introduction

1.1 Background

Interactive proof systems were developed by Goldwasser et al. [15] to prove membership in a language. In such a system, a *prover* and a *verifier* interacts so that at the end of the protocol the prover convinces the verifier (with overwhelming probability) that indeed the input is a member of some prefixed language. A *zero-knowledge* interactive proof between a prover and a verifier was introduced to capture the scenario

S. Dutta is financially supported under the NICT International Exchange Program.

S. Dutta (✉) · K. Sakurai
Faculty of Information Science and Electrical Engineering, Kyushu University, Fukuoka, Japan
e-mail: saby.math@gmail.com

K. Sakurai
e-mail: sakurai@inf.kyushu-u.ac.jp

D. Giri et al. (eds.), *Proceedings of the Fifth International Conference on Mathematics and Computing*, Advances in Intelligent Systems and Computing 1170, https://doi.org/10.1007/978-981-15-5411-7_7

that after the interaction, the verifier is convinced of the fact that the input belongs to a language but gains no more knowledge. Later, zero-knowledge *proof of knowledge* was introduced in which a polynomial-time prover not only proves membership in a zero-knowledge manner but also provides evidence of the fact that he really possesses/knows a *witness* for some predicate about the input. Feige et al. [12] and Tompa et al. [27] gave formal definitions and implementations of a zero-knowledge (interactive) proof of knowledge system. However, what we are interested in this paper is not exactly the knowledge/possession of witness for proving a statement but to "prove the ability to perform a computational task". Such a scenario was first considered by Yung [29] and Koyama [18] independently and later by Bellare et al. [4].

1.2 Related Work and Our Contribution

Initial two works [18, 29] are along the same line. Both of them considered the scenario when someone possesses some extraordinary power to solve a *hard* problem and he wants to "prove this ability" to a verifier. Yung [29] considered the power to solve integer factorization and gave a "zero-knowledge proof of computational ability" protocol by using two zero-knowledge proof of knowledge protocols as subroutines. The author also gave a protocol for proving computational power for "samplable verifiable" problem by using two perfect zero-knowledge proof protocols as subroutines. The techniques are novel which help to extend "proofs of knowledge" to "proofs of proving computational power" but the complexity of the protocols are very high. Later Bellare et al. [4] formalized the idea and provided theoretical constructions.

On the other hand, Koyama [18] proposed a novel idea to prove "directly" the computational power that a prover possesses. He proposed protocols with only two flows of message transfer between a prover and a verifier which rapidly decreased the cost of the protocol. On the downside, the protocols remain no longer zero-knowledge protocols as claimed by the author. However, the idea of reducing communication complexity just by using two flows of message transfer was novel. Similar idea was also considered by Stinson et al. [26] to build an efficient zero-knowledge identification protocol. Their scheme is secure in the random oracle model and is based on non-standard "knowledge of exponent" problem and standard computational Diffie–Hellman problem. Some more works [3, 28] followed the idea of Koyama [18] in different scenarios.

We first revisit Koyama's protocol in Sect. 3 which acts as a motivation to our work. We analyze the protocols and show some flaws. We then introduce notion of *computationally-independent* one-way functions (CI-OWF) which can be useful to construct interactive protocols for proving computational power. Such proposal of computationally-independent one-way functions was originally given by Sakurai et al. [25].[1] We argue that since in a two-flow message transfer protocol achieving "zero-knowledge" is impossible [13], one has to make a trade-off between the level of security and cost of communication. With moderate and practical security, one can build proof of computational ability using CI-OWF. Moreover, all the existing works consider scenarios like "inverting a one-way function" to be a hard problem and the prover has the power to invert. But there can be hard computational problems which may not require inversion of one-way functions, e.g., computational Diffie–Hellman Problem. Studying a pair of one-way functions as a building block for constructing cryptographic protocols is not new. The work of Goldwasser et al. [14] constructs a pair of claw-free one-way functions to build general signature schemes. Later the concept was studied in [8, 9, 19, 24].

On a different but connected note, recently Komargodski [16] considered the following question: given an adaptive/selective one-way leakage function(s) whether one can construct an one-way function which can resist the leakage. He showed that if the leakage function is adaptive then it is impossible to construct a leakage resilient one-way function in both *random oracle* and *standard model*. However, if the leakage function is fixed ahead in time, i.e., *selective* then it is possible to construct such leakage resilient one-way functions.

We make a connection to the concept of leakage resilient one-way functions with our work. We have observed that some amount of leakage is inevitable from the outputs of a pair of computationally-independent functions (*see* Theorems 10 and 11). The relief was that the leakages were also one-way functions which did not reveal information about inputs.

In 2016, Komargodski [16] considered the problem of constructing leakage resilient one-way functions. The paper [16] is a follow up of a work by Dodis et al. [10]. Dodis et al. constructed leakage resilient one-way functions f in presence of leakage function L. More specifically, they constructed one-way functions f such that given any random image $f(x)$ and lossy leakage function $L(x)$ it is still hard to invert $f(x)$. The problem finds its root in an early work of Canetti et al. [6] where it was shown how to obtain a leakage resilient one-way function assuming that the adversary can leak an arbitrary but "small" subsets of bits of the input. Later, [1, 10] showed the existence of one-way functions assuming that the adversary can leak any lossy function of the input.

Komargodski [16] forwarded the research by considering the problem of constructing leakage resilient one-way functions in presence of arbitrary computationally hiding leakage (auxiliary-input). Two types of leakage were considered—adaptive and selective. We state the main results from [16].

[1] (English abstract of SCIS'89) Recently many researchers propose the protocols which is applied to the interactive proof systems and the global network systems, where they use some one-way functions. But there are not a few protocols which they proposed are unsecure. The reason is mainly owing to a pair of one-way functions which has become no longer one-way. In this note, we introduce the notion "Computationally-Independent-CI-" to study a pair of one-way functions and investigate the property of the functions. Here we apply to CI to the cryptographic protocol.

Theorem 1 ([16]) *Suppose* $\mathcal{O} : \{0, 1\}^* \longrightarrow \{0, 1\}^n$ *is a random oracle and* $\mathcal{L}^{\mathcal{O}} = \{L : \{0, 1\}^n \longrightarrow \{0, 1\}^*\}$ *denotes the collection of all oracle-aided one-way functions. Then there is no leakage resilient one-way function family* $\mathcal{F} = \{f : \{0, 1\}^n \longrightarrow \{0, 1\}^*\}$ *for the leakage family* $\mathcal{L}^{\mathcal{O}}$.

Theorem 2 ([16]) *Suppose* $\mathcal{L} = \{L : \{0, 1\}^n \longrightarrow \{0, 1\}^*\}$ *denotes the collection of all one-way functions. Assuming a public-coin multi-bit point obfuscator, there is no leakage resilient one-way function family* $\mathcal{F} = \{f : \{0, 1\}^n \longrightarrow \{0, 1\}^m\}$ *for the leakage family* $\mathcal{L}$.

In addition to the above two negative results one positive result was given in [16].

Theorem 3 ([16]) *Suppose* $L : \{0, 1\}^n \longrightarrow \{0, 1\}^m$ *be a fixed leakage (one-way) function. Assuming that polynomially many hardcore bits can be extracted from L, there is a leakage resilient one-way function* $f : \{0, 1\}^n \longrightarrow \{0, 1\}^*$ *for the leakage family* $\mathcal{L} = \{L\}$.

Is it possible to achieve pair of computationally-independent one-way functions which do not leak any information? We begin with putting a definition of such pair.

Definition 1 (*Strong CI-OWF*) A pair of one-way functions f, g defined over the same domain is called strong computationally-independent one-way functions if the following three conditions are satisfied:

(CI-a) The function $H(x) = (f(x), g(x))$ is still one-way.
(CI-b) Given $f(x_0)$ for randomly chosen x_0, it is hard to compute $g(x_0)$ and given $g(y_0)$ for randomly chosen y_0, it is hard to compute $f(y_0)$.
(CI-c) There is no one-way (leakage) function $L : \{0, 1\}^* \longrightarrow \{0, 1\}^*$ such that $L(x)$ can be efficiently computed given only $f(x)$ and $g(x)$ (and not the input x).

Theorem 4 *It is impossible to construct strong CI-OWF, both in random oracle model and standard model.*

Our definition of CI-OWF (Definition 2) is a weaker version of the above. Theorems 10 and 11 guarantees that given specific form of selective leakage one-way functions L we can construct a pair of CI-OWF based on RSA and square-DH assumptions.

2 Preliminaries

We mention some basic results that will be needed in the paper.

Variants of Diffie–Hellman Problem

Let us recall the some relevant notions required for our work. Let G be a cyclic group of order n and generator g. The two well-known "hard" problems are the following:

The Discrete Logarithm Problem (DLP): On input $g, g^x \in G$, compute $x \in \mathbb{Z}_n$.

The Computational Diffie–Hellman Problem (CDH): On input $g, g^x, g^y \in G$ with $x, y \in \mathbb{Z}_n$, compute g^{xy}.

It is not hard to see that if an adversary is able to solve DLP then he can also solve CDH, i.e., DLP $\Longrightarrow$ CDH. The converse has been shown to hold for some cases [5, 20] but in general, it is yet to be proved that the problems are indeed equivalent. Numerous cryptographic primitives have been developed based on the DLP and CDH assumptions. Chaum et al. [7] gave a "proof-of-possession" of discrete logarithm in a zero-knowledge manner. In their work they have provided a protocol which enables a *prover* to convince a *verifier* that he really possesses discrete log without actually revealing the value to the verifier.

Maurer and Wolf [21] defined various types of CDH *oracles* and proved their equivalence. One of the interesting type is called *square* Diffie–Hellman problem (SDHP) where the corresponding oracle returns g^{x^2} when given inputs g and g^x. Bao et al. [2] proved the equivalence of CDH and SDH when the underlying group is of prime order. Many more variations of CDH and their connections to DLP have been studied in the literature. For example, l-weak DHP [22], *square root* DHP (SRDHP) [17], equivalence between SDHP and SRDHP [23], to name a few.

3 Revisiting the Protocols of Koyama [18]

Koyama [18] considered a scenario where a cryptanalyst wants to convince a potential buyer that the former has indeed invented an efficient code-breaking algorithm without actually revealing the algorithm to the buyer. To be more precise, suppose a cryptanalyst has found an efficient algorithm to break an encryption scheme (Enc, Dec) completely. So, given any ciphertext $c = E(m)$ he is able to find the underlying message m. Now the cryptanalyst acts as a *prover* and a potential buyer acts as a *verifier*. The main goal of the prover is to demonstrate the breaking of the cryptosystem without revealing any knowledge about the algorithm and the message m to the verifier. Therefore, basic requirements of such a *secure* demonstrating protocol are as follow: it must be "non-cheatable" and revealing no knowledge. Formally, the protocol must be

1. (*Complete*) if the proof is correct then the verifier should accept prover's claim with probability 1.
2. (*Sound*) if the proof is *incorrect* then verifier should reject prover's claim with overwhelming probability.

3. (*zero-knowledge*) the proof does not yield knowledge such as value of m and the code-breaking method.

Based on the above requirements described in [18], the author built two protocols to securely and directly demonstrate the possession of code-breaking algorithm by a cryptanalyst to a potential buyer.

First such a generic protocol (Algorithm 1) was given (Protocol 1 in [18]):

Algorithm 1 General form of direct protocol [18]

---

procedure AGREEMENT BETWEEN PROVER & VERIFIER
Prover and Verifier agree on functions Enc, f, g such that $Enc(m) = g(f(m))$.
Prover and Verifier share a *random* input C in the ciphertext space.
procedure COMPUTATION BY PROVER
Prover computes m from C using his efficient attacking method as $m = E^{-1}(C)$.
Prover computes $P = f(m)$ easily and sends P to the Verifier.
procedure VERIFICATION BY THE VERIFIER
Upon receiving P, Verifier checks $g(P) = C$ or not.
If *equality* holds then the Verifier *accepts* the claim that Prover can break the target public-key cryptosystem. Otherwise, the Verifier *rejects*.

Koyama gave the following protocol (Algorithm 2) for the discrete logarithm problem. The *prover* wants to demonstrate the power of solving DLP to a verifier. Let G be a cyclic group of order n with g as one of the generators.

Algorithm 2 Direct protocol for DLP [18]

---

procedure AGREEMENT BETWEEN PROVER & VERIFIER
Prover and Verifier agree on base value g and modulus n.
Prover and Verifier share a *random* input $C \in G$.
procedure RANDOMNESS CHOSEN BY VERIFIER
Verifier randomly generates R relatively prime to $n - 1$.
Computes S such that $RS = 1 \; mod(n - 1)$.
Computes $k = g^R$ and sends k to the Prover.
procedure COMPUTATION BY THE PROVER
Prover uses his efficient algorithm to compute m such that $g^m = C$.
Computes $P = k^m$ and sends P to the Verifier.
procedure VERIFICATION BY THE VERIFIER
Verifier computes P^S and checks $P^S = C$ or not.
If *equality* holds then the Verifier *accepts*. Otherwise, the Verifier *rejects*.

Koyama gave another protocol (Algorithm 3) to demonstrate the power of a prover that he can attack RSA cryptosystem.

Algorithm 3 Direct protocol for RSA [18]

procedure AGREEMENT BETWEEN PROVER & VERIFIER
  Prover and Verifier agree on public key $e = km$, $k, m \geq 3$ and RSA modulus n (whose factorization is unknown to both).
  Prover and Verifier share a *random* input $C \in \{0, 1, \ldots, n-1\}$.
procedure COMPUTATION BY THE PROVER
  Prover uses his efficient algorithm to compute M from (c, e, n) such that $M^e = C$.
  Computes $P = M^k \ mod\ n$ and sends P to the Verifier.
procedure VERIFICATION BY THE VERIFIER
  Verifier computes P^m and checks $P^m = C$ or not.
  If *equality* holds then the Verifier *accepts*. Otherwise, the Verifier *rejects*.

4 Analysis of Koyama's Protocols

We show some flaws present in Koyama's direct protocols. Both the protocols do not satisfy conditions for zero-knowledge. We further claim that the protocol for the DLP is not sound. That is, a cheating prover is always able to make an honest verifier accept. In this regard, we again mention that solving DLP implies solving CDH but the converse is yet to be proved as true. Although there are a number of works [5, 20] that shows evidence to the fact that possibly CDH and DLP are equivalent, but it is not proved for all cyclic groups. An adversary attacking CDH problem is thus less powerful adversary.

Suppose $\mathcal{A}$ is an adversary who is able to solve CDH problem and not DLP. The adversary $\mathcal{A}$ can cheat an honest verifier into making him believe that $\mathcal{A}$ has the power of solving DLP in the following manner:

1. After receiving $k = g^R$ from the verifier (line 7 of Algorithm 2), $\mathcal{A}$ possesses $(g, C = g^m, k = g^R)$.
2. $\mathcal{A}$ computes $g^{mR} \longleftarrow (g, g^m, g^R)$ using his algorithm to solve CDHP.
3. $\mathcal{A}$ sends g^{mR} to the verifier who then computes (line 12 of Algorithm 2) $(g^{mR})^S = g^m = C$.
4. Since the equality holds, the verifier *accepts* (line 13 of Algorithm 2).

We now have the following theorem.

Theorem 5 *The protocol (Algorithm 2) for DLP does not satisfy the soundness property. Hence, the protocol is not secure.*

Remark 1 We want to make a remark at this point that although we have shown that there is a direct attack to Koyama's protocol for DLP, the protocol is also not a zero-knowledge protocol. The verifier comes to know about a Sth root of a randomly chosen element C of G.

Theorem 6 *The protocol (Algorithm 3) for RSA is not secure. More specifically, the protocol lacks zero-knowledge property.*

Proof We simply note that the scheme is not zero-knowledge, as claimed by Koyama [18]. The verifier gets a random mth root of C, which makes the scheme insecure even against an honest verifier.

We conclude that, Koyama's "zero-knowledge" requirement is not exactly the same as standard zero-knowledge requirement. It is but a loose version of the standard one. The main requirement was not to leak the message m and to keep the design of the attacking algorithm completely hidden.

5 Computationally-Independent One-Way Functions

Building cryptographic tools on the basis of existence of two one-way functions is not new in the literature. Goldwasser et al. [14] considered *claw-free* permutations to build signature schemes resisting *chosen-message attack*. Later, Damgard [8] constructed collision-free hash functions based on the claw-free permutations and Russell [24] showed existence of collision-free hash functions is equivalent to the existence of claw-free pairs of pseudo-permutations. Krawczyk et al. [19] provided non-interactive trapdoor commitments based on claw-free permutations and lastly, Dodis et al. [9] showed that claw-free pair of permutations deliver more security than trapdoor permutations when building full-domain hash like signature schemes.

We begin with a proposal to defining computationally-independent one-way functions drawing motivation from Koyama's work and justify the reasons behind it. We refer to the full version [11] for more details and proofs.

5.1 On the Definition of Computationally-Independent One-Way Function

Definition 2 (*adopted from* [25]) A pair of one-way functions f, g is called computationally-independent one-way functions if they satisfy the following two conditions:

- (CI-a) The function $H(x) = (f(x), g(x))$ is still one-way.
- (CI-b) Given $f(x_0)$ for randomly chosen x_0, it is hard to compute $g(x_0)$ and given $g(y_0)$ for randomly chosen y_0, it is hard to compute $f(y_0)$.

In the above definition the first condition viz. (CI-a) ensures the fact that nothing about the input is revealed when both the functional outputs are obtained, which is a basic requirement. This condition is called the one-wayness of the pair. The second condition guarantees the computational independence of the functions, i.e., it is hard to compute one function given the output of the other function. To understand the necessity of the second condition let us consider the following example.

Example 1 Let us choose the functions $f(x) = x^2\ mod\ n$ and $g(x) = x^4\ mod\ n$, where $n = pq$ is a product of two large safe primes so that we are in a RSA group. Now we can see that (CI-a) is automatically satisfied because computing a pre-image would be equivalent to finding integer square root modulo n which is widely believed to be a hard problem. However, computing $g(x)$ from $f(x)$ is easy and when intended to be applied as a proof of computing power protocol, a cheating verifier can easily fool an honest verifier.

Remark 2 We point out that the two conditions (CI-a) and (CI-b) are independent. One does not imply the other. In Example 1 we see that although (CI-a) holds, (CI-b) does not hold true. Therefore, (CI-a) does not imply (CI-b). To see that the reverse implication is false we consider the following Example 2.

Example 2 Let us consider the functions $f(x) = x^2\ mod\ n$ and $g(x) = x^3\ mod\ n$ where n is a product of two large safe primes.

We now give two examples of family of computationally-independent one-way functions based on RSA problem and variants of Diffie–Hellman problems.

Example 3 Let us consider the family of functions with RSA modulus $n = pq$, $\mathcal{F} = \{x^2\ mod\ n, x^3\ mod\ n, \ldots, x^s\ mod\ n, \ldots\}$. Let us choose the functions $f(x) = x^4\ mod\ n$ and $g(x) = x^6\ mod\ n$ from the above family $\mathcal{F}$. Now we can see that (CI-a) is satisfied because from $x^4\ mod\ n$ and $x^6\ mod\ n$ one can compute $x^2\ mod\ n$, but producing a pre-image is as hard as finding a square root in the RSA group. The idea of the proof is as follows:

Suppose there is an adversary $\mathcal{A}$ who can successfully break the one-wayness of $H(x) = (x^4\ mod\ n, x^6\ mod\ n)$. We can now construct an adversary $\mathcal{B}$ having an oracle access to $\mathcal{A}$ to solve the problem of finding square root. $\mathcal{B}$ is given a quadratic-residue y in the RSA group and is asked to find a square root of it. $\mathcal{B}$ computes (y^2, y^3) and passes the tuple to $\mathcal{A}$. $\mathcal{B}$ outputs the number that he receives from $\mathcal{A}$. The success probability of $\mathcal{B}$ is equal to the success probability of $\mathcal{A}$.

Now it is easy to prove that computing $g(x)$ from $f(x)$ is hard. An adversary who can break the above hardness with non-negligible success probability ϵ can be used as an oracle to compute square root an element with success probability roughly $\frac{\epsilon}{4}$. Also, computing $f(x)$ from $g(x)$ is hard because if one can compute $f(x)$ from $g(x)$ then he can actually extract cube root of x^6.

We can now state the following two theorems.

Theorem 7 *Let* $\mathcal{F} = \{x^k\ mod\ n\}_{k\geq 2}$ *denote a family of one-way functions over* $\mathbb{Z}_n^*$ *with RSA modulus* $n = pq$, *where* p, q *are two large safe primes of same size. Let* $f(x) = x^s\ mod\ n$ *and* $g(x) = x^t\ mod\ n$ *be two functions from the family such that* $gcd(s, t) = d > 1$ *and neither* s *divides* t *nor* t *divides* s. *Then the pair* $(f(x), g(x))$ *give computationally-independent one-way functions.*

Theorem 8 *Let* $\mathcal{F} = \{x^k\ mod\ n\}_{k\geq 2}$ *denote a family of one-way functions over* $\mathbb{Z}_n^*$ *with RSA modulus* $n = pq$, *where* p, q *are two large safe primes of same size. Any two functions* $x^a\ mod\ n$ *and* $x^b\ mod\ n$ *from the family such that* a, b *are relatively prime fails to be a pair of computationally-independent one-way functions.*

Another result based on the computational square Diffie–Hellman Problem is stated below.

Theorem 9 *Let G be a cyclic group of order p and generator g. Let f, h be two functions defined by* $f(x) = g^x$ *and* $h(x) = g^{x^2}$ *for all* $x \in \mathbb{Z}_p^*$*. Then the pair* $(f(x), h(x))$ *gives a computationally-independent pair of one-way functions.*

5.2 On Leakage of Information from CI-OWF

So far we have discussed the definition of computationally-independent one-way functions and gave two rich classes of examples. In this section, we discuss on the amount of information leaked (about the input) from the CI-OWF pair $(f(x), g(x))$. From the output values one can compute every efficiently computable function with input $(f(x), g(x))$. We ask the most important question: how much leakage about the input x is observed from the values $f(x)$ and $g(x)$?

From the condition (CI-a) of Definition 2 the leakage cannot be x itself. Thus the leakage can be thought as the output of some one-way leakage function L when evaluated at x, i.e., $L(x)$.

Theorem 10 *The leakage obtained from* $f(x) = x^s \bmod n$ *and* $g(x) = x^t \bmod n$ *satisfying all the conditions of Theorem 7 can be written as* $L(x) = x^d \bmod n$ *where* d *is the greatest common divisor of* s *and* t*.*

Theorem 11 *The leakage obtained from* $f(x) = g^x$ *and* $h(x) = g^{x^2}$ *satisfying all conditions of Theorem 9 can be written as* $L(x) = g^{Q(x)}$*, where* $Q(x)$ *is a quadratic polynomial in* x*.*

6 Protocol for Proving Ability

In this section, we discuss two-flow protocols (one challenge from the verifier and one response from the prover) for proving computational ability with the help of CI-OWF (f, g). One fully secure way to implement a protocol is to follow the footsteps of Yung [29]. However, we are interested in two-flow protocols.

The broad idea for constructing such protocols is as follows: the *verifier* sends a challenge in the form of $f(r)$ to the *prover* and prover responds back with $g(r)$.

This way the prover establishes the fact that he has computational power to solve the "hard" problem of computing $g(r)$ from $f(r)$, where r is chosen randomly by the verifier. We note that it includes the scenario of "the power of inverting f" (which was the original motivation for defining proof of ability [18, 29]) as well as "the power to extract $g(r)$ from $f(r)$ without explicitly computing r" capturing a notion of malleability.

We now describe modifications of Koyama's protocols to convert them into two-flow protocols assuming an *honest* verifier. We modify our requirements for the protocols as follows:

1. (*Completeness*) if the proof is correct then the verifier should accept prover's claim with probability 1.
2. (*Soundness*) if the proof is *incorrect* then verifier should reject prover's claim with overwhelming probability.
3. (*privacy*) the proof does not reveal the input r (to an eavesdropper or a dishonest verifier) and the "powerful" algorithm remains completely hidden.

We have already shown that the protocol of Koyama for DLP (Algorithm 2) is not sound. We propose the following protocol based on the hardness of Square DH (SDH) problem. Note that in a prime order group CDH problem and SDH problem are equivalent [2]. We recall that SDH assumption says that given g and g^x it is hard to compute g^{x^2}, where x is randomly chosen element from $\mathbb{Z}_p^*$. Suppose a prover wants to prove "securely" his ability to solve the SDH problem. The following simple protocol based on the CI-OWF (g^x, g^{x^2}) provides a solution to this problem in presence of an honest verifier.

Algorithm 4 Direct protocol for SDH problem

1: **procedure** AGREEMENT BETWEEN PROVER & VERIFIER
2: Prover and Verifier agree on base value g and modulus n.
3: **procedure** CHALLENGE BY VERIFIER
4: Verifier randomly generates r and computes $g^r = C$.
5: Sends C to the Prover.
6: **procedure** COMPUTATION BY THE PROVER
7: Prover uses his efficient algorithm to compute $g^{r^2} = R$ from C.
8: Sends R to the Verifier.
9: **procedure** VERIFICATION BY THE VERIFIER
10: Verifier computes C^r and checks $C^r = R$ or not.
11: If *equality* holds then the Verifier *accepts*. Otherwise, the Verifier *rejects*.

Conclusion

We study the notion of computationally-independent one-way functions and provide two rich classes of examples based on standard assumptions. We analyzed the two-flow protocols of Koyama [18] and showed some flaws inherent in the protocols. Last we described a concrete two-flow protocol on the basis of newly introduced computationally-independent one-way functions. Finding further examples of such family of functions remain an interesting open issue.

References

1. Alwen, J., Dodis, Y., Wichs, D.: Survey: leakage resilience and the bounded retrieval model. In: ICITS 2009, pp. 1–18 (2009)
2. Bao, F., Deng, R.H., Zhu, H.: Variations of Diffie-Hellman problem. In: ICICS'03, pp. 301–312 (2003)
3. Bao, F., Lee, C.-C., Hwang, M.-S.: Cryptanalysis and improvement on batch verifying multiple RSA digital signatures. Appl. Math. Comput. **172**(2), 1195–1200 (2006)
4. Bellare, M., Goldreich, O.: Proving computational ability. Studies in Complexity and Cryptography 2011, pp. 6–12 (2011)
5. den Boer, B.: Diffie-Hellman is as strong as discrete log for certain primes. In: CRYPTO'88, pp. 530–539 (1988)
6. Canetti, R., Dodis, Y., Halevi, S., Kushilevitz, E., Sahai, A.: Exposure-resilient functions and all-or-nothing transforms. In: EUROCRYPT 2000, pp. 453–469 (2000)
7. Chaum, D., Evertse, J.-H., van de Graaf, J., Peralta, R.: Demonstrating possession of a discrete logarithm without revealing it. In: CRYPTO'86, pp. 200–212 (1986)
8. Damgard, I.: Collision free hash functions and public key signature schemes. In: EUROCRYPT'87, pp. 203–216 (1987)
9. Dodis, Y., Reyzin, L.: On the power of claw-free permutations. In: SCN'02, pp. 55–73 (2002)
10. Dodis, Y., Haralambiev, K., Lpez-Alt, A., Wichs, D.: Cryptography against continuous memory attacks. In: FOCS 2010, pp. 511–520 (2010)
11. Dutta, S., Sakurai, K.: Theory and application of computationally independent one-way functions: interactive proof of ability - revisited. Cryptology ePrint Archive: Report 2019/323 (2019)
12. Feige, U., Fiat, A., Shamir, A.: Zero-knowledge proofs of identity. J. Cryptol. **1**(2), 77–94 (1988)
13. Goldreich, O., Kahan, A.: How to construct constant-round zero-knowledge proof systems for NP. J. Cryptol. **9**(3), 167–190 (1996)
14. Goldwasser, S., Micali, S., Rivest, R.L.: A "paradoxical" solution to the signature problem (extended abstract). In: FOCS'84, pp. 441–448 (1984)
15. Goldwasser, S., Micali, S., Rackoff, C.: The knowledge complexity of interactive proof-systems (extended abstract). In: STOC'85, pp. 291–304 (1985)
16. Komargodski, I.: Leakage resilient one-way functions: the auxiliary-input setting. In: TCC (B1) 2016, pp. 139–158 (2016)
17. Konoma, C., Mambo, M., Shizuya, H.: The computational difficulty of solving cryptographic primitive problems related to the discrete logarithm problem. IEICE Trans. Fundam. Electron. Commun. Comput. Sci. E88-A-1, 81–88 (2005)
18. Koyama, K.: Direct demonstration of the power to break public-key cryptosystems. In: AUSCRYPT 1990, pp. 14–21 (1990). https://link.springer.com/chapter/10.1007/BFb0030346
19. Krawczyk, H., Rabin, T.: Chameleon signatures. In: NDSS'00, pp. 143–154 (2000)
20. Maurer, U.M.: Towards the equivalence of breaking the Diffie-Hellman protocol and computing discrete algorithms. In: CRYPTO'94, pp. 271–281 (1994)
21. Maurer, U., Wolf, S.: Diffie-Hellman oracles. In: CRYPTO'96. LNCS, vol. 1109, pp. 268–282 (1996)
22. Mitsunari, S., Sakai, R., Kasahara, M.: A new traitor tracing. IEICE Trans. Fundam. Electron. Commun. Comput. Sci. E85-A-2, 481–484 (2002)
23. Roh, D., Hahn, S.G.: The square root Diffie-Hellman problem. Des. Codes Cryptogr. **62**(2), 179–187 (2012)
24. Russell, A.: Necessary and sufficient conditions for collision-free hashing. In: CRYPTO'92, pp. 433–441 (1992)
25. Sakurai, K., Inoue, T.: Note on some-one way functions (in Japanese). In: The Symposium of Cryptography and Information Security (SCIS'89), Gotenba, Japan, February 2–4 (1989) [available on request from K. Sakurai]

26. Stinson, D.R., Wu, J.: An efficient and secure two-flow zero-knowledge identification protocol. J. Math. Cryptol. **1**(3), 201–220 (2007)
27. Tompa, M., Woll, H.: Random self-reducibility and zero knowledge interactive proofs of possession of information. In: FOCS'87, pp. 472–482 (1987)
28. Verheul, E.R., Van Tilborg, H.C.A.: Cryptanalysis of 'less short' RSA secret exponents. Appl. Algebra Eng. Commun. Comput. **8**(5), 425–435 (1997)
29. Yung, M.: Zero-knowledge proofs of computational power (extended summary). In: EUROCRYPT'89, pp. 196–207 (1989)

NCASH: Nonlinear Cellular Automata-based Hash function

Tapadyoti Banerjee and Dipanwita Roy Chowdhury

Abstract In this paper, we propose a new keyed one-way double-block-length hash function by using Non-linear Cellular Automata (CA), named as NCASH. The structure of NCASH mainly follows the Wide-Pipe construction, which is a modified Merkle Damgård construction along with the concept of Hirose double-block-length one-way compression function. Our design exploits the random evolution of the CA, as well as a simple regular structure to construct the compression function. The analysis shows that NCASH family is secure against the related known attacks.

Keywords Cellular automata · One-way hash function · Double-block-length hash

1 Introduction

The one-way hash function performs a crucial role in cryptography in almost all information security applications. It converts an arbitrary length input message to a compressed and fixed length output. The primary component of a hash function is compression function. Some of the popular hash constructions which are used to design the modern hash function are Sponge construction [3], Merkle Damgård [8], Wide pipe construction [17], Hirose [14], etc. Based on these constructions, many keyed and unkeyed[1] cryptographic hash functions have been proposed, such as HMAC, Message Digest algorithm [20], Secure Hash Algorithm (SHA) [10], etc.

[1] Keyed hash functions produce the hash value from the message by using the secret key. Whereas, unkeyed hash functions accept only the message.

T. Banerjee (✉) · D. Roy Chowdhury
Crypto Research Lab, Department of Computer Science and Engineering, IIT Kharagpur, Kharagpur 721302, West Bengal, India
e-mail: tapadyoti@gmail.com

D. Roy Chowdhury
e-mail: drc@cse.iitkgp.ac.in

D. Giri et al. (eds.), *Proceedings of the Fifth International Conference on Mathematics and Computing*, Advances in Intelligent Systems and Computing 1170, https://doi.org/10.1007/978-981-15-5411-7_8

After this first wave, due to the extensive need for efficient and low computational overhead hash function, SHA-3 [9] competition has started, where Keccak [2] wins and becomes the hash standard. Here, NIST had emphasized mostly on hardware efficiency. In light of security, it is obvious that highly secure construction may not be efficient from the implementation view. This conveys researchers to the classical trade-off outline. They can try to build up optimal architectures by compromising some of the security aspects. In our research, we are walking on an identical track. We try to exploit the Cellular Automata (CA) as one of the crypto primitives to construct an optimal hash function, instead, of the complex architectures. We also provide theoretical proof which assures that our design remains secure.

Till date, a number of research has been done to explore CA as a hash-primitive. The idea was first proposed by Damgård [8], which was successfully attacked by Daemen et al. [7], and they have proposed Cellhash. Later, Mihaljevic et al. [19] suggested a hash function, which is based on linear CA over Galois field. Then a lot of work [1, 11, 15, 16] was performed. Although they have explored the CA, but some complex design architectures have also been used to make the function secure. Meanwhile, another wave of research has started around 2004, where Hirose has proposed the black-box design of double-block-length hash functions [13], which is later redefined in [14]. We try to design a secure double-block-length hash function, which exploits the CA-evolution by avoiding any other complex function.

In this paper, we propose a secure one-way double-block-length keyed hash function using the elegant structure of the CA. To design the compression function, linear CA is used not only to increase the diffusion rate and randomness property but also to resist the power attack; whereas nonlinear CA boosts the security. The theoretical proofs affirm that the construction is secure.

Our Contributions

- NCASH, the new one-way double-block-length keyed hash function is proposed. This design is built on the straightforward and classic CA-based architecture and applies bitwise calculations. The parallel evolution of CA makes it faster as well as this regular structure results in a cost-effective design.
- A new compression function is designed by exploiting the simple architecture of CA for generating hashtag.
- The security analysis for the CA-based compression function has also been performed. It has been shown analytically that NCASH remains secure against some of the relevant attacks.

The residuum of our paper is constructed as follows. Section 2 depicts the design of NCASH explicitly, including the design rationale. Section 3, describes the security analysis. Finally, our work has been concluded in Sect. 4.

2 Design of NCASH

The basic primitive of the proposed model is the Cellular Automata (CA). Therefore, before describing the construction, a brief configuration of CA is given in this section. CA is discrete lattice of cells remains in a special geometry, which comprises with a flip-flop or a memory element with combinational logic function [5, 22]. The cells are updated concurrently in each clock pulse by using the rule or the transition function. It takes the present values of a cell and its neighborhood cells and performs some logical operations on them to update the value of that cell. The next state of the ith cell of an one-dimensional three-neighborhood CA is

$$S_i^{t+1} = f(S_{i-1}^t, S_i^t, S_{i+1}^t), \text{ where } S_i^t \text{ is the state of } i\text{th cell at time } t \quad (1)$$

Here in our design, we use maximum length[2] [4]. hybrid[3] CA with the rule vector consists of rule 90 ($S_i^{t+1} = S_{i-1}^t \oplus S_{i+1}^t$) and 150 ($S_i^{t+1} = S_{i-1}^t \oplus S_i^t \oplus S_{i+1}^t$).

2.1 Architecture

Here, we introduce the proposed CA-based hash function, NCASH. Before depicting the detail design of this function, briefly outline the basic block diagram of NCASH in the Fig. 1.

The design starts with a message padding stage, where the length of the message is padded with it. This padded message is then used for the compression function to generate the hashtag. The detailed design of this hash function is as follows.

2.2 Message Padding

Consider 'M' is the length of the message and each of the message block length is 'b'. To be capable to add the message to the compression function, M is padded with '10*' that makes the padded message a multiple of b-bit. That is, $M = m_1 m_2 m_3 \ldots m_n$, where each m_i $(i \in 1, \ldots, n)$ is the ith message block of length 'b' and if $|m_n| < b$, then $m_n = m_n \| 10^*$ ('$\|$' denotes concatenation of two bit strings). Henceforth, an additional block of length 'b' is then padded with the modified message, which determines the size of the entire message. This stage is illustrated in Fig. 2.

---

[2] When all the states exempting the all '0's state lie in a single CA cycle then this is called maximum length CA.

[3] If the CA cells evolve with different rules; instead of the same rule, it is called hybrid CA.

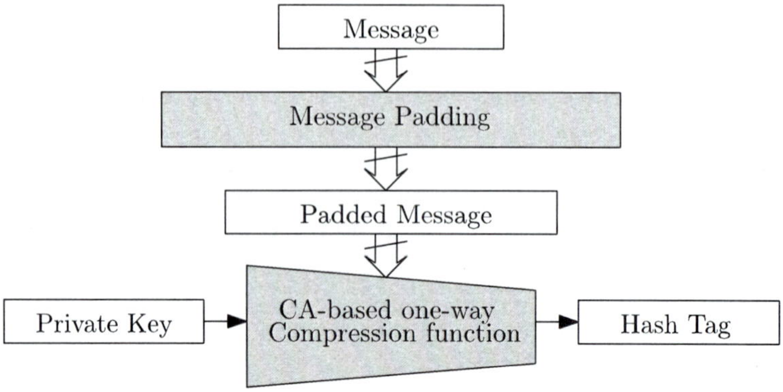

Fig. 1 Design architecture of NCASH

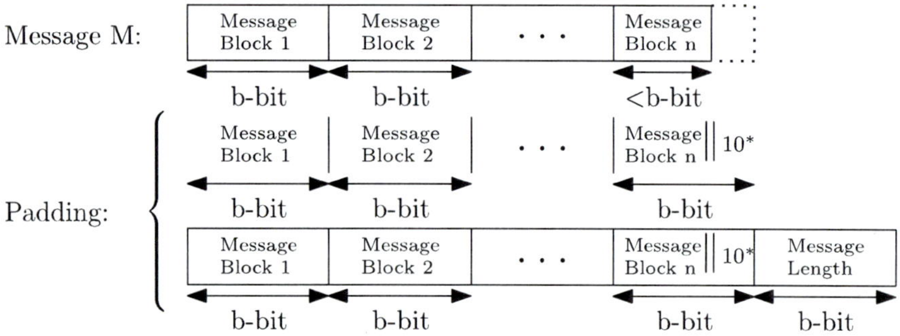

Fig. 2 Pictorial representation of message padding

2.3 Compression Function

In our design, a CA-based one-way compression function is used to produce the hash-tag. The structure of this function mainly follows the Wide-Pipe construction [17], which is a modified Merkle Damgård construction [6], and Hirose double-block-length one-way compression function [14]. We try to generate the hashtag by exploiting the random evolution of CA. The architecture of this function is given in Fig. 3. A combination of Linear Hybrid CA (LHCA)[4] and Nonlinear Hybrid CA (NHCA)[5] are used in every stage of the compression function. In the Fig. 3, for a 'b' block-length message, the ith stage of the function is shown. At the upper half of the compression function, the output of the i–1th stage is inserted into the NHCA and a fixed number

[4]Linear CA consists of only linear operations such as XOR.

[5]Nonlinear CA contains linear rules along with some nonlinear operations such as AND/OR. The linear CA can be converted into nonlinear CA by injecting the nonlinear function at one/more cells of that CA along with the rule-vector [12].

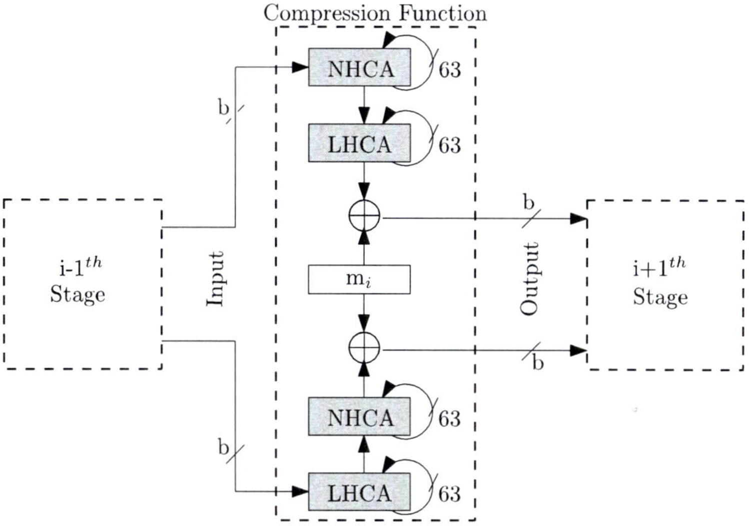

Fig. 3 i^{th} stage of the compression function of NCASH

of pulses (63) are applied. Then the output of the NHCA is turned into the input of the LHCA. To get one of the outputs of this ith stage, the message-block 'm_i' is XORed with the output of the LHCA and it is performed as one of the inputs of the i+1th stage. The lower half of the compression function is originally the reverse order of the upper half, i.e., the sequence of NHCA and LHCA is changed here. The output of the i–1th stage is inserted into the LHCA, and then it is traversed through the NHCA. The second input for the i+1th stage is produced by XORing the output of the NHCA with 'm_i'.

2.4 *Tag Generation*

The hashtag is produced by the output of the n+1th round of the compression function. The higher 'b'-bits from the most significant bit of the tag is actually the output of the NHCA-LHCA chain and the lower 'b'-bits from the least significant bit is the output of the LHCA-NHCA chain. Figure 4 shows the generation of the hashtag.

2.5 *Design of NCASH-l*

The intricate design of the NCASH-*l* is shown in Fig. 5, where '*l*' is the tag-length. For simplicity, a case with only two blocks of the message is shown. In the message padding stage, the length of the message M, i.e., len(M) is padded with the original

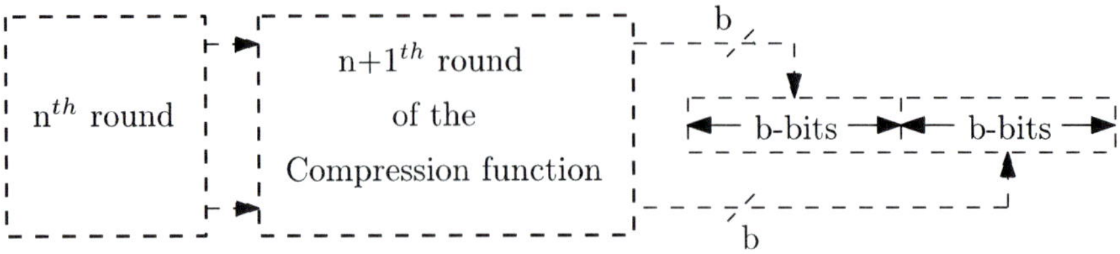

Fig. 4 Schematic diagram of hashtag generation

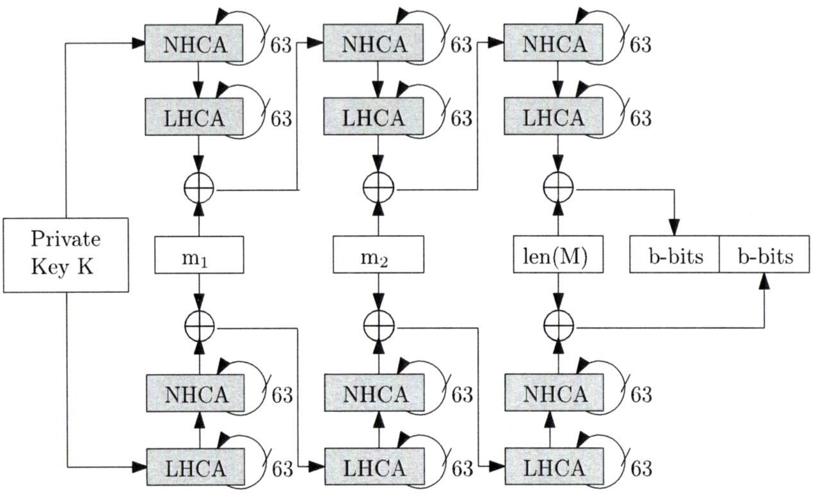

Fig. 5 Illustrated diagram of NCASH-*l*

message and every block of this padded message should be of length ‘b’. Formerly the initial ‘b’-bits of private key, ‘K’ is inserted into the NHCA and LHCA as the initialization vector of these CA (described in Fig. 5). Then these CA are started to evolve and the outputs are fed into the next LHCA and NHCA, respectively. The final outputs of the compression function are produced by XORing the outputs of the CA-chain with the message block. These outputs are considered as the inputs of the next compression function, and so on. Finally, the hashtag is produced as described in Sect. 2.4. The hardware architectural details of NCASH is given in Fig. 6. Here, ‘L-Box’ and ‘NL-Box’ denotes the Linear and Nonlinear function box, respectively, which help to work the CA in dual mode, under the control of ‘NL Controller’ (Nonlinearity injection controller). Whereas round controller controls whether the ‘CA Layers’ will be updated by the ‘b-bit Initial Seed’ (first b-bits of the private key) or the output of the ‘XOR Layer’. The randomness of the tag values generated by this function has successfully passed the NIST Statistical Test Suite for Random and Pseudorandom Number Generators (NIST SP 800-22) [21]. For this test, we have generated 50 lacs of hashtags by using NCASH-256 and apply it on that data-set. Table 1 shows the final analysis report. For this test, each ‘P-Value’ denotes the less probability of the sequence produced by a perfect random number generator than the sequence that was tested, and 'Proportion' denotes the ratio of binary sequences that has passed the corresponding test. As per NIST test suite, for the proportion value, the minimum pass rate for each statistical test except the random excursion (variant) test is approximately = 96 when considering a sample size = 100 binary sequences. Whereas the minimum pass rate for the random excursion (variant) test is approximately = 60 when considering a sample size = 63 binary sequences.

2.6 Design Rationale

The preference of operations and parameters in the construction of NCASH have the theoretical background and not only interrogative. For a better perceptive of the theoretical groundwork behind NCASH, following things are discussed here.

- ***Choice of LHCA***: The maximum-length LHCA is used to frame the compression function. It is used to make the correlation between the two subsequent intermediate values as complex and intricate as possible.
- ***Choice of NHCA with multiple nonlinearity injections***: Here the maximum-length NHCA with multiple nonlinearity injections is also used as a primitive of the compression function. It is a common fact that CA is an excellent random number generator [5] and for NHCA, it is computationally infeasible to find the previous state. So, we use this NHCA to make the function as one-way. Moreover, multiple nonlinearity injections provide better diffusion than the single injection [18].
- ***Value of clock pulse applied on CA***: We establish two constraints on the choice of the number of clock pulses applied on the CA. These are the minimum number of pulses applied and increase the diffusion of the cells. Diffusion is calculated by counting the number of bits affected in the state of the CA after applying the pulse. As we want to get a trade-off value, we consider the IV of the 128-bit NHCA as $0^{63}10^{64}$ (i.e., all '0' and only one '1'), and the nonlinearity inject position is that position where the '1' belongs to. It helps to start diffusion as fast as possible. Analyzing the diffusion values for both LHCA and NHCA, it is observed that at pulse 63, they are totally diffused, i.e., all the bits are affected (127 bits are diffused among 128 bits).
- ***The requirement of the Message Padding***: The message, M is padded with '10*', which makes the padded message a multiple of b-bit. Here 'b' is the message block length. In order to make the construction secure, the length of the original message is also padded with this. Consider two messages 'M' and 'M′', where $m_n = 010^{b-2}$, and $m_n' = 0$. So after padding 10^* at m_n', it will become 010^{b-2}. So, length padding distinguishes 'M' and 'M′'.

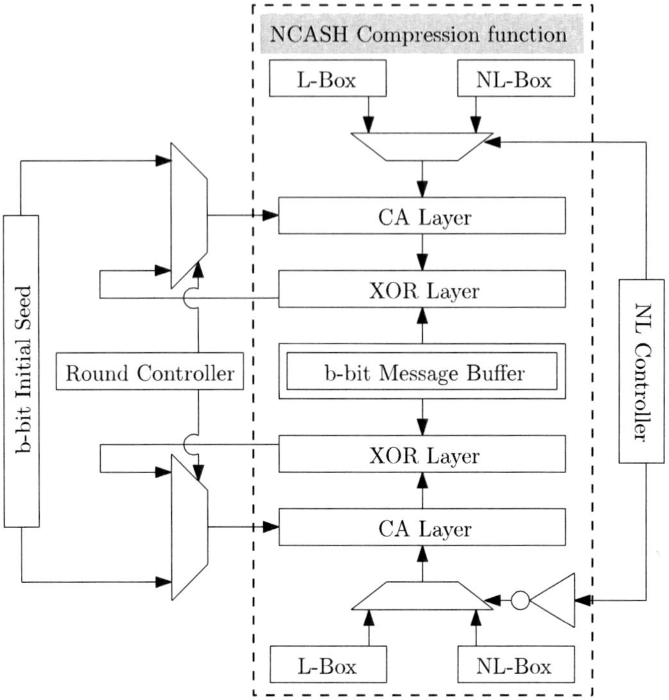

Fig. 6 The hardware architecture of NCASH-*l* family

Table 1 NIST SP 800-22 test suite for NCASH-256

Results for the uniformity of P-values and the proportion of passing sequences			
Test	P-value	Proportion	Interpretation
Frequency (Monobit) test	0.115387	100/100	Success
Frequency test within a block	0.935716	100/100	Success
Runs test	0.334538	100/100	Success
Test for the longest run of ones in a block	0.657933	99/100	Success
Binary matrix rank test	0.162606	99/100	Success
Discrete Fourier transform (Spectral) test	0.401199	99/100	Success
Non-overlapping template matching test (avg.)	0.474438	100/100	Success
Overlapping template matching test	0.657933	97/100	Success
Maurer's "universal statistical" test	0.162606	100/100	Success
Linear complexity test	0.275709	99/100	Success
Serial test (avg.)	0.377337	100/100	Success
Approximate entropy test	0.574903	99/100	Success
Cumulative sums (Cusum) test (avg.)	0.514269	100/100	Success
Random excursions test (avg.)	0.38151	63/63	Success
Random excursions variant test (avg.)	0.467236	63/63	Success

3 Security Analysis

In this section, we show that NCASH is less likely to be susceptible to many attacks and it has a strong compression function. The security analysis of this scheme is done here.

3.1 Key Recovery Attack

A key-recovery attack is an attempt of the adversary to recover the cryptographic key of an encryption scheme. With respect to the keyed hash function, one of the main security aspects is to actively resist this attack.

- ***Exhaustive Key Search***: The exhaustive key search, or brute-force search, is the primary approach to identify the correct key by obtaining all possible valid keys. If we consider NCASH-256, exhaustive research key attack requires 2^{128} possible entities. Thus, it is computationally infeasible.
- ***Tag Sensitivity to One-bit Changes***: To evaluate the tag sensitivity level of our suggested work, the following experiment is executed with respect to the avalanche effect notion. The avalanche effect is the desirable property of cryptographic algorithms, which indicates small changes in input (e.g., flipping a single bit) causes a drastic change in the output (e.g., half the output bits flip). Here in this experiment, we flip the bits of the key one at a time and calculate the Hamming distance[6] for all obtained output tags for NCASH-256. Figure 7 shows the resultant graph of this analysis, where the Hamming distances are plotted in the verticle axis and positions of the changed bits are plotted in the horizontal axis. It is clearly visible that at least half of the bits are different in each case.

3.2 Strength of the Compression Function

NCASH follows an iterative structure (see Fig. 5). Now, if the iterated compression function is collision resistant, then so is the resultant iterated hash function. Here, we try to find the probability for getting the same output from the compression function. Consider Fig. 8, which consists of only the upper half of the desired function (described in Fig. 3).

Here in this figure, if we unfold the 'W_i' then it will be as follows:

$$W_i = LHCA(NHCA(W_{i-1})) \oplus m_i \tag{2}$$

[6]The Hamming distance between two binary strings which has an equal length is the number of locations at which the analogous bits are disparate.

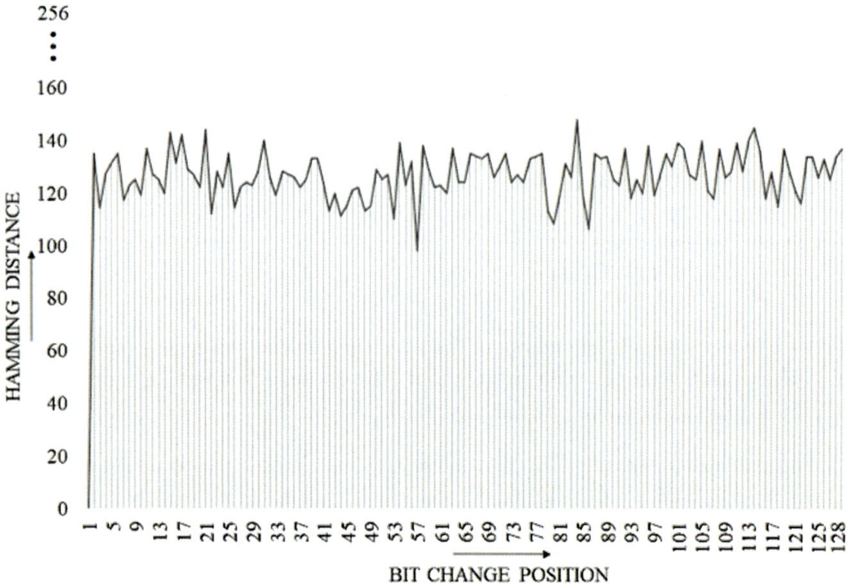

Fig. 7 Graphical representation of the avalanche effect

Fig. 8 Compression function in partial form

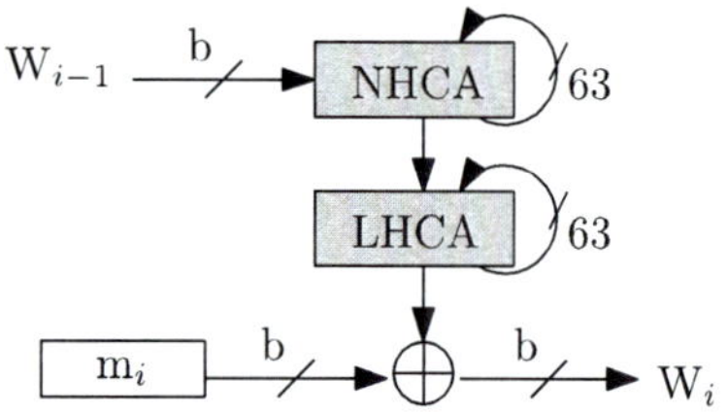

In Eq. 2, NHCA(W_{i-1}) has $2^b - 1$ possible values. Moreover, it passes through another LHCA, which is a good random number generator [5]. So, we can consider it as a b-bit random number. Then it XORed with another b-bit message block 'm_i', which is also another random number. Therefore, 'W_i' is nothing but the XOR output of two b-bit strings. Now, we need to check the probability of $W_i = W_j$, where $i \neq j$. We know that

$$probability = \frac{n(E)}{n(S)}$$

where $n(E)$: Number of outcomes favorable for the event,

$n(S)$: Total number of possible outcomes.

Here, 'W' is produced by XORing two b-bit strings. So, 2^b number of two b-bit combination can produce the same 'W'. Hence, $n(E) = 2^b$. To find the value of $n(S)$, fixed the value of the first component of Eq. 2, and take the all possible combinations of 'm'. As there are such 2^b number of 'm' values for each 'LHCA(NHCA($W_{previous}$))' values, and also there are 2^b possible values for this CA-chain. Hence, $n(S) = 2^b \times 2^b = 2^{b+b} = 2^{2b}$.

$$\begin{aligned} \therefore probability &= \frac{2^b}{2^{2b}} \\ &= \frac{1}{2^{2b-b}} \\ &= \frac{1}{2^b} \end{aligned}$$

Now this probability is same for the another b-bit output of the compression function, which consists of only the reverse CA-chain with respect to Fig. 8; i.e., LHCA-NHCA chain. As these two values are mutually exclusive to each other, hence the total probability to get the same output-set from this function is $\frac{1}{2^b} \times \frac{1}{2^b} = \frac{1}{2^{b+b}} = \frac{1}{2^{2b}}$.

Remark 1 For NCASH-256, the probability to get the same output-set from the compression function is $\frac{1}{2^{256}}$.

Note 1 The randomness of the 'W_i' (described in the Fig. 8) has also successfully passed the NIST Statistical Test Suite for Random and Pseudorandom Number Generators (NIST SP 800-22) [21], and the result set is similar to Table 1.

3.3 *Preimage, 2nd Preimage and Collision Security*

In case of the preimage attack or the second preimage attack, an adversary tries to find a message such that the hash of that message equals to a given hash value. Hence the adversary can apply exhaustive key search or brute-force search. He should randomly pick values from the message set and examine each value unless and until a collision occurs. So, the level of effort for an l-bit hash value is 2^l.

In case of a collision-resistant attack, here an adversary tries to find a pair of messages from the message set, that should give the same hash value. The level of effort is calculated by birthday paradox.[7] In essence, from a uniform distribution in the range, 0 through N–1 and choice is made uniformly and randomly, then the probability to get repeated element exceeds 0.5 after $\sqrt{N}$. Thus, for a l-bit hash value, we can expect to find two messages, which is chosen uniformly and randomly will give same hash value within $\sqrt{2^l}$ or $2^{l/2}$ attempts.

The specific values of the security parameters are given in Table 2.

[7] In probability theory, the birthday problem or birthday paradox concerns the probability that, in a set of N randomly chosen people, some pair of them will have the same birthday.

Table 2 Specification of NCASH family

NCASH family	Preimage security	2nd preimage security	Collision security
NCASH-80	2^{80}	2^{80}	2^{40}
NCASH-128	2^{128}	2^{128}	2^{64}
NCASH-160	2^{160}	2^{160}	2^{80}
NCASH-256	2^{256}	2^{256}	2^{128}

4 Conclusion

This paper proposes a new one-way keyed hash function using Cellular Automata, called NCASH. The randomness of the hashtag depends only on the randomness of the proper CA-based structure, instead of any other complex logic as used in the other hash functions. Here, linear and nonlinear CA are engaged to construct the compression function with a simple and elegant structure to produce the one-way double-block-length keyed hash. NCASH family is constructed and implemented with different hash length and cryptanalysis. The architecture and security analysis of this function signify that it is less likely to be vulnerable to many attacks.

References

1. Belfedhal, A.E., Faraoun, K.M.: Building secure and fast cryptographic hash functions using programmable cellular automata. J. Comput. Inf. Technol. **23**(4), 317–328 (2015)
2. Bertoni, G., Daemen, J., Peeters, M., Van Assche, G.: KECCAK sponge function family main document (2010)
3. Bertoni, G., Daemen, J., Peeters, M., Van Assche, G.: Duplexing the sponge: single-pass authenticated encryption and other applications. In: International Workshop on Selected Areas in Cryptography, pp. 320–337. Springer (2011)
4. Chang, T., Song, I., Bae, J., Kim, K.S.: Maximum length cellular automaton sequences and its application. Signal Process. **56**(2), 199–203 (1997)
5. Pal Chaudhuri,P., Roy Chowdhury, D., Nandi, S., Chattopadhyay, S.: Additive Cellular Automata: Theory and Applications, vol. 1. Wiley, New York (1997)
6. Coron, J.S., Dodis, Y., Malinaud, C., Puniya, P.: Merkle-Damgård revisited: how to construct a hash function. In: Annual International Cryptology Conference, pp. 430–448. Springer (2005)
7. Daemen, J., Govaerts, R., Vandewalle, J.: A framework for the design of one-way hash functions including cryptanalysis of Damgård's one-way function based on a cellular automaton. In: International Conference on the Theory and Application of Cryptology, pp. 82–96. Springer (1991)
8. Damgård, I.B.: A design principle for hash functions. In: Conference on the Theory and Application of Cryptology, pp. 416–427. Springer (1989)
9. Dworkin, M.J.: SHA-3 standard: permutation-based hash and extendable-output functions. Technical report (2015)
10. Eastlake 3rd, D., Jones, P.: US secure hash algorithm 1 (SHA1). Technical report (2001)
11. Echandouri, B., Hanin, C., Omary, F., Elbernoussi, S.: Keyed-CAHASH: a new fast keyed hash function based on cellular automata for authentication. Int. J. Comput. Sci. Appl. **14**(2) (2017)

12. Ghosh, S., Sengupta, A., Saha, D., Roy Chowdhury, D.: A scalable method for constructing nonlinear cellular automata with period 2^n-1. In: International Conference on Cellular Automata, pp. 65–74. Springer (2014)
13. Hirose, S.: Provably secure double-block-length hash functions in a black-box model. In: International Conference on Information Security and Cryptology, pp. 330–342. Springer (2004)
14. Hirose, S.: Some plausible constructions of double-block-length hash functions. In: International Workshop on Fast Software Encryption, pp. 210–225. Springer (2006)
15. Jamil, N., Mahmood, R., Muhammad, R.: A new cryptographic hash function based on cellular automata rules 30, 134 and omega-flip network. In: Proceedings of the 2012 International Conference on Information and Computer Networks (ICICN 2012), Singapore, vol. 27, pp. 163–169 (2012)
16. Kuila, S., Saha, D., Pal, M., Roy Chowdhury, D.: CASH: cellular automata based parameterized hash. In: International Conference on Security, Privacy, and Applied Cryptography Engineering, pp. 59–75. Springer (2014)
17. Lucks, S.: Design principles for iterated hash functions. IACR Cryptology ePrint Archive 2004, p. 253 (2004)
18. Maiti, S., Roy Chowdhury, D.: Achieving better security using nonlinear cellular automata as a cryptographic primitive. In: International Conference on Mathematics and Computing, pp. 3–15. Springer (2018)
19. Mihaljevic, M., Zheng, Y., Imai, H.: A fast cryptographic hash function based on linear cellular automata over GF(q) (1998)
20. Rivest, R.: The MD5 message-digest algorithm. Technical report (1992)
21. Rukhin, A., Soto, J., Nechvatal, J., Smid, M., Barker, E.: A statistical test suite for random and pseudorandom number generators for cryptographic applications, NIST Special Publication 800-22. Technical report, Booz-Allen and Hamilton Inc., Mclean Va (2001)
22. Shiffman, D.: The Nature of Code: Simulating Natural Systems with Processing. Daniel Shiffman (2012)

Conditions for Feedback Shift Registers to Generate Permutations

Vamshi Krishna Kammadanam, Virendra Sule, and Yi Hong

Abstract The paper explores conditions to be satisfied by feedback shift registers (FSRs) to generate a permutation (alternatively a nonsingular map) of the state space over various fields. Such a condition is well known over the binary field $\mathbb{F}_2$. This is extended to small finite fields and a simple sufficient condition is stated for larger fields. Permutations with cycles or periodic sequences generated by their action of lengths which are of exponential order of the register length are attractive for construction of pseudorandom sequences which have many applications.

Keywords Feedback shift registers · Boolean functions · Boole–Shannon expansion · Permutations

1 Introduction

Feedback shift registers (FSRs) are necessary in many areas such as stream ciphers for encryption and pseudorandom sequence generation for cryptography, design of counters, sequences with good correlation for communications, message authentication, etc. In most of these applications it is desired to generate exponentially long (compared to register length) periodic sequences which satisfy randomness properties. Problems arising from such applications have been studied since long for which texts such as [3, 4, 9] are well-known classics. One of the applications of FSRs (of register length n) is also to primarily generate a permutation which has a cycle of an

V. K. Kammadanam (✉)
IITB-Monash Academy, IIT-Bombay, Powai, Mumbai 400076, Maharashtra, India
e-mail: vamshi.iitkgp@gmail.com

V. Sule (✉)
IIT-Bombay, Powai, Mumbai 400076, Maharashtra, India
e-mail: vrs@ee.iitb.ac.in

Y. Hong
Monash University, Clayton, Melbourne, VIC, Australia
e-mail: yi.hong@monash.edu

D. Giri et al. (eds.), *Proceedings of the Fifth International Conference on Mathematics and Computing*, Advances in Intelligent Systems and Computing 1170, https://doi.org/10.1007/978-981-15-5411-7_9

exponential size 2^n (or p^n for p prime) by its action in a vector space of dimension n over $\mathbb{F}_2$ (respectively $\mathbb{F}_p$). Such a permutation necessarily has at least one long periodic orbit (or a cycle) of exponential size. The cyclic orbit is an attractive set for extending the public key exchange scheme analogous to the well-known Diffie–Hellman protocol [2]. Application of FSRs for alternative public key schemes has also been proposed by [8].

This paper is aimed at investigating conditions under which FSRs over finite fields $\mathbb{F}_q$ for small q generate permutations of $\mathbb{F}_q^n$ (a property also known as nonsingularity of the FSR map). Such a condition has been well known in [3] for the binary field $q = 2$. We extend this condition to other finite fields. From a practical viewpoint such conditions are useful only for nonlinear FSRs. For linear FSRs the condition is very simple and well known. Moreover even if sequences of period $2^n - 1$ can be generated with required randomness properties using permutations generated by linear feedback shift registers (LFSRs), these have very small rank (or linear complexity) equal to n. The well-known Berlekamp–Massey [1, 6] algorithm can be used to compromise such LFSR sequences using only $2n$ samples of the output. Hence the problem of generating FSRs with nonlinear feedback functions assumes significance since sequences generated by such FSRs have at least one exponential size orbit and an output sequence with large rank profile. Due to these reasons study of FSRs continues to be an active area of research over last several decades and several of its problems are still unsettled. In [5], the author gives the condition for nonsingularity of FSR in general finite fields in terms of linear equations however it is not clear whether the condition is computationally constructive or is useful for construction of such FSRs. In [7], the author gives the conditions for nonsingularity of FSRs for a degree of at most two. This paper gives conditions for nonsingularity of FSRs in terms of inequations and gives necessary and sufficient conditions for restricted degree nonlinear functions as feedback function of the FSR as well as for over fields of small characteristic. These results show that the general case of these conditions over finite fields is computationally too complex to be of any practical use.

2 Conditions for FSRs to Be a Nonsingular Map

Let $\mathbb{F}$ be a finite field. An FSR of length n over $\mathbb{F}$ generates a map $F : \mathbb{F}^n \to \mathbb{F}^n$ given by

$$(x_1, x_2, \ldots, x_n) \to (x_2, x_3, \ldots, x_n, f(x_1, x_2, \ldots, x_n)) \tag{1}$$

where $f(x_1, \ldots, x_n)$ is known as the feedback function of the FSR. The problem we address is to determine conditions on the feedback function f such that the map F is a permutation of set of all strings $\mathbb{F}^n$. As a short hand notation we denote F as

$$F(x_1, y) = (y, f(x_1, y)) \tag{2}$$

where $y \in \mathbb{F}^{n-1}$.

2.1 Well-Known Condition over Binary Field

The above problem was solved in [3] for the binary field $\mathbb{F}_2$.

Proposition 1 *When* $\mathbb{F} = \mathbb{F}_2$, $n > 1$, *the map* F *is a permutation if and only if*

$$f(x_1, y) = f_0(y) \oplus x_1$$

where f_0 *is an arbitrary Boolean function in variables* $y = (x_2 \ldots x_n)$.

A proof is given in [3]. A slight variation of it is as presented below as ideas in this proof shall be utilised in later developments.

Proof The map F is a permutation iff $F(x_1, y)$ is one to one. Due to the shift structure of the map F it follows that the map is one to one iff the feedback function f is such that for $x_1 \neq w$, $f(x_1, y) \neq f(w, y)$ for any y in $\mathbb{F}_2^{n-1}$. Then it follows that this condition is equivalent to saying for any x_1 in $\mathbb{F}_2$ and $w = x_1 \oplus 1 = x'$ the complement,

$$f(x_1 \oplus 1, y) \oplus f(x_1, y) \neq 0 \tag{3}$$

for all $y \in \mathbb{F}_2^{n-1}$. Expanding f

$$f(x_1, y) = f(0, y) \oplus x_1 f(1, y) \tag{4}$$

where $f_0(y) = f(0, y)$ and $f_1(y) = f(1, y)$ are arbitrary Boolean functions in variable y. As

$$\begin{aligned} f(x_1 + 1, y) \oplus f(x_1, y) &= ((x_1 \oplus 1) \oplus x) f_1 \\ &= f_1 \end{aligned}$$

$f_1(y) \neq 0$ for any y. Hence, $f_1 = 1$. This proves the theorem.

In the next sections we extend to this proof to other finite fields. We shall extend the above theorem to $\mathbb{F}_3$, $\mathbb{F}_5$, $\mathbb{F}_4$ and give a sufficient condition to generalise it to $\mathbb{F}_q$.

2.2 Conditions for FSRs to Be a Permutations in Non-binary Fields

We now develop conditions for FSR for various fields $\mathbb{F}_q$ to generate permutations other than the binary field. These conditions are dependent on the nature of functions of several variables in respective fields. The number of these conditions grow proportional to q hence we consider the problem of developing these conditions over small fields such as $\mathbb{F}_3$, $\mathbb{F}_5$ and $\mathbb{F}_{2^2}$. To the author's knowledge these conditions are not known for general finite fields. Another feature we wish to emphasise here is that these conditions are formulated as computational problems whose solutions allow us to verify whether or not the map is a permutation.

2.3 *Permutation Condition for* $\mathbb{F}_3$

Proposition 2 *When* $\mathbb{F} = \mathbb{F}_3$, $n > 1$, *the map* F *in (2) is a permutation if and only if*

$$f(x, y) = f_0(y) + x f_1(y)$$

where $f_0(y)$ *is an arbitrary function while* $f_1(y) \neq 0$ *for all* $y \in \mathbb{F}_3^{n-1}$.

Proof The map F is a permutation iff $F(x, y)$ is one to one. Due to the shift structure of the map F it follows that the map is one to one iff the feedback function f is such that for $x \neq w$, $f(x, y) \neq f(w, y)$ for any y in $\mathbb{F}_3^{n-1}$. So, let $b \neq 0$ in $\mathbb{F}_3$ and $w = x + b$. Then it follows that this condition is equivalent to saying for all x in $\mathbb{F}_3$ and for all $b \neq 0$

$$f(x + b, y) - f(x, y) \neq 0 \tag{5}$$

for all $y \in \mathbb{F}_3^{n-1}$. Since f is a function whose arguments x, y take values in $\mathbb{F}_3$, it follows that f has a unique representation as

$$f(x, y) = f(0, y) + x f(1, y) + x^2 f(2, y) \tag{6}$$

where $f_0(y) = f(0, y)$, $f_1(y) = f(1, y)$ and $f_2(y) = f(2, y)$ are arbitrary functions in variables y. Using this expansion the above condition is

$$\begin{aligned} f(x + b, y) - f(x, y) &= ((x + b) - x) f_1 + ((x + b)^2 - x^2) f_2 \\ &= (b f_1 + b^2 f_2) + (2b f_2) x \neq 0 \end{aligned}$$

Letting x take all values in $\mathbb{F}_3$ and writing the above polynomials in following form it follows that the condition Eq. (5) is equivalent to

$$\begin{aligned} b f_1 + b^2 f_2 &\neq 0 \\ b f_1 + (b^2 + 2b) f_2 &\neq 0 \\ b f_1 + (b^2 - 2b) f_2 &\neq 0 \end{aligned}$$

for all $b \neq 0$. Substituting values of $b \in \mathbb{F}_3$, we get following equations.

$$f_1 + f_2 \neq 0 \tag{7}$$
$$f_1 \neq 0 \tag{8}$$
$$f_1 + 2f_2 \neq 0 \tag{9}$$
$$f_1 + 2f_2 \neq 0 \tag{10}$$
$$f_1 + f_2 \neq 0 \tag{11}$$
$$f_1 \neq 0 \tag{12}$$

However, the Eqs. (10), (11) and (12) are repetitions of Eqs. (7), (8) and (9). Hence, the condition for permutation Eq. (5) is equivalent to the condition that

$$f_1 \neq 0 \tag{13}$$
$$f_2 = 0 \tag{14}$$

This proves the theorem.

We may call functions $f(y)$ which do not vanish for any point y in $\mathbb{F}^n$ as *nowhere vanishing functions* in n variables. These are thus not the same as constant 1 as in the binary field since a product of irreducible polynomials over $\mathbb{F}$ of degree less than 3 is such a function over $\mathbb{F}_3$.

2.4 Permutation Condition over $\mathbb{F}_5$

The above procedure to determine conditions for the FSR to be permutation can be repeated over the next prime field $\mathbb{F}_5$. However as noticed in the above section the number of conditions increase with the field size and even for $\mathbb{F}_5$ these are formidable to simplify into a compact condition as in $\mathbb{F}_3$ case. Hence we shall only derive a sufficient condition which is simple enough to verify for practical purposes.

Proposition 3 *When $\mathbb{F} = \mathbb{F}_5$, $n > 1$, the map F (2) is a permutation if*

$$f(x, y) = \begin{cases} f_0(y) + (x \pm (\gamma^2 + \gamma)x^2 + \gamma x^3) f_1(y) \text{ or} \\ f_0(y) + x^3 f_3(y) \end{cases}$$

where $f_0(y)$ is an arbitrary function of y while $f_1(y)$ and $f_3(y)$ are both nowhere vanishing for $y \in \mathbb{F}_5^{n-1}$ and $\gamma = \{0, 2, 3\}$.

Proof The map F is a permutation iff $F(x, y)$ is one to one. Due to the shift structure of the map F it follows that the map is one to one iff the feedback function f is such that for $x \neq w$, $f(x, y) \neq f(w, y)$ for any y in $\mathbb{F}_5^{n-1}$. So, let $b \neq 0$ in $\mathbb{F}_5$ and $w = x + b$. Then it follows that this condition is equivalent to saying for all x in $\mathbb{F}_5$ and for all $b \neq 0$

$$f(x + b, y) - f(x, y) \neq 0 \tag{15}$$

for all $y \in \mathbb{F}_5^{n-1}$. Expanding the function f in x and noting that x takes values in $\mathbb{F}_5$

$$f(x, y) = f(0, y) + xf(1, y) + x^2 f(2, y) + x^3 f(3, y) + x^4 f(4, y) \tag{16}$$

where $f(i, y)$, $0 \le i \le 4$ are arbitrary functions in variable y. Let these be denoted by $f_i(y)$, respectively.

$$\begin{aligned} f(x+b, y) - f(x, y) &= ((x+b) - x) f_1 \\ &+ ((x+b)^2 - x^2) f_2 \\ &+ ((x+b)^3 - x^3) f_3 \\ &+ ((x+b)^4 - x^4) f_4 \\ &= (bf_1 + b^2 f_2 + b^3 f_3 + b^4 f_4) \\ &+ (2bf_2 + 3b^2 f_3 + 4b^3 f_4)x \\ &+ (3bf_3 + 6b^2 f_4)x^2 + 4bf_4 x^3 \\ &= (bf_1 + b^2 f_2 + b^3 f_3 + b^4 f_4) \\ &+ (2bf_2 + 3b^2 f_3 + 4b^3 f_4)x \\ &+ (3bf_3 + b^2 f_4)x^2 + 4bf_4 x^3 \end{aligned} \tag{17}$$

Letting x and b take all values in $\mathbb{F}_5$ and writing the above polynomials in following form it follows that the condition Eq. (15) is satisfied iff

$$f_1 + f_2 + f_3 + f_4 \neq 0 \tag{18}$$
$$f_1 + 2f_2 + 4f_3 + 3f_4 \neq 0 \tag{19}$$
$$f_1 + 3f_2 + 4f_3 + 2f_4 \neq 0 \tag{20}$$
$$f_1 + 4f_2 + f_3 + 4f_4 \neq 0 \tag{21}$$
$$f_1 + 3f_2 + 2f_3 \neq 0 \tag{22}$$
$$f_1 + 4f_2 + 3f_3 \neq 0 \tag{23}$$
$$f_1 + 1f_2 + 3f_3 \neq 0 \tag{24}$$
$$f_1 + 2f_2 + 2f_3 \neq 0 \tag{25}$$
$$f_1 + f_3 \neq 0 \tag{26}$$
$$f_1 + 4f_3 \neq 0 \tag{27}$$

We consider two cases for solving the above equations. First case: Let $f_1 \neq 0$. From Eqs. (26), (27) we get $f_3 \neq \{f_1, 4f_1\}$. So, let $f_3 = \gamma f_1$, where $\gamma = \{0, 2, 3\}$. Substituting α, γ in Eqs. (22), (23), (24), (25), we get

$$f_2 \neq \{\pm(1 + 3\gamma)\alpha, \pm 2(1 + 2\gamma)\alpha\} \tag{28}$$

Substituting $\gamma = \{0, 2, 3\}$ in Eq. (28) we get,

$$f_2 \neq \begin{cases} \{\alpha, 2\alpha, 3\alpha, 4\alpha\} & \text{if } \gamma = 0 \\ \{0, 2\alpha, 3\alpha\} & \text{if } \gamma = 2 \\ \{0, \alpha, 4\alpha\} & \text{if } \gamma = 3 \end{cases}$$

Implies

$$f_2 = \begin{cases} 0 & \text{if } \gamma = 0 \\ \alpha \text{ or } 4\alpha & \text{if } \gamma = 2 \\ 2\alpha \text{ or } 3\alpha & \text{if } \gamma = 3 \end{cases} \tag{29}$$

Equation (29) can be rewritten as

$$f_2 = \begin{cases} (\gamma^2 + \gamma)\alpha & \text{or} \\ 4(\gamma^2 + \gamma)\alpha \end{cases} \tag{30}$$

where $\gamma = \{0, 2, 3\}$. Similarly substituting f_1, f_2, f_3 in Eqs. (18), (19), (20), (21) we get the following:

$$f_4 \neq \begin{cases} \{4(1+\gamma^2+2\gamma)), 3(1+2\gamma^2+\gamma), \\ 2(1+3\gamma^2+2\gamma), (1+4\gamma^2)\} & \text{or} \\ \{4(1+4\gamma^2), 3(1+3\gamma^2+2\gamma), \\ 2(1+2\gamma^2+\gamma), (1+\gamma^2+2\gamma)\} \end{cases}$$

Substituting $\gamma = \{0, 2, 3\}$ above we get

$$f_4 \neq \begin{cases} \{\alpha, 2\alpha, 3\alpha, 4\alpha\} & \text{if } \gamma = 0 \\ \{\alpha, 2\alpha, 3\alpha, 4\alpha\} & \text{if } \gamma = 2 \\ \{\alpha, 2\alpha, 3\alpha, 4\alpha\} & \text{if } \gamma = 3 \end{cases}$$

So, $f_4 = 0$ for $\gamma = \{0, 2, 3\}$ and we get the following result:

$$f(x, y) = f_0(y) + xf_1(y) \pm (\gamma^2 + \gamma)x^2 f_1(y) + \gamma x^3 f_1(y)$$

where $f_3(y) \neq 0$ for all values of $y \in \mathbb{F}_5^{n-1}$ and $\gamma = \{0, 2, 3\}$. Second case: Let $f_1 = 0$. Then from Eqs. (26), (27) we get $f_3 \neq 0$. Let $f_3 = \alpha \in \mathbb{F}_5$ and substitute in Eqs. (22), (23), (24), (25), we get $f_2 \neq= \{\alpha, 2\alpha, 3\alpha, 4\alpha\}$. Implying $f_2 = 0$. Similarly we get $f_4 = 0$ from Eqs. (18), (19), (20), (21) and we get the following result:

$$f(x, y) = f_0(y) + x^3 f_3(y)$$

where $f_3(y) \neq 0$ for all values of $y \in \mathbb{F}_5^{n-1}$.

2.5 *Permutation Condition for* $\mathbb{F}_4$

Next we take the case of small extension field $\mathbb{F}_{2^4}$.

Proposition 4 *When* $\mathbb{F} = \mathbb{F}_4$, $n > 1$, *the map* F *in (2) is a permutation if and only if*

$$f(x, y) = f_0(y) + \gamma x f_1(y) + (\gamma \oplus 1)x^2 f_2(y)$$

where $f_0(y)$ *is arbitrary function of* y *while* $f_1(y)$ *and* $f_2(y)$ *are nowhere vanishing for* $y \in \mathbb{F}_4^{n-1}$ *and* $\gamma \in \mathbb{F}_2$.

Proof The condition required for permutation of the FSR map $(x, y) \to (y, f(x, y))$ to be a permutation of $\mathbb{F}_4^n$ where (x, y) is an n-tuple over $\mathbb{F}_4$, with y an $n-1$-tuple and $f : \mathbb{F}_4^n \to \mathbb{F}_4$ is a function. The structure indicates that the map is a permutation iff f is such that for $x \neq x'$, $f(x, y) \neq f(x', y)$ for any y in $\mathbb{F}_4^{n-1}$. So let $b \neq 0$ in $\mathbb{F}_4$. Then it follows that this condition is equivalent to saying for any x in $\mathbb{F}_4$ and $b \neq 0$

$$f(x + b, y) - f(x, y) \neq 0 \tag{31}$$

consider the polynomial function representation of $f(x, y)$

$$f(x, y) = f_0 + x f_1 + x^2 f_2 + x^3 f_3 \tag{32}$$

as x is to take values in $\mathbb{F}_4$ and where $f_i(y)$ are only functions of y and denoted as f_i. Then

$$\begin{aligned} f(x + b, y) - f(x, y) &= ((x + b) - x) f_1 + ((x + b)^2 - x^2) f_2 \\ &\quad + ((x + b)^3 - x^3) f_3 \\ f(x + b, y) - f(x, y) &= (b f_1 + b^2 f_2 + b^3 f_3) + (b^2 f_3)x + (b f_3)x^2 \end{aligned} \tag{33}$$

Letting x and b take all values in $\mathbb{F}_4$ and writing the above polynomials in following form it follows that the condition Eq. (31) is satisfied iff

$$f_1 + f_2 + f_3 \neq 0 \tag{34}$$
$$f_1 + \theta f_2 + \theta^2 f_3 \neq 0 \tag{35}$$
$$f_1 + \theta^2 f_2 + \theta f_3 \neq 0 \tag{36}$$
$$f_1 + \theta f_2 \neq 0 \tag{37}$$
$$f_1 + \theta^2 f_2 \neq 0 \tag{38}$$
$$f_1 + f_2 \neq 0 \tag{39}$$

The solution of the above equations are divided into two cases. First case: Let $f_1 = \alpha \in \mathbb{F}_4^*$. From Eqs. (37), (38) and (39) we get

$$f_2 \neq \{\alpha, \theta\alpha, \theta^2\alpha\}$$

Implying $f_2 = 0$. Substituting f_1, f_2 in Eqs. (34), (35) and (36) we get

$$f_3 \neq \{\alpha, \theta\alpha, \theta^2\alpha\}$$

Implies $f_3 = 0$ and we get the following result:

$$f(x, y) = f_0(y) + xf_1(y) \tag{40}$$

Second Case: Let $f_1 = 0$. From Eqs. (37), (38) and (39) we get $f_2 \neq 0$. So, let $f_2 = \alpha \in \mathbb{F}_4^*$. Substituting f_1, f_2 in Eqs. (34), (35) and (36) we get

$$f_3 \neq \{\alpha, \theta\alpha, \theta^2\alpha\}$$

Implies $f_3 = 0$ and we get the following result:

$$f(x, y) = f_0(y) + x^2 f_2(y) \tag{41}$$

Combining Eqs. (40) and (41) we get the below result

$$f(x, y) = f_0(y) + \gamma x f_1(y) + (\gamma \oplus 1)x^2 f_2(y)$$

where $f_1(y) \neq 0$ and $f_2(y) \neq 0$ for all values of $y \in \mathbb{F}_4^{n-1}$ and $\gamma \in \mathbb{F}_2$.

3 Conditions for Permutation Under Restrictions on f

Next we develop necessary and sufficient conditions under restrictions on function f in Eq. (2). A simplifying restriction is on degree of f in x_1. Following conditions give these results for $f(x_1, y)$ in n variables described in Eqs. (1) and (2) restricted in degree of x_1.

Proposition 5 *When $\mathbb{F} = \mathbb{F}_q$, $q = p^r$, where p is prime and when the degree of f in x_1 is restricted to 1, the map F is a permutation if and only if*

$$f(x_1, y) = f_0(y) + x_1 f_1(y)$$

where f_0 is an arbitrary function in variable $y = (x_2 \dots x_n)$ and $f_1(y)$ is a nowhere vanishing function for all $y = (x_2 \dots x_n) \in \mathbb{F}_q^{n-1}$.

Proof The map F in Eq. (2) is a permutation iff $F(x, y)$ is one to one. Due to the shift structure of the map F it follows that the map is one to one iff the feedback function f is such that for $x \neq w$, $f(x, y) \neq f(w, y)$ for any y in $\mathbb{F}_q^{n-1}$. So, let $b \neq 0$ in $\mathbb{F}_q$ and $w = x + b$. Then it follows that this condition is equivalent to saying for all x in $\mathbb{F}_q$ and for all $b \neq 0$

$$f(x + b, y) - f(x, y) \neq 0 \tag{42}$$

for all $y \in \mathbb{F}_q^{n-1}$. Since f is a function whose arguments x, y take values in $\mathbb{F}_q$, it follows that f has a unique representation with degree restricted to 1 in x_1 as

$$f(x, y) = f_0(y) + x_1 f_1(y) \tag{43}$$

where $f_0(y)$ and $f_1(y)$ are arbitrary functions in variables y. Using this expansion the above condition is

$$\begin{aligned} f(x + b, y) - f(x, y) &= ((x + b) - x) f_1 \\ &= b f_1 \neq 0 \end{aligned}$$

Since $b \neq 0$ this shows that f_1 is a nowhere vanishing function of y. This proves the theorem.

Proposition 6 *When* $\mathbb{F} = \mathbb{F}_{2^r}$, $r > 1$, $n > 1$ *and when the degree of* f *in* x_1 *is restricted to* 2, *the map* F *is a permutation if and only if*

$$f(x_1, y) = f_0(y) + \gamma x_1 f_1(y) + (\gamma \oplus 1) x_1^2 f_1(y)$$

where f_0 *is an arbitrary Boolean function in variables* $y = (x_2 \dots x_n)$ *and* $f_1(y)$ *and* $f_2(y)$ *are nowhere vanishing functions for all* $y = (x_2 \dots x_n) \in \mathbb{F}_{2^r}^{n-1}$ *and for arbitrary* $\gamma \in \mathbb{F}_2$.

Proposition 7 *When* $\mathbb{F} = \mathbb{F}_{2^r}$, $r > 1$, $n > 1$ *and when the degree of* f *in* x_1 *is restricted to* 3, *the map* F *is a permutation if and only if*

$$f(x_1, y) = f_0(y) + \gamma x_1 f_1(y) + (\gamma \oplus 1) x_1^2 f_2(y)$$

where f_0 *is an arbitrary Boolean function in variables* $y = (x_2 \dots x_n)$ *and* $f_1(y)$ *and* $f_2(y)$ *are nowhere vanishing functions for all* $y = (x_2 \dots x_n) \in \mathbb{F}_{2^r}^{n-1}$ *and for arbitrary* $\gamma \in \mathbb{F}_2$.

For restriction on f to second degree in x_1, the condition becomes different from Proposition (6) when p is odd.

Proposition 8 *When* $\mathbb{F} = \mathbb{F}_{p^r}$, *p is an odd prime* $r > 0$, $n > 1$ *and when the degree of* f *in* x_1 *is restricted to* 2, *the map* F *is a permutation if and only if*

$$f(x_1, y) = f_0(y) + x_1 f_1(y) + x_1^2 f_2(y)$$

where f_0 *is an arbitrary Boolean function in variables* $y = (x_2 \dots x_n)$ *and* $f_1(y)$ *is a nowhere vanishing functions and* $f_2(y)$ *is a everywhere vanishing function for all* $y = (x_2 \dots x_n) \in \mathbb{F}_{p^r}^{n-1}$.

4 Number of Inequations

For a general field $\mathbb{F}_q$, the number of inequations is $q(q-1)/2$. This comes from the counting argument from the inequation $f(x+b) - f(x) \neq 0$, where $b \in \mathbb{F}_q^*$ and $x \in \mathbb{F}_q$. When $x = 0$, b can take $q-1$ values resulting in $q-1$ inequations and when $x = 1$, b can take $q-1$ values, however, removing the inequation repeated when $x = 0$, results in $q-2$ inequations. Similarly when $x = 2$ results in $q-3$ inequations and so on. Adding them we get $q - 1 + q - 2 + \cdots + 2 + 1 = q(q-1)/2$ inequations.

4.1 Conditions for Nowhere Vanishing Functions

Conditions for FSRs derived in this paper always involve functions in several variables over finite fields which are nowhere vanishing. Although a general characterisation of such functions does not seem to be well known in the literature, it is easy to state sufficient conditions for a function to be nowhere vanishing. These are developed below.

First let $\mathbb{F}_q$ be a finite field and $f(x)$ be a polynomial function of one variable over $\mathbb{F}_q$ then f is nowhere vanishing iff it has no factor of the type $(x-a)$ for any a in $\mathbb{F}_q$. Hence f must be a residue of the product of powers of irreducible polynomials in $\mathbb{F}_q[X]$ modulo $X^q - X$. Next we consider the problem in several variables.

Let f be a $\mathbb{F}_q$ valued function on $\mathbb{F}_q^n$, i.e., a function of n variables. Then by unique factorisation over $\mathbb{F}_q[X_1, \dots, X_n]$ each of the irreducible factors must be a nowhere vanishing function. Hence, a simple but highly restrictive sufficient condition is that f is a product of residues of powers of irreducible polynomials of one variables in $\mathbb{F}_q[X_i]$ modulo $X_i^q - X_i$. Such a condition is practically useful to construct FSRs to form permutations. However, the general mathematical problem of characterising nowhere vanishing functions over finite fields is beyond the scope of this paper.

5 Conclusion

The well-known condition for an FSR to generate a permutation over binary field is extended to conditions over small fields $\mathbb{F}_3$ and $\mathbb{F}_4$. Similarly, sufficient conditions are derived for $\mathbb{F}_5$ and general finite field $\mathbb{F}_q$. The construction of such permutations involves construction of non vanishing functions over finite fields which is a computational problem not fully well explored in literature. In general, conditions for permutation involve solution of systems of inequations as opposed to equations. Hence, as a general problem, the problem of characterising varieties defined by polynomial equations and inequations over finite fields should turn out to be a relevant and interesting problem of Computational Mathematics.

References

1. Berlekamp, E.: Algebraic Coding Theory. McGraw-Hill Series in Systems Science. McGraw-Hill, New York (1968). https://books.google.co.in/books?id=nIIpAQAAMAAJ
2. Diffie, W., Hellman, M.: New directions in cryptography. IEEE Trans. Inf. Theory **22**(6), 644–654 (1976)
3. Golomb, S.W., et al.: Shift register sequences. Aegean Park Press, Laguna Hills (1982)
4. Golomb, S., Gong, G.: Signal Design for Good Correlation: For Wireless Communication, Cryptography, and Radar. Cambridge University Press, Cambridge (2005). https://books.google.co.in/books?id=KZamDAEACAAJ
5. Lai, X.: Condition for the nonsingularity of a feedback shift-register over a general finite field (corresp.). IEEE Trans. Inf. Theory **33**(5), 747–749 (1987)
6. Massey, J.: Shift-register synthesis and BCH decoding. IEEE Trans. Inf. Theory **15**(1), 122–127 (1969)
7. Mullen, G.L.: Permutation polynomials and nonsingular feedback shift registers over finite fields. IEEE Trans. Inf. Theory **35**(4), 900–902 (1989)
8. Niederreiter, H.: A public-key cryptosystem based on shift register sequences. In: Pichler, F. (ed.) Advances in Cryptology – EUROCRYPT'85, pp. 35–39. Springer, Berlin (1986)
9. Rueppel, R.: Analysis and Design of Stream Ciphers. Communications and Control Engineering. Springer, Berlin (2012). https://books.google.co.in/books?id=86FHnwEACAAJ

A Decision-Making Framework for Green Supplier Selection Using FAHP and Taguchi Quality Loss Functions

Ashoke Kumar Bera, Dipak Kumar Jana, Debamalya Banerjee, and Titas Nandy

Abstract Due to ever-increasing consciousness and considerable pressures from various parties, organizations have compelled to implement the green practices in their daily activities. However, green issues are novel, evolving constantly and require a regular study in this area to understand the problems properly. Presently, manufacturing sectors are facing some major challenges like environmental impacts, cost optimization, employee management, project management, corporate-social responsibility, global economy crisis, etc. Moreover, the supplier selection (SS) depends on the consideration of several conflicting subjective and objective factors. Considering the above facts, this paper proposes an integrated model using fuzzy analytical hierarchy process (FAHP) and Taguchi loss function for selecting suitable green suppliers based on crucial economic and environmental favourable factors (FF) and unfavourable factors (UF). The proposed method measures the loss of supplier's performance due to the relevant factors and identifies the best supplier to perform effectively for the organization. The triangular fuzzy numbers (TFN) are used in this study to handle the imprecision involved in the supplier evaluation process.

Keywords Group decision-making · Green supplier selection · Fuzzy AHP · Taguchi quality loss function · Favourable factors · Unfavorable factors

A. K. Bera (✉) · D. K. Jana
Haldia Institute of Technology, MAKAUT, Haldia 721657, India
e-mail: beraashoke123@gmail.com

D. K. Jana
e-mail: dipakjana@gmail.com

D. Banerjee · T. Nandy
Jadavpur University, Kolkata 700032, India
e-mail: dbanerjee@production.jdvu.ac.in

T. Nandy
e-mail: titas_nandy@gmail.co.in

D. Giri et al. (eds.), *Proceedings of the Fifth International Conference on Mathematics and Computing*, Advances in Intelligent Systems and Computing 1170, https://doi.org/10.1007/978-981-15-5411-7_10

1 Introduction

Evaluation and selection of suppliers for an organization is recognized as a crucial issue because it directly affects the competitive position as well as profitability of the organization. The process is further complicated due to large number of influencing quantitative and qualitative factors [24] and the involvement of large number of suppliers [1]. Present business trends, ever-increasing regulations and conflicting organizational objectives have further increased the complexity of SS decisions. Now, the challenge is to identify the supplier management as well as develop the strategies to handle SS effectively.

To manage the environmental degradation, the organizations are focusing more on greening the internal operations as well as concentrating on inter-organizational activities and look into the performance of their suppliers. The objective is to develop and maintain the eco-friendly supply chain. It involves the consideration of environmental issues into the traditional SS factors [6]. Price, quality, lead time performance and service level are the main conventional SS factors, while harmful emissions, carbon footprint, efficient use of energy, water and recycling of used and end of life products have been the more common environmental measures [3]. Thus, the organizations can use the suitable multicriteria decision-making (MCDM) methods for selecting suppliers considering both conventional and environmental factors. Very few studies found in this area that focus on the development of suitable MCDM tools to integrate both kinds of factors. The recent study proposed by Govindan et al. [6] pointed out that most of the modelings in this particular area are dependent on the inclusion of fuzzy theory into the conventional MCDM to handle uncertainty in input data and in particular the vagueness or imprecision associated with human judgments.

In this work, an integrated model is proposed using FAHP and Taguchi loss function for selecting appropriate green suppliers based on economical and environmental FF and UF for an Indian manufacturing company. The developed method measures the loss of supplier performance because of the relevant factors and identifies the best supplier to perform effectively for the organization. The fuzzy set theory (FST) is integrated with the proposed model to consider the uncertainty of the input parameters. Due to the growing environmental pressures, it is required to improve the environmental performance of suppliers. However, despite the importance of this area, less number of studies have been investigated in industrial sectors as compared to those of other sectors [6]. Thus, this paper is developed to investigate the suitability of the model in an industry that is a major source to the environmental pollution.

The main contributions are summarized as follows.

(1) An integrated FAHP method with Taguchi loss function is developed for effective SS.
(2) The importance of various factors of SS are evaluated by FAHP method.
(3) Taguchi loss function is introduced to calculate the loss of supplier's performance corresponding to relevant FF and UF and identify the best supplier to perform effectively for the organization.

(4) A case study from an Indian manufacturing organization is proposed to illustrate the applicability of the model in an imprecise environment.

2 Related Work

2.1 Supplier Evaluation Factors

A number of works on the use of relevant factors and decision models for SS have been found in [6]. Some researchers have developed the tradeoffs amongst these factors to select the suitable suppliers for solving specific problem along with various constraints. Several researchers [5, 10, 28] have identified various factors for SS such as price, delivery, quality, capacity, demand and manufacturing capability, etc. Due to the imposition of environmental controls and practices, some of the recent studies consider a set of conventional and environmental factors to handle green SS problems [6].

Govindan et al. [6] found Environmental Management Systems (EMS) as one of the important environmental criteria for green SS. Nielsen et al. [19] analyzed several related papers and identified EMS as the most important factor amongst over 90 identified factors. EMS is a chain of practices and processes that improves the environmental performance and operational efficiency of the organization. It includes the management of energy usage, waste produced in manufacturing [19] and establishing standards [29]. Many organizations use one or more EMS practices for green SS [29]. Using these reviews a combination of conventional and green SS factors for the organization are selected by the decision-makers. Tables 2 and 3 describe various conventional and environmental FF and UF used for this study.

2.2 Methods Used for SS

Several techniques for SS have been proposed for last few decades ranging from single-objective optimization model to complex multi-objective models. The uses of integrated models are more famous due to their flexibility [18]. Chai et al. [2] and Govindan et al. [6] reviewed various published methods related to traditional and green SS. For effective SS both quantitative and qualitative factors are to be considered. Qualitative factors cannot be simply expressed in numerical values [7]. Thus, these factors are expressed by using appropriate linguistic variables. As the area of green supply chain management (SCM) continues to grow, the use of qualitative methods is also increasing to deal with numerous environmental and social factors [27].

In conventional systems, the losses occur due to the deviation of products from their specification limits, and hence unacceptable [22]. As per Taguchi any shift from the set target results in a loss and higher the quality means the lesser the shift from target value. Thus, when the loss is zero, the characteristics measurement exactly matches with the target value. The loss due to the deviation is generally measured with the help of quadratic functions. The companies use these loss functions to estimate and reduce the losses systematically which occur due to the variation from the target [12]. These functions are recently used for various non-manufacturing applications. Most of the SS methods involve numerous data collected through sampling, surveys, investigation or opinion of experts. Such data are not easily obtainable in time or in many practical cases data collected may be insufficient, vague or incomplete. In these situations the decisions are made considering qualitative factors which are often influenced by the subjective judgment of the decision-makers [21]. The uncertainties in green SS problems are more prominent due to the involvement of sustainability factors because they are often expressed in qualitative terms. Also the model based on subjective evaluation by a decision-maker is always not as accurate as it requires more expertise, knowledge and experience [16]. Incorporating FST in SS problem can produce more practical solutions because they transform human judgments into a mathematical formula [30].

3 Preliminaries

3.1 *Fuzzy Sets*

In many circumstances, available information are inadequate to model practical problems due to imprecision, vagueness and subjective judgment of DMs [20]. Also, as the weights and performance ratings of suppliers cannot be assessed precisely, the fuzzy set theory is used to consider the uncertainty involved in human judgments. Zadeh [30] developed fuzzy sets as an extension to the classical set. The membership grades of elements of classical set is either 1 or 0. Hsieh et al. [8] developed the mathematical approach of fuzzy sets. A fuzzy number $\tilde{A}$ on R to be a TFN if its membership grade $\mu_{\tilde{A}}(x) : R \to [0, 1]$ and expressed by Eq. (1). The membership functions hold the following conditions: (i) a_1 to a_2 is an increasing function, (ii) a_2 to a_3 is a decreasing function and (iii) $a_1 \leq a_2 \leq a_3$; where a_1, a_2, a_3 are the parameters of triangular fuzzy number.

$$\mu_{\tilde{A}}(x) = \begin{cases} 0 & \text{for } x < a_1;\ x > a_3 \\ \dfrac{x - a_1}{a_2 - a_1} & \text{for } a_1 \leq x \leq a_2 \\ 1 & \text{for } x = a_2 \\ \dfrac{a_3 - x}{a_3 - a_2} & \text{for } a_2 \leq x \leq a_3 \end{cases} \tag{1}$$

Operations on Fuzzy Numbers

(a) **Addition**

$$\tilde{A} \oplus \tilde{B} = (a_1, a_2, a_3) \oplus (b_1, b_2, b_3) = (a_1 + b_1, a_2 + b_2, a_3 + b_3) \quad (2)$$

(b) **Subtraction**

$$\tilde{A} \ominus \tilde{B} = (a_1, a_2, a_3) \ominus (b_1, b_2, b_3) = (a_1 - b_3, a_2 - b_2, a_3 - b_1) \quad (3)$$

(c) **Multiplication**

$$\tilde{A} \otimes \tilde{B} = (a_1, a_2, a_3) \otimes (b_1, b_2, b_3) = (a_1 b_1, a_2 b_2, a_3 b_3) \quad (4)$$

(d) **Division**

$$\tilde{A} \oslash \tilde{B} = (a_1, a_2, a_3) \oslash (b_1, b_2, b_3) = \left(\frac{a_1}{b_3}, \frac{a_2}{b_2}, \frac{a_3}{b_1}\right) \quad (5)$$

(e) **Reciprocal of fuzzy number**

$$\tilde{A}^{-1} = (a_1, a_2, a_3)^{-1} = \left(\frac{1}{a_3}, \frac{1}{a_2}, \frac{1}{a_1}\right) \quad (6)$$

3.2 *Fuzzy AHP*

AHP is a multicriteria decision-making method was proposed and propound by Saaty [23]. In this method, the problem is arranged hierarchically at different levels. Every level consists of a limited number of elements [13]. There are at least three levels in the hierarchy: goal at the top, decision criteria in the middle and the alternatives are placed at the bottom. Laarhoven and Pedrycz [15] combined fuzzy logic with AHP principle and developed FAHP. This method is used to model complex problems in various sectors such as performance evaluation of shipping company [4], best restaurant site selection for restaurant [11], selection of suitable suppliers in pharmaceutical industry [14], etc. In this work, the pairwise comparison of qualitative factors is achieved by using linguistic scale given in Table 1. The methodological steps are described as

Step 1: Pairwise comparison matrix of factors is constructed as shown in Eq. (7) for FF and UF using the linguistic perceptions of DMs. Then the linguistic perceptions is converted into equivalent fuzzy numbers with the help of linguistic scale (Table 1).

$$\tilde{A} = \begin{matrix} & C_1 & C_2 & \dots & C_n \\ C_1 \\ C_2 \\ \vdots \\ C_n \end{matrix} \begin{pmatrix} 1 & \tilde{a}_{12} & \cdots & \tilde{a}_{1n} \\ \tilde{a}_{21} & 1 & \cdots & \tilde{a}_{2n} \\ \dots & \dots & \dots & \dots \\ \tilde{a}_{n1} & \tilde{a}_{n2} & \dots & 1 \end{pmatrix} \tag{7}$$

Step 2: Geometric mean of all the factors of the comparison matrix is evaluated by using Eq. (8).

$$\tilde{r}_i = (\tilde{a}_{i1} \otimes \cdots \tilde{a}_{ij} \otimes \cdots \tilde{a}_{in})^{\frac{1}{n}} \tag{8}$$

where $\tilde{a}_{ij}$ is fuzzy comparison value of factor i over factor j and $\tilde{r}_i$ is the geometric mean of fuzzy comparison value of factor i.
Step 3: Fuzzy weight of factors is obtained by using Eq. (9) [8].

$$\tilde{w}_i = \tilde{r}_i \otimes (\tilde{r}_1 \oplus \cdots \tilde{r}_i \oplus \cdots \tilde{r}_n)^{-1} \tag{9}$$

where $\tilde{w}_i = (a_{1i}, a_{2i}, a_{4i})$ is the fuzzy weight of the ith factor. The eigenvalue λ_{max} can be calculated by $A_W = \lambda_{max} W$, where $\tilde{A}$ denotes the pairwise comparison matrix, $\tilde{W}$ denotes the importance of the factors and λ_{max} is the maximum eigen value. After computation of weights and eigenvalue of factors, the consistency ratio of comparison matrix is calculated by using the eigenvalue λ_{max}, which is used for computation of consistency index. The consistency index (CI) is expressed by Eq. (10).

$$CI = (\lambda_{max} - n)/(n - 1) \tag{10}$$

The consistency ratio (CR) is used for estimation of consistency of pairwise comparisons. The CR is computed by Eq. (11). Where RI obtained from the table and known as Random Consistency Index (RI) developed by Saaty [23] shown in Table 4.

$$CR = CI/RI \tag{11}$$

If $CR < 10\%$, the comparisons are acceptable, otherwise adjusted. RI represents the average index for randomly generated weights. As λ_{max} is a TFN, it is defuzzified into a crisp number to compute the CI. Here central value method is used to obtain corresponding crisp number.

3.3 *Taguchi Loss Function*

Taguchi quality loss function (QFL) is used to measure losses due to the variation of product quality or service from the target [25]. The losses are calculated to quantify the losses caused by the product due to quality variation [17]. Taguchi's QFL can be

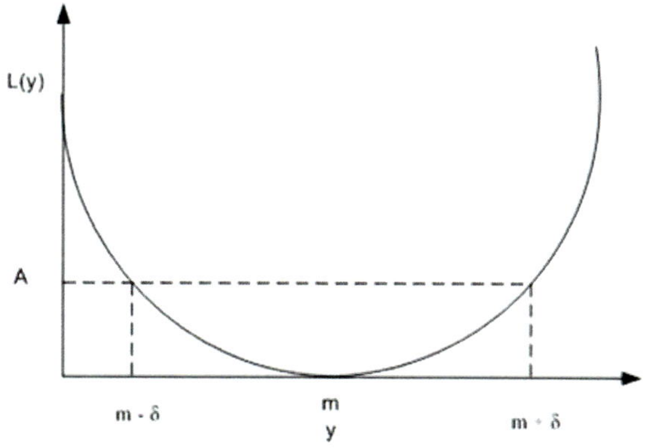

Fig. 1 Loss function: nominal-is-best

successfully used in many areas due to its flexibility such as health care applications [26], manufacturing sector [9], etc. The companies use these functions to quantify amount of losses incurred by company due to cost of failure, cost of maintenance or adverse environmental effects such as pollution or very high production cost. Three types of QLF functions were proposed by Taguchi are

(i) Nominal is the best: By using this loss function the characteristic values are measured either positive or negative. The losses are measured from preset target value. The target can be at the centre or some shift in both the side of centre. Taguchi loss function is shown in Fig. 1 and mathematically given as:

$$L(y) = C(y - m)^2 \tag{12}$$

where $L(y)$ denotes the loss due to particular quality parameter y; m denotes the average value of the specification; C is the average loss coefficient, its value remains constant and depends on the cost of specification limit and the width (e.g., $m \pm \delta$) of the specification, where δ is the customer's tolerance.

(ii) Smaller-is-better: This is positive measurable quality parameter and the ideal value is zero. The value nearer to zero means the better is the quality. This is known as one-sided specification limit (minimum) shown in Fig. 2 and it is given by Eq. (13), where A denotes the average quality loss.

$$L(y) = C(y)^2,\ C = A/\delta^2 \tag{13}$$

(iii) Higher-is-better: This is non-negative measurable quality parameter and the ideal value is infinity. The value closer to infinity means the better is the quality. This is also known as one-sided specification limit (maximum) and expressed by Eq. (14) and shown in Fig. 3, where A denotes the average quality loss.

$$L(y) = C/(y)^2,\ C = A/\delta^2 \tag{14}$$

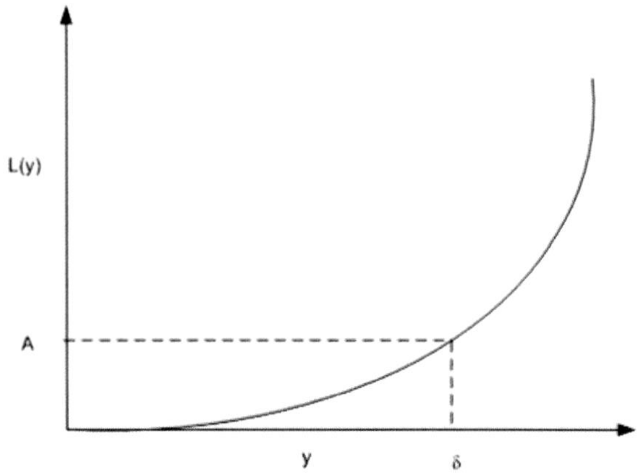

Fig. 2 Loss function: smaller-is-better

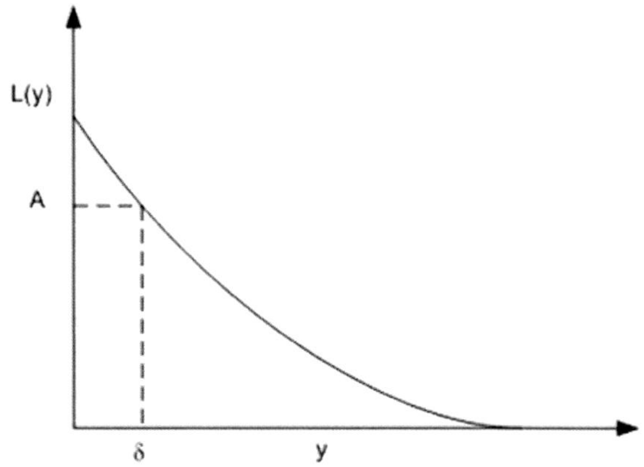

Fig. 3 Loss function: higher-is-better

4 Proposed Methodology

In this section, the proposed decision framework is explained. Firstly, the importance of factors for green suppliers evaluation is determined by using FAHP and then Taguchi loss functions are used to find the ranking order of suppliers and select the suitable supplier for the company. The flow diagram of proposed framework is shown in Fig. 4. The following steps summaries the procedure of the proposed model:

(1) Develop the list of the (FF) and (UF) required for evaluation of supplier for the company.
(2) Perform pairwise comparisons of factors using linguistic preferences given by the DMs.
(3) Evaluate the relative importance of individual FF (W_{FFi}) and UF (W_{UFj}) using FAHP approach.

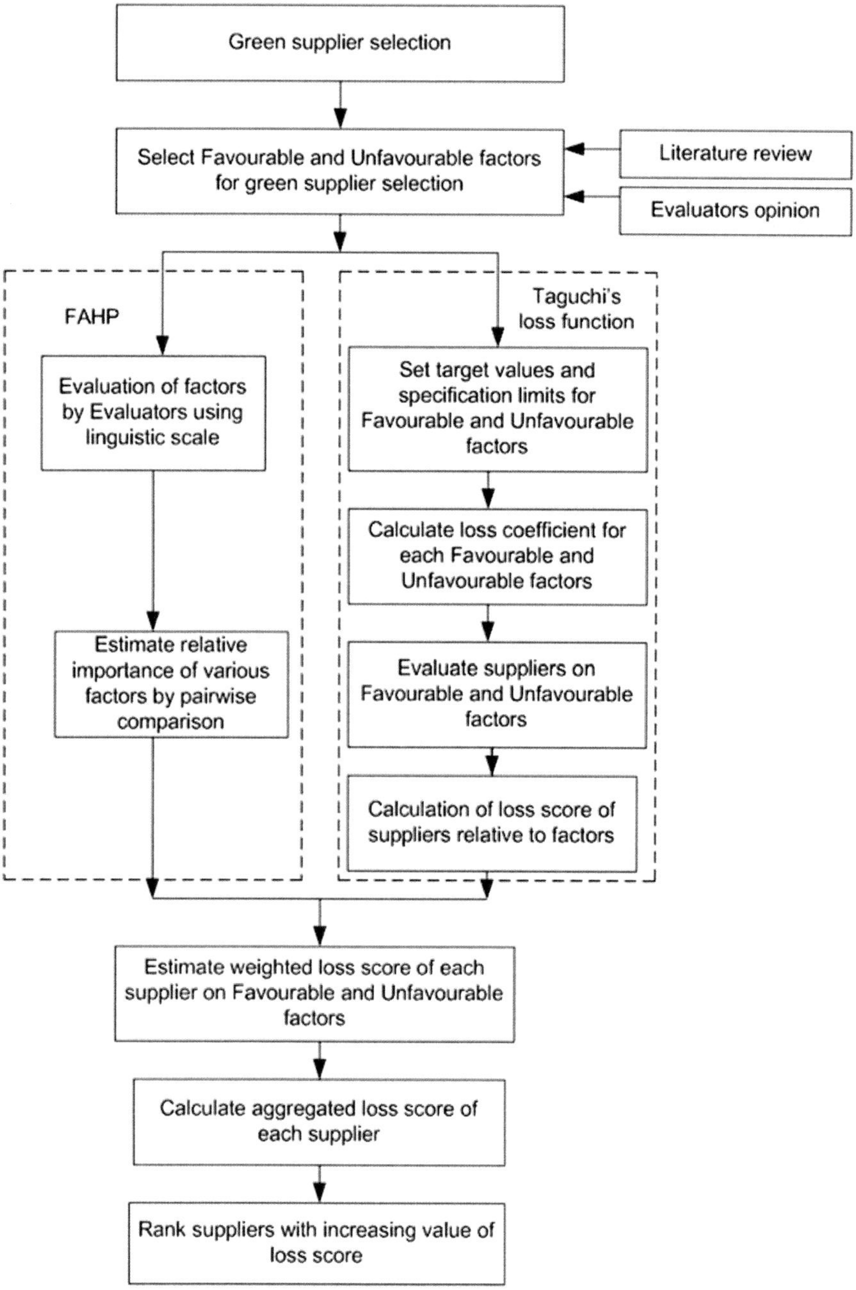

Fig. 4 Integrated framework for supplier selection

(4) Specify the minimum acceptable quality loss $\%(X_i)$ for each FF based as per the requirements of said company.
(5) Estimate the loss coefficient (C_i) for FF with the help of one-sided higher the better loss function of Taguchi as explained in [22].
(6) Specify the upper limit of acceptable quality loss $\%(Y_j)$ for each UF as per the requirement of said company.
(7) Estimate the loss coefficient (C_j) of UF with the help of one-sided lower the better loss function of Taguchi as explained in Eq. (14).
(8) Identify the suppliers who can perform within the specification limits for purchasing activity. The suppliers are denoted by s.
(9) Estimate the performance (x_{is}) of each supplier with respect to FF.
(10) Compute Taguchi LSs, $b_{is} = C_i(\frac{1}{x^2},)$ where b_{is} denotes the loss incurred due to performance of supplier s.
(11) Evaluate the weighted loss score (LS) of each vendor, $B_S = \sum W_{Bi} b_{is}$
(12) Compute the performance of each supplier with respect to UF. Let y_{js} denotes the performance of each supplier s.
(13) Compute the Taguchi's LSs, $r_{js} = C_j y_{js}^2$, where r_{js} is the loss incurred by the supplier s.
(14) Evaluate weighted LS of each supplier, $UF_s = \sum W_{UFj} r_{js}$, UF_s denotes the supplier's weighted LS due to all UF.
(15) Find out the aggregate LS of each supplier by summing weighted LSs for FF and UF. Let $(LS)s = FF_s + UF_s$, where $(LS)_s$ denotes the final aggregate LS for supplier, s. The suppliers are ranked based on the aggregated LSs and the most appropriate supplier is selected for the organization by the lowest LS. Where i= number of FF $= 1, 2, \ldots, BF$; j= number of UF $= 1, 2, \ldots, UF$ and s= number of suppliers $= 1, 2, \ldots, S$.

5 A Case Study

In this section, the proposed model is used to solve green SS problem for a reputed manufacturing company working in Eastern part of India desires to select suitable green suppliers for supplying appropriate material required for its new product line. To meet this specific goal, the company has set up a decision making committee (DMC) consist of three experts. Expert 1, a senior manager of the company looking after supply chain, expert 2, a manager in supply chain department of the organization and expert 3 is selected from academia with vast experience in SCM, innovations, manufacturing. All the DMs are selected due to their vast experience (more than 10 years) in the field of green SS and innovations. It is assumed that the committee is homogeneous in spite of their age, experience and sex. After preliminary screening, the committee has shortlisted four feasible suppliers (S_1, S_2, S_3 and S_4) for further assessment. The committee has prepared a set of questionnaires for determination of related FF and UF for green supplier evaluation after consulting with the company executives and researchers in this domain. The committee has identified various

Table 1 Triangular fuzzy variables for linguistic comparison

Intensity of fuzzy scale	Definition of linguistic values	Triangular fuzzy scale
$\tilde{1}$	Equally important	(1, 1, 1)
$\tilde{3}$	Very weakly important	(2, 3, 4)
$\tilde{5}$	Moderately important	(4, 5, 6)
$\tilde{7}$	Strongly important	(6, 7, 8)
$\tilde{9}$	Absolutely important	(7, 8, 9)
$\tilde{2}, \tilde{4}, \tilde{6}, \tilde{8}$	Intermediate values	(x−1, x, x+1)
$1/\tilde{x}$	Between two adjacent judgements	$(1/(x+1), 1/x, 1/(x-1))$

FF and UF for green supplier evaluation as shown in Tables 2 and 3. Using the steps described in Sect. 4, the evaluation and selection of supplier are carried out as follows:

(1) Pairwise comparison of FF and UF is carried out by using linguistic scale shown in Table 1. The perception of DMC is summarized in Tables 4 and 7. The relative importance of FF and UF is shown in Tables 6 and 8. The consistency ratio of comparison matrix is determined as $CR = 0.067$ and 0.087 for FF and UF matrix, respectively (Tables 6 and 8).
(2) The relative weight of each FF and UF are calculated using FAHP and the results of relative weights of FF are $W_{FF_i} = (0.074, 0.148, 0.041, 0.505, 0.233)$, and for UF are $W_{UF_j} = (0.0635, 0.1417, 0.2431, 0.0935, 0.2023, 0.1386, 0.1172)$.
(3) The target and specification limit of each FF and UF according to company's requirements are set by DMC to minimize the quality loss due to supplier's performance.
(4) The loss coefficient (C_i) for FF is obtained by using Eq. (14) and the results is shown in Table 9.
(5) The loss coefficient (C_j) for UF is determined by using Eq. (13) and the results are shown in Table 10.
(6) Decision-maker's perception of four suppliers in each FF and UF is depicted in Tables 11 and 12 respectively.
(7) Individual LSs of suppliers are calculated using performance value and loss coefficient C. supplier's individual LSs and relative importance with respect to FF and UF are determined and depicted in Tables 13 and 14, respectively. The weighted LS of each supplier is calculated and shown in the same tables.
(8) The aggregated LSs of each supplier is obtained by summing the weighted LSs for FF and UF. The suppliers are then ranked according to aggregated LS and is presented in Table 15.

Results and Analysis

The supplier S_3 with minimum LS 90.85 is selected as the best supplier for the organization. The DMC believes that supplier S_3 will perform optimally for the organization with respect to the selected factors. The result shows that supplier S_3 is

Table 2 FF considered for green SS

Factors	Description
Flexibility (FF_1)	Degree of flexibility to maintain good relationship with the organization
Adoption of green technology (FF_2)	Use of green equipment for manufacturing, reduction of energy utilization and control of green house gas emission
R & D capability of supplier (FF_3)	Availability of skilled experts and technologies for further development in manufacturing process and innovations
Reputation of supplier (FF_4)	Ability to produce green products, compliance with standards, financial and management capability and risk minimization
Cost reduction (FF_5)	Ability to reduce cost associated with manufacturing

Table 3 UF considered for green SS

Factors	Description
Demand risk (UF_1)	Enable to maintain demand fluctuations
Environmental training (UF_2)	Lack of environmental awareness training to employees
OHS policy (UF_3)	Lack of OHS policy adoption
Environmental risk (UF_4)	Enable to control environmentally dangerous activities
Performance risk (UF_5)	Inability to meet the desired performance level
Process risk (UF_6)	Enable to meet the required quantity or quality
Logistics risk (UF_7)	Inability to meet the delivery schedule

Table 4 Random consistency (Saaty, 1980)

Size of the matrix	1	2	3	4	5	6	7	8	9	10
Random consistency	0	0	0.58	0.9	1.12	1.24	1.32	1.41	1.45	1.49

Table 5 Pairwise comparison of FF

	FF_1	FF_2	FF_3	FF_4	FF_5
FF_1	(1, 1, 1)	(1/4, 1/3, 1/2)	(2, 3, 4)	(1/7, 1/6, 1/5)	(1/5, 1/4, 1/3)
FF_2	(2, 3, 4)	(1, 1, 1)	(3, 4, 5)	(1/6, 1/5, 1/4)	(1/3, 1/2, 1)
FF_3	(1/4, 1/3, 1/2)	1/5, 1/4, 1/3	(1, 1, 1)	(1/9, 1/8, 1/7)	(1/6, 1/5, 1/4)
FF_4	(5, 6, 7)	(4, 5, 6)	(7, 8, 9)	(1, 1, 1)	(2, 3, 4)
FF_5	(3, 4, 5)	(1, 2, 3)	(4, 5, 6)	1/4, 1/3, 1/2	(1, 1, 1)

Table 6 Fuzzy weight of FF

	Fuzzy weight		Priority weight
FF_1	(0.0470, 0.0740, 0.1134)	0.2344	0.074
FF_2	(0.0881, 0.1452, 0.2346)	0.4679	0.148
FF_3	(0.0272, 0.0407, 0.0609)	0.1288	0.041
FF_4	(0.3395, 0.5219, 0.7351)	1.5965	0.505
FF_5	(0.1370, 0.2350, 0.3640)	0.7360	0.233
		3.1637	

$\lambda_{max} = 5.301, CI = 0.075, RCI = 1.12, CR = 0.067 < 0.10$

Table 7 Pairwise comparison of UF

	UF_1	UF_2	UF_3	UF_4	UF_5	UF_6	UF_7
UF_1	(1, 1, 1)	(1/4, 1/3, 1/2)	(1/6, 1/5, 1/4)	(1, 1, 1)	(1/4, 1/3, 1/2)	(1/3, 1/2, 1)	(1/3, 1/2, 1)
UF_2	(2, 3, 4)	(1, 1, 1)	(1/3, 1/2, 1)	(1/4, 1/3, 1/2)	(1, 1, 1)	(1, 2, 3)	(1, 2, 3)
UF_3	(4, 5, 6)	(1, 2, 3)	(1, 1, 1)	(2, 3, 4)	(1, 1, 1)	(1, 2, 3)	(1, 2, 3)
UF_4	(1, 1, 1)	(2, 3, 4)	(1/4, 1/3, 1/2)	(1, 1, 1)	(1/4, 1/3, 1/2)	(1/3, 1/2, 1)	(1/3, 1/2, 1)
UF_5	(2, 3, 4)	(1, 1, 1)	(1, 1, 1)	(2, 3, 4)	(1, 1, 1)	(1, 2, 3)	(1, 2, 3)
UF_6	(1, 2, 3)	(1/3, 1/2, 1)	(1/3, 1/2, 1)	(1, 2, 3)	(1/3, 1/2, 1)	(1, 1, 1)	(1, 2, 3)
UF_7	(1, 2, 3)	(1/3, 1/2, 1)	(1/3, 1/2, 1)	(1, 2, 3)	(1/3, 1/2, 1)	(1/3, 1/2, 1)	(1, 1, 1)

Table 8 Fuzzy weight of UF

	Fuzzy weight		Priority weight
UF_1	(0.0354, 0.0614, 0.1238)	0.2207	0.0635
UF_2	(0.0720, 0.1424, 0.2781)	0.4925	0.1417
UF_3	(0.1252, 0.2556, 0.4640)	0.8448	0.2431
UF_4	(0.0505, 0.0905, 0.1840)	0.3249	0.0935
UF_5	(0.1134, 0.2152, 0.3742)	0.7029	0.2023
UF_6	(0.0581, 0.1290, 0.2946)	0.4817	0.1386
UF_7	(0.0496, 0.1058, 0.2518)	0.4073	0.1172
		3.4747	

$\lambda_{max} = 7.691, CI = 0.115, RCI = 1.32, CR = 0.087 < 0.10$

Table 9 Specification limit and range for FF

Factors	Relative weights	Target value (%)	Specification limit (%)	Range (%)	Loss coefficient (C)
FF_1	0.0740	100	55	100 – 55	30.25
FF_2	0.1480	100	70	100 – 70	49.00
FF_3	0.0410	100	60	100 – 60	36.00
FF_4	0.5050	100	85	100 – 85	72.25
FF_5	0.2330	100	82	100 – 82	67.24

Table 10 Specification limit and range of UF

Factors	Relative weights	Target value (%)	Specification limit (%)	Range (%)	Loss coefficient (C)
UF_1	0.0585	0	10	0 – 10	10000
UF_2	0.1567	0	8	0 – 8	15625
UF_3	0.2393	0	2	0 – 2	250000
UF_4	0.1090	0	5	0 – 5	40000
UF_5	0.2053	0	7	0 – 7	20408
UF_6	0.1270	0	8	0 – 8	15625
UF_7	0.1043	0	7	0 – 7	20408

Table 11 Estimation of supplier's performance on FF

Supplier	FF_1	FF_2	FF_3	FF_4	FF_5
S_1	82	76	70	95	85
S_2	75	92	65	90	88
S_3	70	87	75	94	92
S_4	68	82	74	92	83

Table 12 Estimation of supplier's performance on UF

Supplier	UF_1	UF_2	UF_3	UF_4	UF_5	UF_6	UF_7
S_1	8	4	0.9	3	3	7	2
S_2	7	6	1.2	2	5	5	4
S_3	5	3	0.5	1	4	2	3
S_4	6	4	1.4	4	2	6	7

Table 13 Calculation of LSs of suppliers on FF

Supplier	FF_1 RI	Losses	FF_2 RI	Losses	FF_3 RI	Losses	FF_4 RI	Losses	FF_5 RI	Loss	Weighted LS
S_1	0.074	44.99	0.148	84.83	0.041	73.47	0.505	80.06	0.233	88.913	80.042
S_2	0.074	53.78	0.148	57.89	0.041	85.21	0.505	89.20	0.233	82.954	80.414
S_3	0.074	61.73	0.148	64.74	0.041	64.00	0.505	81.77	0.233	75.897	75.750
S_4	0.074	65.42	0.148	72.87	0.041	65.74	0.505	85.36	0.233	93.250	83.157

Table 14 Calculation of LS of suppliers on UF

Supplier	UF_1 RI	Loss	UF_2 RI	Losses	UF_3 RI	Losses	UF_4 RI	Losses
S_1	0.0635	64	0.1417	25.00	0.2431	20.25	0.0935	36.00
S_2	0.0635	49	0.1417	56.25	0.2431	36.00	0.0935	16.00
S_3	0.0635	25	0.1417	14.06	0.2431	6.25	0.0935	4.00
S_4	0.0635	36	0.1417	25.00	0.2431	49.00	0.0935	64.00
	UF_5 RI	Loss	UF_6 RI	Losses	UF_7 RI	Losses	Weighted LS	
S_1	0.2023	18.37	0.1386	76.563	0.1172	8.1633	31.18	
S_2	0.2023	51.02	0.1386	39.0625	0.1172	32.6531	40.89	
S_3	0.2023	32.65	0.1386	6.25	0.1172	18.3673	15.10	
S_4	0.2023	8.16	0.1386	56.25	0.1172	100.00	44.89	

where RI = Relative Importance

Table 15 Aggregated LS of suppliers and ranking

Supplier	Weighted LS for FFs	Weighted LS for UFs	Aggregated LS	Rank
S_1	80.042	31.18	111.22	2
S_2	80.414	40.89	121.31	3
S_3	75.750	15.10	90.85	1
S_4	83.157	44.89	128.05	4

performing consistently in both FF and UF and the aggregated LS of supplier S_3 is 22% lower than that of supplier S_1; 34% lower than supplier S_2 and 41% lower than supplier S_4.

The sensitivity analysis of the result is carried out by changing the relative weights of FF and UF as (0.05, 0.10, 0.15, 0.25, 0.045); (0.20, 0.20, 0.20, 0.20, 0.20); (0.45, 0.25, 0.15, 0.10, 0.05) for FF and (0.15, 0.15, 0.15, 0.15, 0.15, 0.15, 0.10); (0.10, 0.15, 0.15, 0.15, 0.15, 0.15, 0.15); (0.20, 0.20, 0.20, 0.10, 0.10, 0.10, 0.10) for UF and shown graphically in Figs. 5 and 6, respectively. From this analysis, it is clear that,

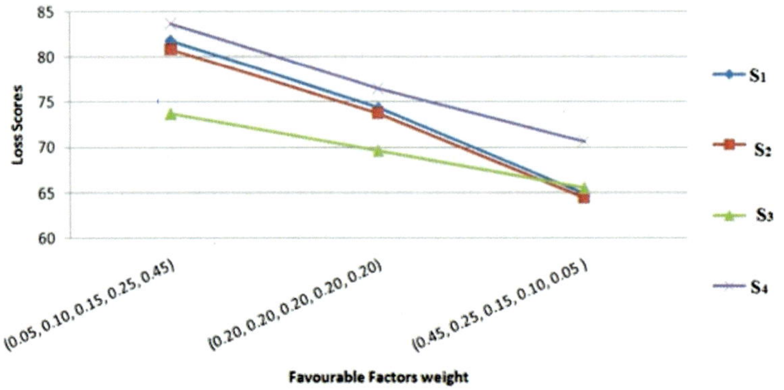

Fig. 5 Sensitivity analysis on FF

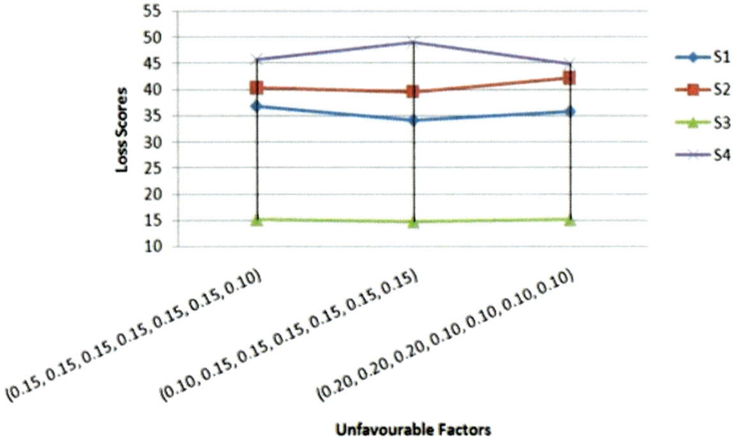

Fig. 6 Sensitivity analysis on UF

changes in relative weights did not affect the ranking order of suppliers; hence, it means that the DMs perceptions on the performance level of suppliers are consistent. Also, in all sets of relative weight, the loss incurred by supplier S_3 is consistent and thus, vendor S_3 can perform consistently considering above-mentioned FF and UF.

6 Managerial Insights

The results of this work have yielded some managerial suggestions as follows:

(a) Manufacturing industries are facing practical problems related to the decision-making for selection of suitable green suppliers to improve their efficiency and reputation. This work helps to develop a systematic integrated model for selecting appropriate green suppliers to fulfill their requirements.

(b) This study not only selects the best green suppliers for the manufacturing organization, but also provides a clear idea of the FF and UF for green supplier evaluation.

(c) As the manufacturing industries presently focus on green strategies, it is essential to find out suitable green suppliers with high capability in greening the supply chain and guide the suppliers to develop their capability against the benchmark set by the organization.

(d) Moreover, other managerial decision-making problems can be solved easily by using this integrated method.

7 Conclusions

The main purpose of this work is to select most appropriate green suppliers for a manufacturing company located in Eastern part of India considering FF and UF. Generally, there is a gap between the available information and performances of the alternatives (suppliers) at the time of decision-making and thus the information are incomplete and uncertain. Due to involvement of several factors of qualitative and quantitative in nature and large number of suppliers, the decision-making becomes a challenging task. This study evaluates the suppliers by measuring the losses incurred by suppliers because of their performance for selected FF and UF. Vendor S_3 with Aggregated LS 90.85 is selected as a suitable supplier for the manufacturing company. Some limitations of the proposed model are: (1) Limited numbers of factors (environmental) are considered in favourable and unfavourable categories for this study and (2) Various decision making techniques can be applied in future for the SS process in the manufacturing industries.

References

1. Bai, C., Sarkis, J.: Integrating sustainability into supplier selection with grey system and rough set methodologies. Int. J. Prod. Econ. **124**(1), 252–64 (2010)
2. Chai, J., Liu, J.N.K., Ngai, E.W.T.: Application of decision-making techniques in supplier selection: a systematic review of literature. Expert Syst. Appl. **40**(1), 3872–3885 (2013)
3. Choi, T.M.: Optimal apparel supplier selection with forecast updates under carbon emission taxation scheme. Comput. Oper. Res. **40**(11), 2646–2655 (2013)
4. Chou, T.Y., Liang, G.S.: Application of a fuzzy multi-criteria decision making model for shipping company performance evaluation. Marit. Policy Manag. **28**(1), 375–392 (2001)
5. Dikson, G.W.: An analysis of vendor selection: systems and decisions. J. Purch. **1**(2), 5–17 (1966)
6. Govindan, K., Rajendran, S., Sarkis, J., Murugesan, P.: Multi criteria decision making approaches for green supplier evaluation and selection: a literature review. J. Clean. Prod. **98**(1), 66–83 (2013)

7. Hancock, B., Ockleford, E., Windridge, K.: An Introduction to Qualitative Research. Research Design Service Yorkshire and Humber (2009)
8. Hsieh, T.Y., Lu, S.T., Tzeng, G.H.: Fuzzy MCDM approach for planning and design tenders selection in public office buildings. Int. J. Proj. Manag. **22**(7), 573–584 (2004)
9. Hui, Y.V., Leung, L.C.: Optimal economic tool regrinding with Taguchi's quality loss function. Eng. Econ. **39**(4), 313–331 (1994)
10. Ho, W., Xu, X., Dey, P.K.: Multi-criteria decision making approaches for supplier evaluation and selection: a literature review. Eur. J. Oper. Res. **202**(1), 16–24 (2010)
11. Hwang, H.S., Ko, W.H.: A restaurant planning model based on fuzzy-AHP method. In: Proceedings of ISAHP, pp. 1–14 (2003)
12. Kethley, R.B., Waller, T.A.: Improving customer service in the real estate industry: a property selection model using Taguchi loss functions. Total Qual. Manag. **13**(6), 739–748 (2002)
13. Khajeeh, M.: Water conservation in Kuwait: a fuzzy analysis approach. Int. J. Ind. Eng. **6**(1), 90–105 (2010)
14. Khorasani, O., Bafruei, M.K.: A fuzzy AHP approach for evaluating and selecting supplier in pharmaceutical industry. Int. J. Acad. Res. **3**(1), 346–352 (2011)
15. Laarhoven, P.J.M.V., Pedrycz, W.: A fuzzy extension of Saaty's priority theory. Fuzzy Sets Syst. **11**(1), 199–227 (1983)
16. Li, G.D., Yamaguchi, D., Nagai, M.A.: Grey-based rough decision-making approach to supplier selection. Int. J. Adv. Manuf. Technol. **36**(2), 1032–40 (2008)
17. Magdalena, R.: Supplier selection for food industry: a combination of Taguchi loss function and fuzzy analytical hierarchy process. Asian J. Technol. Manag. **5**(1), 13–22 (2012)
18. Nguyen, H.T., Dawal, S.Z.M., Nukman, Y., Aoyama, H.: A hybrid approach for fuzzy multi-attribute decision making in machine tool selection with consideration of the interactions of attributes. Expert Syst. Appl. **41**(6), 3078–90 (2014)
19. Nielsen, I.E., Banaeian, N., Golinska, P., Mobli, H., Omid, M.: Green supplier selection criteria: from a litreture review to a flexible framework for determination of suitable criteria. In: Golinska, P. (ed.) Eco Production Book Series: Logistic Operations, Supply Chain Management and Sustainability, pp. 79–100. Springer, Berlin (2014)
20. Olfat, A., Govindan, K., Khodaverdi, R.: A fuzzy multi criteria approach for evaluating green supplier's performance in green supply chain with linguistic preferences. Resour. Conserv. Recycl. (working paper) (2012)
21. Oztaysi, B.: A decision model for information technology selection using AHP integrated TOPSIS-grey: the case of content management systems. Knowl.-Based Syst. **59**(1), 48–57 (2014)
22. Pi, W.N., Low, C.: Supplier evaluation and selection via Taguchi loss functions and an AHP. Int. J. Adv. Manuf. Technol. **27**(5–6), 625–630 (2006)
23. Saaty, T.L.: The Analytic Hierarchy Process. McGraw-Hill, New York (1980)
24. Sarkis, J., Talluri, S.: A model for strategic supplier selection. J. Supply Chain Manag. **38**(4), 18–28 (2002)
25. Taguchi, G., Elsayed, E.D., Hsiang, T.: Quality Engineering in Production Systems. McGraw-Hill, New York (1989)
26. Taner, T., Antony, J.: Comparing public and private hospital care service quality in Turkey. Leadersh. Health Serv. **19**(2), 1–10 (2006)
27. Varsei, M., Soosay, C., Fahimnia, B., Sarkis, J.: Framing sustainability performance of supply chains with multidimensional indicators. Supply Chain Manag.: Int. J. **19**(3), 242–57 (2014)
28. Weber, C., Current, J., Benton, W.: Vendor selection criteria and methods. Eur. J. Oper. Res. **50**(1), 2–18 (1991)
29. Wong, T.N., Lee, L.H., Sun, Z.: CSR and environmental criteria in supplier selection. In: Kachitvichyanukul, V., Luong, H.T., Pitakaso, R. (eds.) Proceedings of the Asia Pacific Industrial Engineering & Management Systems Conference, vol. 140(1), pp. 283–294 (2012)
30. Zadeh, L.A.: Fuzzy sets. Inf. Control **8**(3), 338–53 (1965)

Genetic Algorithm-Based Fuzzy Programming Method for Multi-objective Stochastic Transportation Problem Involving Three-Parameter Weibull Distribution

Adane Abebaw Gessesse and Rajashree Mishra

Abstract In real-life situations, it is difficult to handle multi-objective stochastic transportation problems. It can't be solved directly using traditional mathematical programming approaches. In this paper, we proposed a solution procedure to handle the above problem. The proposed solution procedure is a hybridization of the evolutionary algorithm called a genetic algorithm and a classical mathematical programming technique called fuzzy programming method. This hybrid approach is called a genetic algorithm-based fuzzy programming method. The supply and demand parameters of the constraints follow a three-parameter Weibull distribution. To complete the proposed problem a total of three steps are required. Initially, the probabilistic constraints are handled using stochastic simulation. Then, we checked the feasibility of probability constraints by the stochastic programming with the genetic algorithm without deriving the deterministic equivalents. Then, the genetic algorithm-based fuzzy programming method is considered to generate non-dominated solutions for the given problem. Finally, a numerical case study is presented to illustrate the methodology.

Keywords Multi-objective programming · Stochastic transportation problem · Weibull distribution · Fuzzy programming · Genetic algorithm

1 Introduction

The aim of transportation problem is to minimize the objective functions which includes minimization of cost, minimization of damage, and minimization of time

A. Abebaw Gessesse (✉) · R. Mishra
Department of Mathematics, School of Applied Sciences, KIIT Deemed to be University, Bhubaneswar 751024, India
e-mail: adaneab@gmail.com

R. Mishra
e-mail: rajashreemishra011@gmail.com

D. Giri et al. (eds.), *Proceedings of the Fifth International Conference on Mathematics and Computing*, Advances in Intelligent Systems and Computing 1170, https://doi.org/10.1007/978-981-15-5411-7_11

during transporting of items from source to destination. Hitchcock [5] was the first who developed transportation problem.

In real-life decision-making problems, the decision-maker faces the challenges to minimize more than one factors, for example, transportation cost, transportation time, etc. which leads to multi-objective transportation problem. The objective functions are conflicting with each other and non-commensurable. The typical issue faced in tackling such type of mathematical problem is to propose methodology to obtain a compromise solution among a large number of non-dominated solutions.

The parameters associated with transportation problem are not deterministic or fixed value always. Due to several reasons, namely, weather during transportation, condition of transportation medium, seasonal demand of customers at destination, etc. the parameters may be uncertain in nature. In a mathematical programming model, uncertainties are addressed using fuzzy program set theory or probability theory. In the present paper, we deal the parameters that address uncertainty using the probability theory. The presence of probability in a mathematical programming problem leads to stochastic programming.

In our proposed work, attention has been given to solve stochastic transportation problem where the supply and demand parameters follow three-parameter Weibull distribution having more than one objective function. The mathematical model is known as multi-objective stochastic transportation problem (MOSTP). We suggest a GA-based fuzzy programming method of solution procedure for handling MOSTP involving random variable which follows three-parameter Weibull distribution. To establish the solution procedures of the proposed problem, the probabilistic constraints are handled by stochastic simulation-based GA. The feasibility of probability constraints is checked by the stochastic programming with the GA without deriving the deterministic equivalents and it is maintained throughout the problem. Then, the genetic algorithm-based fuzzy programming method is considered to generate Pareto optimal solutions for the given problem.

This paper is organized as follows. Following the introduction section, a literature review has been presented in Sect. 2. The deterministic equivalent model is presented in Sect. 3. Case study, and results and discussion have been presented in Sects. 4 and 5, respectively. The concluding remarks are given in Sect. 6 followed by references.

2 Literature Survey

Different models have been discussed on stochastic transportation problems by several researchers. Mahapatra [9] have discussed the solution procedures and applied fuzzy programming method for solving the multi-objective deterministic transportation problem. Biswal and Samal [4] presented a mathematical programming model which is either Cauchy random variables or multi-choice type for source and demand parameters of the stochastic transportation problem. Roy [12] established a multi-choice stochastic transportation problem, where the supply and demand parameters of the constraints follow Weibull distribution. Quddoos [11] developed a multi-choice stochastic transportation problem and obtained an equivalent deterministic model, where accessibility and request parameters follow general form of distribution. Under the light of conic scalarizing function, Roy et al. [13] investigated the study of multi-choice MOTP. Recently, Roy et al. [14] under intuitionistic fuzzy environment examined MOTP.

Genetic algorithms (GA) developed by Holland [6] are adaptive procedures that find solutions to such multi-objective problems by an evolutionary process based on natural selection. Since GA, because of its population-based nature, can obtain multiple Pareto optimal solutions in a single simulation run, it is very popular in solving MOSTP. It has the following three advantages rather than classical methods: (1) to find a set of solutions as close as possible to the Pareto front (PF) (known as convergence); (2) to find a well-distributed set of solutions (known as diversity); and (3) to cover the entire PF (known as coverage) [16].

Many researchers have studied on evolutionary computing and its application for solving transportation problems. Zaki et al. [17] proposed an efficient GA by incorporating the benefits of both GA and local search (LS) scheme for solving MOTP and others. Mousa et al. [10] designed an efficient evolutionary algorithm for solving MOTP coordinating the benefits of both GA and LS. Bharathi and Vijayalakshmi [3] presented an application of evolutionary algorithms to the MOTP. Sanjay et al. [15] presented a solution procedure of a MOTP with fuzzy stochastic simulation-based GA. Ardjmand et al. [1] applied GA to a new bi-objective stochastic model for transportation, location, and allocation of hazardous materials. Recently, Karthy and Ganesan [7] applied GA for solving MOTP.

3 Basic Preliminaries

Bounded Random Number (BRN): One way to generate random numbers in C++ is to use the function rand() which produces an integer between 0 and RAND_MAX, where RAND_MAX is defined in #stdlib as $(2^{15} - 1)$. Therefore, a random number on an interval [0,1] can be generated as
Step 1: m = rand().
Step 2: m $\leftarrow (m/RAND\_MAX)$.

Weibull distribution: It is broadly used to investigate the combined loss of execution of a mind-boggling framework in frameworks building. It is also used to depict the information on holding up time until the point when an occasion happens. Moreover, the Weibull distribution has numerous applications in medicinal, natural, and earth sciences [8]. Weibull distribution is defined by the following parameters [2]: α: scale parameters, θ: shape parameter, and μ: location parameter. Its probability density function (pdf) is expressed as follows:

$$f(x) = \frac{\theta}{\alpha}\left(\frac{x-\mu}{\alpha}\right)^{\theta-1} e^{-\left(\frac{x-\mu}{\alpha}\right)^{\theta}}, \tag{1}$$

where $x \geq \mu$, θ, and $\alpha > 0$.
It can be generated as follows:
Step 1: Use location μ, scale α, shape θ.
Step 2: Generate m from BRN(0,1).
Step 3: Return $n = (log(1/(1-m)))^{1/\theta}$.
Step 4: Return $\mu + \alpha n$.

4 Mathematical Model of MOSTP

Mathematically, a MOSTP where randomness is considered in the right-hand side constraints is expressed as

$$min : Z^k(x) = \sum_{s=1}^{m} \sum_{t=1}^{n} c_{st}^k x_{st}, k = 1, 2, \ldots, K. \tag{2}$$

Subject to

$$P\left(\sum_{t=1}^{n} x_{st} \leq a_s\right) \geq 1 - \gamma_s, s = 1, 2, \ldots, m. \tag{3}$$

$$P\left(\sum_{t=1}^{n} x_{st} \geq b_t\right) \geq 1 - \delta_t, t = 1, 2, \ldots, n. \tag{4}$$

$$x_{st} \geq 0, \forall s, t, \tag{5}$$

where $0 < \gamma_s < 1$ and $0 < \delta_t < 1, \forall s, t$.

Let $a_s (s = 1, 2, \ldots, m)$ and $b_t (t = 1, 2, \ldots, n)$ are Weibull random variables. $c_{st}^k (s = 1, 2, \ldots, m; t = 1, 2, \ldots, n; k = 1, 2, \ldots, K)$ represents the coefficients associated with the decision variables in the objective function, and γ_s and δ_t are the given confidence level in which the sth and tth constraint violations are accepted.

5 GA-Based Fuzzy Programming Method for MOSTP

The GA-based fuzzy programming method is designed to solve the MOSTP. The algorithmic steps are described as follows:

Step 1: Fix GA parameters and termination criteria.
Step 2: The parameter for distributions is generated.

Step 3: The GA population is initialized for the first objective function with the given constraints.
Step 4: Initialization of generation (gen = 1).
Step 5: Apply the bounds on the population and calculate constraints for each objective function.
Step 6: Check the feasibility condition, if satisfied then move to Step 7 or else go to Step 3.
Step 7: Probability criteria is checked, if satisfied then move to Step 8 or else go to Step 3.
Step 8: Calculate the functional value, i.e., the objective function.
Step 9: Apply selection, crossover, and mutation, respectively.
Step 10: Again calculate the functional value, i.e., the objective function.
Step 11: Again probability criteria is checked, if satisfied then move to Step 12 or else go to Step 3.
Step 12: Apply Elitism.
Step 13: Check the stopping criteria. If reached, the present population is the best population else gen = gen + 1 and go to Step 5.
Step 14: Ideal solution is obtained.
Step 15: Repeat the steps from Step 2 to Step 14 for the other objective function till all the objective functions are done.
Step 16: Construct a payoff matrix as shown in Table 1.
Step 17: From the constructed payoff matrix, determine the bounds for kth objective function $Z^k(x)$, $k = 1, 2, \ldots, K$. Find the best lower bound L_b^k and worst upper bound U_w^k, $k = 1, 2, \ldots, K$ for minimization type.
Step 18: To the k(th) objective function $Z^k(x)$, define a membership function $\mu_{Z^k}(x)$ which is evaluated as follows:

$$\mu_{Z^k}(x) = \begin{cases} 1, & if\, Z^k(x) \le L_b^k \\ \frac{U_w^k - Z^k(x)}{U_w^k - L_b^k}, & if\, L_b^k < Z^k(x) < U_w^k, k = 1, 2, \ldots, K \\ 0, & if\, Z^k(x) \ge U_w^k \end{cases} . \tag{6}$$

Step 19: Formulate a single objective mathematical programming (SOMP) problem using max-min operator with augmented variable λ as

$$max : \lambda. \tag{7}$$

Subject to

$$\lambda \le \mu_{Z^k}(x), k = 1, 2, \ldots, K \tag{8}$$

$x \in R$, where R is the attainable region.
Step 20: Finally, solve the SOMP problem (7)–(8) using GA to obtain the Pareto solutions.

The flow diagram of GA-based fuzzy programming method is shown in Fig. 1.

Table 1 Payoff matrix

Ideal solutions	Objective functions			
	$Z^1(x)$	$Z^2(x)$	...	$Z^K(x)$
$x^{(1)}$	$Z^1(x^{(1)})$	$Z^2(x^{(1)})$	...	$Z^K(x^{(1)})$
$x^{(2)}$	$Z^1(x^{(2)})$	$Z^2(x^{(2)})$	...	$Z^K(x^{(2)})$
$\vdots$	$\vdots$	$\vdots$	...	$\vdots$
$x^{(K)}$	$Z^1(x^{(K)})$	$Z^2(x^{(K)})$	...	$Z^K(x^{(K)})$

6 Case Study

A case study is considered by taking the data egg transportation from Andhra Pradesh to Maharashtra in India. A group of suppliers supply eggs by collecting from Andhra Pradesh in the cities like Hyderabad, Kareem Nagar, and Nizamabad to Aurangabad, Nanded, and Latur Districts of Maharashtra selected randomly through the nine routes.

The objective is to find the optimum solution for minimizing the percentage breakages and overall time. Due to unpredictable factors the supply and demand points may not be known previously. However, by investigating the reviewed information, it is found that the supply and the demand parameters follow Weibull random variables. To minimize the total cost, MOSTP is used.

Let time (T_{st}), percentage breakages (P_{st}), and allocation (x_{st}) are nonnegative real variables. The sources are $A_s(s = 1, 2, 3)$ where eggs are accessible, $A_1 =$ Hyderabad, $A_2 =$ Nizamabad and $A_3 =$ Karimnagar, and $R_t(t = 1, 2, 3)$ are requests, where eggs are required, $R_1 =$ Aurangabad, $R_2 =$ Nanded, and $R_3 =$ Latur.

The time and percentage of breakage matrices with an average speed of 37.5 kilometers per hour (km/h) are presented in Tables 2 and 3 from each supply to each destination points in nine different routes.

The mathematical programming model is expressed in Eqs. (9)–(12).

$$\begin{aligned} min: Z^2 = {} & 15.03x_{11} + 8.56x_{12} + 7.64x_{13} + 14.5x_{21} + 7.28x_{22} \\ & +9.3x_{23} + 10.82x_{31} + 3.88x_{32} + 5.26x_{33} \end{aligned} \tag{9}$$

$$\begin{aligned} min: Z^3 = {} & 5.75x_{11} + 3.28x_{12} + 2.92x_{13} + 5.45x_{21} + 2.78x_{22} \\ & + 3.56x_{23} + 16.36x_{31} + 1.48x_{32} + 2.02x_{33} \end{aligned} . \tag{10}$$

Subject to

$$P\left(\sum_{t=1}^{n} x_{st} \le R_s\right) \ge 1 - \gamma_s, s = 1, 2, 3. \tag{11}$$

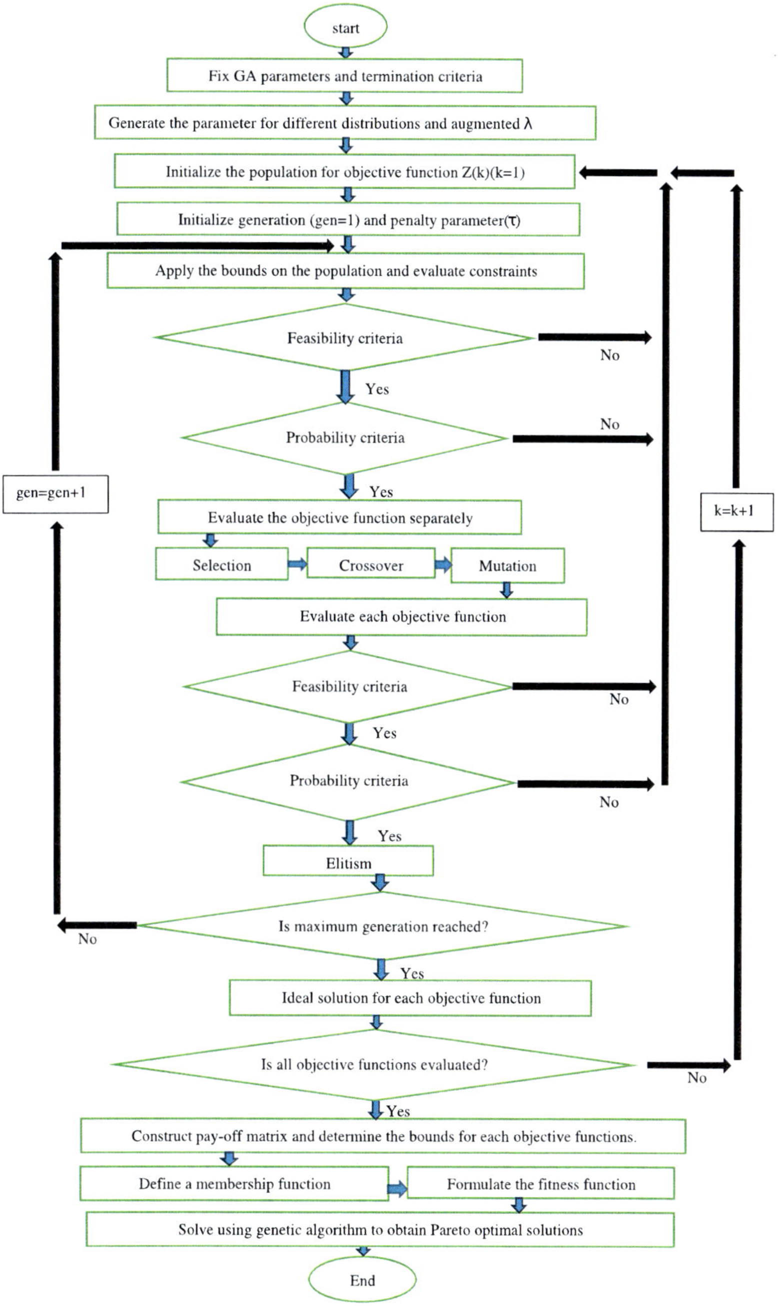

Fig. 1 Flow diagram of GA-based fuzzy programming approach to solve MOSTP

Table 2 Time matrix (T_{st})

Source	Destination		
	R_1	R_2	R_3
A_1	15.03	8.56	7.64
A_2	14.5	7.28	9.3
A_3	10.82	3.88	5.26

Table 3 Percentage breakage matrix (P_{st})

Source	Destination		
	R_1	R_2	R_3
A_1	5.75	3.28	2.92
A_2	5.45	2.78	3.56
A_3	16.36	1.48	2.02

Table 4 Parameters for suppliers

Location	Shape	Scale	DLL (γ_s)
$\mu_1 = 3$	$\theta_1 = 5.8$	$\alpha_1 = 3.0$	$\gamma_1 = 0.01$
$\mu_2 = 2$	$\theta_2 = 6.5$	$\alpha_2 = 4.0$	$\gamma_2 = 0.02$
$\mu_3 = 2.5$	$\theta_3 = 7.2$	$\alpha_3 = 2.5$	$\gamma_3 = 0.03$

Table 5 Parameters for demands

Location	Shape	Scale	DLL (γ_s)
$\mu_1^{'} = 0.5$	$\theta_1^{'} = 6.2$	$\alpha_1^{'} = 5.0$	$\delta_1 = 0.7$
$\mu_2^{'} = 1$	$\theta_2^{'} = 6.9$	$\alpha_2^{'} = 4.0$	$\delta_2 = 0.8$
$\mu_3^{'} = 1.5$	$\theta_3^{'} = 7.3$	$\alpha_3^{'} = 3.0$	$\delta_3 = 0.9$

$$P\left(\sum_{t=1}^{n} x_{st} \geq A_t\right) \geq 1 - \delta_t, t = 1, 2, 3, \quad (12)$$

where $x_{st} \geq 0, s = 1, 2, 3; t = 1, 2, 3;$ and $0 < \gamma_s < 1, 0 < \delta_t < 1, \forall s, t.$

Three known parameters of Weibull distribution with determined likelihood level (DLL) of suppliers and demands are taken, respectively, and are represented in Tables 4 and 5.

Using the data provided in Tables 4 and 5, the above MOSTP is solved using GA-based fuzzy programming method. Applying the steps of GA-based fuzzy programming method, we obtain two ideal solutions as

$$X^{(1)} = [0.542522, 2.09677, 2.19941, 0.674487, 0.777126, 2.06256, 0.801564, 0.293255, 0.410557]^T$$
$$X^{(2)} = [0.899316, 1.76442, 2.80547, 0.386119, 0.513196, 0.391007, 0.312805, 0.097752, 0.483871]^T$$

Table 6 Payoff matrix for the case study

Solutions	Objective functions	
	$Z^1(x)$	$Z^2(x)$
$X^{(1)}$	89.4957	43.9752
$X^{(2)}$	69.3341	30.313

with the value of $Z^1 = 89.4957$ and $Z^2 = 30.313$, respectively. Using the two ideal solutions a payoff matrix is formulated in Table 6. The upper and lower bounds of the objective functions are obtained from the payoff matrix. A best lower bound and a worst upper bound recorded underneath for all of the objective functions are $L_b^1 = 69.3341, U_w^1 = 89.4957, L_b^2 = 30.313$, and $U_w^2 = 43.9752$. Hence, using min-max operator with augmented λ, the MOSTP reduces to single objective stochastic transportation problem (SOSTP).

$$max : \lambda. \tag{13}$$

Subject to

$$\begin{aligned} &15.03x_{11} + 8.56x_{12} + 7.64x_{13} + 1.45x_{21} + 7.28x_{22} + 9.3x_{23} + 10.82x_{31} \\ &+3.88x_{32} + 5.6x_{33} + 20.1616\lambda \leq 89.4957. \end{aligned} \tag{14}$$

$$\begin{aligned} &5.75x_{11} + 3.28x_{12} + 2.92x_{13} + 5.45x_{21} + 2.78x_{22} + 3.56x_{23} + 16.36x_{31} \\ &+1.48x_{32} + 2.02x_{33} + 13.6622\lambda \leq 43.9752. \end{aligned} \tag{15}$$

$$P\left(\sum_{t=1}^{3} x_{1t} \leq a_1\right) \geq 1 - \gamma_1 \tag{16}$$

$$P\left(\sum_{t=1}^{3} x_{2t} \leq a_2\right) \geq 1 - \gamma_2 \tag{17}$$

$$P\left(\sum_{t=1}^{3} x_{3t} \leq a_3\right) \geq 1 - \gamma_3 \tag{18}$$

$$P\left(\sum_{s=1}^{3} x_{s1} \geq b_1\right) \geq 1 - \delta_1 \tag{19}$$

$$P\left(\sum_{s=1}^{3} x_{s2} \geq b_2\right) \geq 1 - \delta_2 \tag{20}$$

Table 7 Pareto optimal solutions for Pc = 0.9

Pm	x_{11}	x_{12}	x_{13}	x_{21}	x_{22}	x_{23}	x_{31}	x_{32}	x_{33}	Z^1	Z^2
0.01	0.449658	3.17693	1.39296	0.322581	2.1652	0.68915	0.0733138	0.698925	0.464321	77.3917	30.4758
0.01	0.948192	0.488759	0.669599	0.273705	2.50244	0.97263	0.376344	0.874878	4.49658	85.9014	37.4564
0.01	0.948192	0.488759	0.669599	0.273705	2.54154	1.01173	0.337243	0.874878	4.49658	86.1266	37.0646
0.01	2.19941	0.488759	0.664712	0.273705	0	0.97263	0.376344	0.874878	4.50635	86.5035	37.6996
0.01	2.18964	0.498534	0.674487	0.293255	0.058651	0.982405	0.337243	0.874878	4.5259	86.9961	37.4082
0.01	0.410557	3.17693	1.39296	0.322581	4.66764	0.0635386	0.0733138	0.698925	0.464321	89.2036	34.9805
0.01	1.33431	0.400782	1.48094	0.532747	3.08407	0.513196	0.576735	1.05572	1.83284	89.7265	41.3155
0.01	0.449658	3.17693	1.39296	0.322581	4.66764	0.68915	0.0244379	0.698925	0.151515	93.4353	36.0011
0.01	2.19453	0.058651	0.957967	0.224829	2.96676	0.86999	0.430108	1.01173	2.26784	94.2617	41.2932
0.01	0.410557	3.17693	1.39296	0.322581	4.66764	0.68915	0.0733138	0.698925	0.464321	95.0218	37.2077
0.05	0.0439883	2.41935	0.00977517	0.933529	0.723363	2.2434	0.791789	0.913978	0.0293255	73.379	37.6677
0.05	0.923754	0.791789	1.83284	0.899316	0.083089	0.0439883	0.513196	0.14174	4.63343	79.1933	36.5146
0.05	0.0439883	2.10655	0.713587	0.620723	1.97458	2.2434	0.791789	0.904203	0.0293255	80.6138	40.456
0.05	0.0439883	2.41935	0.00977517	1.24633	1.97458	1.9306	0.791789	0.913978	0.26393	85.3485	42.2112
0.05	1.11437	1.5738	0.249267	0.234604	0.518084	2.47312	1.06061	0.278592	2.6393	88.7379	46.9159
0.05	0.0391007	2.41447	0.0146628	1.21212	0.723363	2.23851	0.786901	0.904203	2.52199	90.3158	44.0796
0.05	0.498534	0.156403	0.542522	2.2043	1.99902	0.635386	0.288368	3.31867	1.93548	91.5782	38.3354
0.05	0.0391007	2.38025	0.00977517	1.24633	1.04594	2.2434	0.791789	0.913978	2.57087	93.2235	45.2469
0.05	0.0439883	2.41935	0.00977517	1.24633	0.0977517	4.74585	0.791789	0.913978	0.0293255	96.633	46.542
0.05	1.21701	0.801564	2.41447	0.889541	0.102639	2.67351	1.24633	0.102639	0.483871	98.5375	52.8476
0.1	1.01662	0.26393	1.12903	0.146628	0.532747	1.6911	0.0244379	1.6129	0.596285	57.5555	22.2999
0.1	0.361681	1.08993	2.45846	0.874878	0.0635386	0.215054	0.166178	2.6784	0.0146628	60.9641	25.256
0.1	1.06549	0.459433	2.45846	0.249267	0.200391	1.31965	0.0146628	1.42717	0.640274	65.1396	25.0712
0.1	0.943304	0.410557	2.49756	0.30303	0.14174	1.4565	0.259042	1.76931	0.83089	69.783	29.8291
0.1	0.987292	0.503421	1.18768	0.249267	2.74194	1.4174	0.112414	1.27077	0.210166	72.2321	28.9676
0.1	0.117302	1.29521	1.98436	1.40762	1.48583	0.630499	0.322581	0.0733138	0.738025	72.7584	31.6406
0.1	0.938416	0.464321	1.20723	0.190616	2.70283	1.47605	0.0146628	1.42717	0.835777	73.5624	28.2918
0.1	0.977517	0.620723	2.42913	0.083089	2.74194	0.156403	0.0977517	1.72043	1.28055	75.6533	30.1141
0.1	0.85044	0.454545	2.35582	0.249267	2.70283	1.5347	0.0146628	0.185728	0.640274	76.4823	29.404
0.1	0.317693	0.107527	2.37537	0.224829	3.4262	1.3783	0.0244379	2.33627	0.596285	77.3297	29.8344
0.1	0.987292	1.39296	2.45846	0.239492	3.01564	0.205279	0.0146628	0.332356	0.645161	77.7227	29.879

Table 8 Pareto optimal solutions for Pc = 0.8

Pm	x_{11}	x_{12}	x_{13}	x_{21}	x_{22}	x_{23}	x_{31}	x_{32}	x_{33}	Z^1	Z^2
0.01	0.0391007	0.86999	1.58847	0.44477	0.391007	3.02053	0.195503	0.72825	2.48778	75.584	31.2824
0.01	0.483871	1.01173	1.99902	0.522972	0.0537634	3.11339	0.371457	0.747801	2.03812	85.7757	37.322
0.01	0.542522	0.806452	0.816227	0.620723	3.37243	0.782014	0.136852	3.19159	4.11046	97.603	38.9559
0.01	0.542522	0.806452	0.659824	0.620723	3.39687	0.156403	1.38807	3.19159	2.85924	97.7246	54.2825
0.01	0.112414	0.0635386	0.439883	3.21114	2.62952	1.24145	0.0293255	3.64125	0.816227	101.583	38.8872
0.01	0.229717	0.806452	0.190616	0.620723	3.37243	0.782014	1.38807	3.19159	4.11046	101.66	55.8005
0.01	0.542522	0.806452	0.816227	0.620723	3.37243	0.860215	0.449658	3.19159	4.11046	101.715	44.3518
0.01	0.542522	0.806452	0.816227	0.464321	3.37243	0.156403	1.38807	3.19159	4.13001	103.158	56.3858
0.01	0.542522	0.180841	0.816227	0.620723	3.37243	0.782014	1.38807	2.56598	4.11046	103.359	56.448
0.01	0.28348	2.42913	1.88172	0.180841	3.74389	1.15347	0.259042	2.73705	2.25318	105.31	43.4323
0.05	0.278592	4.00782	1.10948	0.953079	0.0391007	0.361681	0.518084	2.39003	1.1437	85.3334	38.9012
0.05	1.54936	2.44379	0.796676	0.948192	0.351906	0.439883	0.518084	1.21701	1.16325	87.1403	39.5895
0.05	0.00977517	3.25024	2.00391	0.175953	0.259042	1.63245	1.38319	0.469208	3.26002	96.8322	53.9676
0.05	1.47116	4.00782	1.10948	0.948192	0.0391007	0.361681	0.518084	1.21701	1.13392	98.5841	43.976
0.05	1.54936	4.32063	1.14858	0.967742	0.0391007	0.0488759	0.518084	1.21212	1.1437	100.143	44.5713
0.05	1.54936	4.32063	1.10948	0.948192	0.0391007	0.351906	0.518084	1.21701	1.1437	102.398	45.4366
0.05	1.54936	4.30108	1.10948	0.948192	0.0391007	0.439883	0.478983	1.21701	1.14858	102.651	45.0559
0.05	0.288368	0.786901	4.54545	0.215054	0.923754	1.05083	1.53959	0.508309	3.57283	102.837	58.15
0.05	0.610948	2.9912	1.10948	2.90323	0.0391007	0.361681	0.518084	1.21701	1.1437	105.352	46.3699
0.05	2.77615	0.83089	3.49462	0.0439883	0.146628	0.835777	1.15347	1.04106	0.884653	106.188	54.7137
0.1	0.884653	0.478983	2.22385	0.669599	2.14565	0.376344	0.938416	1.34897	0.83089	82.9743	43.1328
0.1	0.215054	1.20723	2.66862	1.39296	1.46139	1.38807	0.42522	0.738025	0.00977517	85.2162	37.6531
0.1	0.953079	0.591398	0.493646	1.27077	2.11632	2.02835	0.0635386	0.938416	1.09971	85.9683	33.5412
0.1	0.0782014	1.33431	1.9697	0.7087	1.99413	0.361681	0.12219	1.01173	4.91202	86.8875	34.6901
0.1	1.97458	0.327468	1.63245	0.630499	1.35386	0.376344	0.938416	1.64223	0.791789	88.1417	45.1169
0.1	0.953079	0.591398	0.215054	1.28055	2.11632	1.28543	1.39296	1.09971	0.483871	88.8434	50.8803
0.1	1.42717	0.278592	0.503421	1.94037	1.41251	1.96481	0.102639	0.97263	0.0733138	89.6425	35.3533
0.1	2.13587	0.171065	2.21896	0.679374	0.738025	0.337243	0.957967	1.64223	0.909091	90.3984	46.2158
0.5	0.508309	0.0928641	1.20723	1.38319	2.33138	0.205279	1.37341	0.381232	3.65103	92.1397	51.9112
0.5	0.860215	0.469208	0.816227	1.38807	1.33431	0.957967	0.909091	3.52884	1.29032	92.2467	46.2552

$$P\left(\sum_{s=1}^{3} x_{s3} \geq b_3\right) \geq 1 - \delta_3 \tag{21}$$

$$x_{st} \geq 0, \forall s = 1, 2, 3; t = 1, 2, 3. \tag{22}$$

7 Results and Discussion

The proposed GA-based fuzzy programming approach is coded in C++ Code:: Blocks 16.01 compiler. The size of population is taken as 100. The number of generations taken is 100. From the 100 generations, best 10 Pareto optimal solutions are shown in Tables 7 and 8 for different values of probability of crossover and mutation.

8 Conclusion

In this study, we propose a solution procedure for solving MOSTP by considering the random variables as free market activity which follows a three-parameter Weibull distribution. The proposed mathematical model is solved using a genetic algorithm (GA)-based fuzzy programming method without deriving the deterministic equivalent. GA-based fuzzy programming method is superior in comparison to classical methods which help the decision-maker in generating Pareto optimal solutions, which, in turn, will help the decision-maker in taking a better decision by considering all the possible directions of the problem. The Pareto optimal solutions represent the situations in the solution space of the problem. A contextual investigation for transporting of the egg is demonstrated to illustrate the technique when both the supply and demand points follow a three-parameter Weibull distribution.

References

1. Ardjmand, E., Young II, W.A., Weckman, G.R., Bajgiran, O.S., Aminipour, B., Park, N.: Applying genetic algorithm to a new bi-objective stochastic model for transportation, location, and allocation of hazardous materials. Expert Syst. Appl. **51**, 49–58 (2016)
2. Bartkute, V., Sakalauskas, L.: The method of three-parameter Weibull distribution estimation. ACUTM **12**, 65–78 (2008)
3. Bharathi, K., Vijayalakshmi, C.: Optimization of multi-objective transportation problem using evolutionary algorithms. Glob. J. Pure Appl. Math. **12**(2), 1387–1396 (2016)
4. Biswal, M., Samal, H.: Stochastic transportation problem with Cauchy random variables and multi choice parameters. J. Phys. Sci. **17**(11), 117–130 (2013)
5. Hitchcock, F.L.: The distribution of a product from several sources to numerous localities. J. Math. Phys. **20**, 224–230 (1941)

6. Holland, J.: Adaptation in Natural and Artificial Systems: An Introductory Analysis with Application to Biology. Control and Artificial Intelligence. University of Michigan Press, Ann Arbor (1975)
7. Karthy, T., Ganesan, K.: Multi-objective transportation problem - genetic Algorithm approach. Int. J. Pure Appl. Math. **119**(9), 343–350 (2018)
8. Krishnamoorthy, K.: Handbook of sTatistical Distributions with Applications. CRC Press, Boca Raton (2016)
9. Mahapatra, D.R., Roy, S.K., Biswal, M.: Stochastic based on multi-objective transportation problems involving normal randomness. Adv. Model. Optim. **12**(2), 205–223 (2010)
10. Mousa, A.A., Geneedy, H.M., Elmekawy, A.Y.: Efficient evolutionary algorithm for solving multiobjective transportation problem. J. Nat. Sci. Math. **4**(1), 77–102 (2010)
11. Quddoos, A., ull Hasan, M.G., Khalid, M.M.: Multi-choice stochastic transportation problem involving general form of distributions. SpringerPlus **3**(1), 565–574 (2014)
12. Roy, S.K.: Multi-choice stochastic transportation problem involving weibull distribution. IJOR **21**(1), 38–58 (2014)
13. Roy, K., Maity, G., Weber, G.W., Gök, S.Z.A.: Conic scalarization approach to solve multichoice multi-objective transportation problem with interval goal. Ann. Oper. Res. **253**(1), 599–620 (2017)
14. Roy, S.K., Ebrahimnejad, A., Verdegay, J.L., Das, S.: New approach for solving intuitionistic fuzzy multi-objective transportation problem. Sādhanā, **43**(1), 3 (2018)
15. Sanjay, D., Acharya, S., Rajashree, M.: Genetic algorithm based fuzzy stochastic transportation programming problem with continuous random variables. Opsearch **53**(4), 835–872 (2016)
16. Trivedi, A., Srinivasan, D., Sanyal, K., Ghosh, A.: A survey of multiobjective evolutionary algorithms based on decomposition. IEEE Trans. Evol. Comput. **21**(3), 440–462 (2017)
17. Zaki, S.A., Mousa, A.A.A., Geneedi, H.M., Elmekawy, A.Y.: Efficient multiobjective genetic algorithm for solving transportation, assignment, and transshipment problems. Appl. Math. **3**(1), 92–99 (2012)

Multi-choice Multi-objective Fractional Probabilistic Programming Problem

Berhanu Belay and Srikumar Acharya

Abstract In this paper, we present a multi-objective fractional probabilistic programming problem where the parameters in the objective functions and right-hand side constraints are multi-choice in nature. Multi-choice multi-objective objective fractional probabilistic programming problem is formulated. The multi-choice parameters in the right-hand side of constraints are random variables following Cauchy distribution. Multi-choice multi-objective fractional probabilistic programming problem is transformed to deterministic equivalent multi-choice multi-objective fractional programming problem using chance-constrained programming problem. Lagrange interpolating polynomial approach is proposed to deal with multi-choice parameters. The multi-choice multi-objective fractional programming problem is transformed to non-linear mixed integer multi-objective programming problem. An $\epsilon-$ constrained method is applied to find the compromise solutions of non-linear mixed integer multi-objective programming problem using different values of ϵ. LINGO software is applied to find the optimal solution of single objective non-linear mixed integer programming problem. Finally, a numerical example is given to illustrate the proposed programming problem.

Keywords Multi-choice programming problem · Multi-objective programming problem · Probabilistic programming problem · Fractional programming problem · ϵ-constraint method

---

B. Belay (✉) · S. Acharya
Department of Mathematics School of Applied Sciences, KIIT Deemed to be University, Bhubaneswar 751024, India
e-mail: berhanubelay2@gmail.com

S. Acharya
e-mail: sacharyamfa@kiit.ac.in

D. Giri et al. (eds.), *Proceedings of the Fifth International Conference on Mathematics and Computing*, Advances in Intelligent Systems and Computing 1170, https://doi.org/10.1007/978-981-15-5411-7_12

1 Introduction

In some real-life decision-making problems, decision-makers face a challenge to optimize multiple ratio objectives which are conflicting in nature. For example, a production planning may require maximization of profit-to-cost ratio and minimization of overtime-to-regular time ratio. Such type of problem is called multi-objective fractional programming (MFP) problem. MFP problem is applicable in different problems, like planning of agricultural problem, health care and hospital planning, production planning problem, inventory problem, financial planning, etc. Many methodologies are designed to solve MFP problem. The simplex method [9] is one method to find the efficient solution of multi-objective linear fractional programming problem. Luhandjula [10] solved MFP using a fuzzy approach. By applying techniques used in [7] for suitable transformation, [3] solved MFP problem based on set theoretic. A special MFP problem having identical denominators is investigated by Dutta et. al. [8]. Again [11] proposed a method to obtain a compromise solution for MFP problem. In an MFP problem, the parameters may be multi-choice in nature. From a set of values of parameters, decisions are made to choose best value. In this case, the problem is called multi-choice multi-objective programming (MMFP) problem. Chang [4, 5] proposed a method to solve multi-choice goal programming model. Biswal and Acarya [1] presented a general mathematical programming problem to solve multi-choice multi-objective programming (MMP) problem by considering the parameters in the right-hand side constraints as multi-choice parameters. Latter in [2] they used interpolating polynomial for multi-choice parameter to convert the problem in to non-linear mixed integer multi-objective programming problem. In some real-life decision-making problems, multi-choice parameters may be uncertain due to incomplete information or vague information or uncertain information. Such uncertainties can be addressed by fuzzy programming problem or stochastic programming problem. The mathematical models are said to be stochastic programming problems, if the parameters are characterized by random variables following known distributions. There are two popular techniques of stochastic programming problems to handle random variables, namely, chance-constrained programming problem and two stage programming problem. In chance-constrained programming technique which is proposed in [6], the violation of constraints up to a fixed probability level is allowed. But in case of two stage programming technique constraint violation is not allowed. Linear programming model where multi-choice parameters are random variables following normal distribution is presented by Pradhan and Biswal [13]. Recently, in [12], they presented multi-choice programming problem by considering multi-choice parameters as fuzzy numbers.

In this paper, we attempt to solve the multi-choice multi-objective fractional probabilistic programming (MMFPP) problem by considering the coefficient of objective functions as multi-choice parameters and the right-hand side of the constraints is multi-choice random variables following Cauchy distribution.

This paper is organized into five sections followed by references. The first section presents the brief introduction of MMFPP problem and related works. Section 2 presents the general model of MMFPP problem. Section 3 presents the solution procedure of the formulated model. Numerical example for the problem is presented in Sect. 4 followed by conclusion in Sect. 5.

2 Formulation of Mathematical Model

We present a MMFPP model having Q number of objective functions as

$$\max : Z_q = Z_q(x), q = 1, 2, \ldots, Q \tag{1}$$

subject to

$$P\left(\sum_{j=1}^{n} a_{ij}x_j \leq b_i\right) \geq \eta_i, b_i \in \left\{b_i^{(1)}, b_i^{(2)}, \ldots, b_i^{(r_i)}\right\}, i = 1, 2, \ldots, m \tag{2}$$

$$0 < \eta_i < 1, i = 1, 2, \ldots, m \tag{3}$$

$$x_j \geq 0, j = 1, 2, \ldots, n, \tag{4}$$

where

$$Z_q(x) = \frac{N_q(x)}{D_q(x)} = \frac{\sum_{j=1}^{n} c_{qj}x_j + c_{q0}}{\sum_{j=1}^{n} d_{qj}x_j + d_{q0}}, q = 1, 2, \ldots, Q \tag{5}$$

$$D_q(x) > 0, q = 1, 2, \ldots, Q, c_{qj} \in \left\{c_{qj}^{(1)}, c_{qj}^{(2)}, \ldots, c_{qj}^{(k_j)}\right\} \tag{6}$$

$$d_{qj} \in \left\{d_{qj}^{(1)}, d_{qj}^{(2)}, \ldots, d_{qj}^{(s_j)}\right\}. \tag{7}$$

The mathematical model (1–4) cannot be solved directly due to presence of random variables, multiple objective functions, fractional objective functions, and multi-choice parameters. To find the efficient solution, we have to find the deterministic equivalent of multi-choice probabilistic programming problem which can be solved by using appropriate method. The solution procedure is discussed in the following section.

3 Solution Procedure

To solve the proposed mathematical model, first, we deal with random variables by chance-constrained method. In the next step, we transform MMFP problem in to equivalent MFP problem. Then we transform MFP to equivalent multi-objective programming problem. Finally, we apply $\epsilon-$ constraint method to find the compromise solution of the resulting multi-objective programming problem.

3.1 Formulation of Deterministic Mathematical Model

We find the deterministic equivalent of the problem (1–4) by applying chance-constrained programming technique. We consider that multi-choice random variables b_i in (2) follow Cauchy distribution with known parameters α_i and β_i, where $\alpha_i \in \left\{\alpha_i^{(1)}, \alpha_i^{(2)}, \ldots, \alpha_i^{(r_i)}\right\}$ and $\beta_i \in \left\{\beta_i^{(1)}, \beta_i^{(2)}, \ldots, \beta_i^{(r_i)}\right\}$. The parameters α_i and β_i denote median and scales of the random variables b_i.

A random variable b_i has Cauchy distribution having parameters α_i and β_i, if its probability density function is given by

$$f(b_i) = \frac{\alpha_i}{\pi({\alpha_i}^2 + (b_i - \beta_i)^2)}, -\infty < b_i < \infty, \alpha_i > 0. \tag{8}$$

Now, from the probabilistic problem (2), we have

$$P\left(\sum_{j=1}^{n} a_{ij}x_j \le b_i\right) \ge \eta_i, b_i \in \left\{b_i^{(1)}, b_i^{(2)}, \ldots, b_i^{(r_i)}\right\}, i = 1, 2, \ldots, m \tag{9}$$

$$\Rightarrow p(b_i \ge \sum_{j=1}^{n} a_{ij}x_j) \ge \eta_i. \tag{10}$$

Put $y_i = \sum_{j=1}^{n} a_{ij}x_j$, then the probabilistic constraint (10) is written as

$$\int_{y_i}^{\infty} \frac{\alpha_i}{\pi({\alpha_i}^2 + (b_i - \beta_i)^2)} db_i \ge \eta_i, i = 1, 2, \ldots, m, \tag{11}$$

and after integration, we have

$$\tan^{-1} \frac{y_i - \beta_i}{\alpha_i} \le \frac{\pi}{2} - \eta_i \pi. \tag{12}$$

Solving for y_i, we have

$$y_i \le \beta_i + \alpha_i \tan\left(\frac{\pi}{2} - \eta_i \pi\right) \tag{13}$$

$$\Rightarrow \sum_{j=1}^{n} a_{ij}x_j \le \beta_i + \alpha_i \tan\left(\frac{\pi}{2} - \eta_i \pi\right). \tag{14}$$

Substituting (14) in (2), the deterministic equivalent of the MMFPP problem which is given by (1–4) is expressed as follows:

$$\max : Z_q = Z_q(x), q = 1, 2, \ldots, Q \tag{15}$$

subject to

$$\sum_{j=1}^{n} a_{ij} x_j \leq \beta_i + \alpha_i \tan\left(\frac{\pi}{2} - \eta_i \pi\right) \tag{16}$$

$$x_j \geq 0, j = 1, 2, \ldots, n, \tag{17}$$

where

$$Z_q(x) = \frac{N_q(x)}{D_q(x)} = \frac{\sum_{j=1}^{n} c_{qj} x_j + c_{q0}}{\sum_{j=1}^{n} d_{qj} x_j + d_{q0}}, q = 1, 2, \ldots, Q. \tag{18}$$

3.2 *Conversion of MMFP Problem into MMP Problem*

In this paper, we use Charnes & Cooper's [6] transformation technique and Dutta [8] method to reduce MMFP into MMP by using the transformation

$$y_j = x_j t, t = \frac{1}{D_q(x)}, t > 0. \tag{19}$$

Using this transformation the programming problem (15–17) is transformed to

$$\max : h_q = N_q(y_j/t) \tag{20}$$

subject to

$$D_q(y_j/t) = 1 \tag{21}$$

$$\sum_{j=1}^{n} a_{ij} y_j \leq t\left(\beta_i + \alpha_i \tan\left(\frac{\pi}{2} - \eta_i \pi\right)\right) \tag{22}$$

$$y_j \geq 0, j = 1, 2, \ldots, n, t > 0, \tag{23}$$

where

$$N_q(y_j/t) = \sum_{j=1}^{n} c_{qj} y_j + t c_{q0}, c_{qj} \in \left\{c_{qj}^{(1)}, c_{qj}^{(2)}, \ldots, c_{qj}^{(k_j)}\right\}, \tag{24}$$

$$D_q(y_j/t) = \sum_{j=1}^{n} d_{qj} y_j + t d_{q0}, d_{qj} \in \left\{d_{qj}^{(1)}, d_{qj}^{(2)}, \ldots, d_{qj}^{(s_j)}\right\}. \tag{25}$$

If (t^*, y_j^*) is the efficient solution of (20–23), then x_j^* is the efficient solution of (15–17).

3.3 Lagrange Interpolating Polynomial for Multi-Choice Parameters

We consider that the multi-choice parameters are found in each objective function and the right-hand side of the constraints. To transform MMP problem to multi-objective mathematical programming, we formulate Lagrange interpolating polynomials by announcing an integer variable analogous to every multi-choice parameter. If the parameter has r number of choices, the corresponding integer variable takes r number of nodal points. Every nodal point links to functional value of polynomial at a given multi-choice parameter. Using Lagrange's formula, we substitute multi-choice parameters by interpolating polynomial. For a given multi-choice parameter c_{qj}, $d_{qj} (j = 1, 2, \ldots, n, q = 1, 2, \ldots, Q)$, and $b_i (i = 1, 2, \ldots, m)$ in the denominator, numerator of multi-objective function, and right-hand side constraints, respectively, we introduce an integer variable $u_{qj}(0, 1, \ldots, k_{qj} - 1)$, $v_{qj}(0, 1, \ldots, s_{qj} - 1))$, and $w_i(0, 1, \ldots, r_i - 1))$ which takes k_{qj}, s_{qj}, and r_i number of values, respectively.

Using Lagrange formula, an interpolating polynomial $f_{c_{qj}}(u_{qj})$ that passes through k_{qj} number of points is given by

$$
\begin{aligned}
f_{c_{qj}}(u_{qj}, c_{qj}) = {} & \frac{(u_{qj} - 1)(u_{qj} - 2) \cdots (u_{qj} - k_{qj} + 1)}{(-1)^{(k_{qj}-1)}(k_{qj} - 1)!} c_{qj}^1 \\
& + \frac{u_{qj}(u_{qj} - 2) \cdots (u_{qj} - k_{qj} + 1)}{(-1)^{(k_{qj}-2)}(k_{qj} - 2)!} c_{qj}^2 \\
& + \frac{u_{qj}(u_{qj} - 1)(u_{qj} - 3) \cdots (u_{qj} - k_{qj} + 1)}{(-1)^{(k_{qj}-3)}(k_{qj} - 3)!2!} c_{qj}^3 + \cdots + \\
& \frac{u_{qj}(u_{qj} - 1) \cdots (u_{qj} - k_{qj} + 2)}{(k_{qj} - 1)!} c_{qj}^{k_{qj}}, \; i = 1, 2, \ldots, m.
\end{aligned}
$$

Similarly, an interpolating polynomial $f_{d_{qj}}(v_{qj})$ that passes through all the s_j number of points is formulated by using Lagrange formula as follows:

$$
\begin{aligned}
f_{d_{qj}}(v_{qj}, d_{qj}) = {} & \frac{(v_{qj} - 1)(v_{qj} - 2) \cdots (v_{qj} - s_{qj} + 1)}{(-1)^{(s_{qj}-1)}(s_{qj} - 1)!} d_{qj}^1 \\
& + \frac{v_{qj}(v_{qj} - 2) \cdots (v_{qj} - s_{qj} + 1)}{(-1)^{(s_{qj}-2)}(s_{qj} - 2)!} d_{qj}^2 \\
& + \frac{v_{qj}(v_{qj} - 1)(v_{qj} - 3) \cdots (v_{qj} - s_{qj} + 1)}{(-1)^{(s_{qj}-3)}(s_{qj} - 3)!2!} d_{qj}^3 + \cdots + \\
& \frac{v_{qj}(v_{qj} - 1) \cdots (v_{qj} - s_{qj} + 2)}{(s_{qj} - 1)!} d_{qj}^{s_{qj}}, \; i = 1, 2, \ldots, m.
\end{aligned}
$$

Similarly, using Lagrange formula an interpolating polynomial $f_{b_i}(w_i)$ that passes through on r_i number of points is formulated as

$$f_{b_i}(w_i, b_i) = \frac{(w_i - 1)(w_i - 2)\cdots(w_i - r_i + 1)}{(-1)^{(r_i-1)}(r_i - 1)!} b_i^1 + \frac{w_i(w_i - 2)\cdots(w_i - r_i + 1)}{(-1)^{(r_i-2)}(r_i - 2)!} b_i^2$$
$$+ \frac{w_i(w_i - 1)(w_i - 3)\cdots(w_i - r_i + 1)}{(-1)^{(r_i-3)}(r_i - 3)!2!} b_i^3 + \cdots +$$
$$\frac{w_i(w_i - 1)\cdots(w_i - r_i + 2)}{(r_i - 1)!} b_i^{r_i}, i = 1, 2, \ldots, m.$$

Substituting $f_{c_{qj}}(u_{qj}, c_{qj})$, $f_{d_{qj}}(v_{qj}, d_{qj})$, and $f_{b_i}(w_i, b_i)$ in (20), (21), and (22), respectively, we obtain the equivalent multi-objective mathematical programming problem which is given by

$$\max : h_q = \sum_{j=1}^{n} f_{c_{qj}}(u_{qj}, c_{qj})y_j + tc_{q0} \tag{26}$$

subject to

$$\sum_{j=1}^{n} f_{d_{qj}}(v_{qj}, d_{qj})y_j + td_{q0} = 1 \tag{27}$$

$$\sum_{j=1}^{n} a_{ij}y_j \leq t(f_{b_i}(w_i, b_i) \tag{28}$$

$$y_j \geq 0, j = 1, 2, \ldots, n, t > 0. \tag{29}$$
$$0 \leq u_{qj} \leq k_{qj} - 1 \tag{30}$$
$$0 \leq v_{qj} \leq s_{qj} - 1 \tag{31}$$
$$0 \leq w_i \leq r_i - 1 \tag{32}$$

$$u_{qj}, v_{qj}, w_i \in \mathbb{Z}, i = 1, 2, ..., m, j = 1, 2, ..., n. \tag{33}$$

The resulting non-linear mixed integer multi-objective programming problem (26–33) can be solved by different approaches. In this paper, we use $\epsilon-$constrained method to solve the mathematical problem. To apply the method for maximization problem, define Q number of single objective programming problem using $\epsilon_q, q = 1, 2, ..., Q$ as follows:

step 1: Select the first objective function $h_q(q = 1)$ and solve it as a single objective mathematical programming problem subject to the original constraints and new constraint as $h_q \geq \epsilon_q, q = 2, 3, \ldots, Q(q \neq 1)$.

step 2: Again select the second objective function $h_q(q = 2)$ and optimize the programming problem subject to the original constraints and new constraint as $h_q \geq \epsilon_q, q = 1, 3, \ldots, Q(q \neq 2)$.

step 3: Repeat the procedure Q times for Q objective functions.

The main difficulty of ϵ-constraint method is to determine the bounds of $\epsilon_q, q = 1, 2, \ldots, Q$. To estimate the bounds of ϵ, we use the following procedure:
i: Select the first objective function $h_q(q = 1)$. Then optimize h_1 subject to the given constraints. Next select the second objective function $h_q(q = 2)$ and optimize h_2. Repeat the procedure Q times for Q objective functions. Suppose that $X^{(1)}, X^{(2)}, \ldots, X^{(Q)}$ are the individual optimal solutions of the objective functions $h_1, h_2, \ldots, h_q, q = 1, 2, \ldots, Q$, respectively.
ii: With the help of ideal solutions found in *i*, construct Q by Q payoff matrix. From the payoff matrix, estimate the upper and lower bounds of the objective functions from the payoff matrix as $L_q \leq h_q \leq U_q, q = 1, 2, \ldots, Q$
iii: Determine the bounds of $\epsilon_q, q = 1, 2, \ldots, Q$ which is the point in the range of h_q.

The single objective mathematical model is solved by using suitable mathematical method or software to obtain optimal solution for different values of $\epsilon_q, q = 1, 2, \ldots, Q$.

4 Numerical Example

We now consider a MMFPP problem where the multi-choice parameters in the right-hand side of constraints follow Cauchy distribution and some parameters in the objective function are deterministic multi-choice parameters.

$$\max : Z_1 = \frac{\{2, 5, 8\}x_1 + \{3, 6, 9\}x_2 - 5}{\{1, 3\}x_2 + 4} \tag{34}$$

$$\max : Z_2 = \frac{-\{1, 4\}x_1 + \{3, 5\}x_2 + 4}{\{1, 3\}x_2 + 4} \tag{35}$$

subject to

$$P(-x_1 + 4x_2 \leq b_1) \geq 0.95, b_1 \in \left\{b_1^{(1)}, b_1^{(2)}, b_1^{(3)}\right\} \tag{36}$$

$$P(x_1 + 2x_2 \leq b_2) \geq 0.7, b_2 \in \left\{b_2^{(1)}, b_2^{(2)}, b_2^{(3)}, b_2^{(4)}\right\} \tag{37}$$

$$x_1, x_2 \geq 0. \tag{38}$$

Here the multi-choice parameters in the right-hand side of the constraint are random variables following Cauchy distribution having parameter values α and β given by Table 1. To solve the given problem, we find the deterministic equivalent of the problem using chance constraint programming problem. Since the multi-choice parameters in the right-hand side of the constraint are random variables following Cauchy distribution, we use the model given by (15–17). Hence, the resulting deterministic

Table 1 Value of parameters for Cauchy distributed random variables

Parameters	$b_1^{(1)}$	$b_1^{(2)}$	$b_1^{(3)}$	$b_2^{(1)}$	$b_2^{(2)}$	$b_2^{(3)}$	$b_2^{(4)}$
α	2	3	4	2	3	4	5
β	10	12	14	8	10	12	15

equivalent of the programming problem is expressed as

$$\max : Z_1 = \frac{\{2, 5, 8\}x_1 + \{3, 6, 9\}x_2 - 5}{\{1, 3\}x_2 + 4} \tag{39}$$

$$\max : Z_2 = \frac{-\{1, 4\}x_1 + \{3, 5\}x_2 + 4}{\{1, 3\}x_2 + 4} \tag{40}$$

subject to

$$-x_1 + 4x_2 \leq \{9.95, 11.93, 13.9\} \tag{41}$$

$$x_1 + 2x_2 \leq \{7.98, 9.97, 11.96, 14.96\} \tag{42}$$

$$x_1, x_2 \geq 0. \tag{43}$$

Using Dutta [8] method, the MMFP (39–43) is converted to MMP problem by using change of variable $y = t * x, t > 0$ as

$$\max : h_1 = \{2, 5, 8\}y_1 + \{3, 6, 9\}y_2 - 5t \tag{44}$$

$$\max : h_2 = -\{1, 4\}y_1 + \{3, 5\}y_2 + 4t \tag{45}$$

subject to

$$\{1, 3\}y_2 + 4t = 1 \tag{46}$$

$$-y_1 + 4y_2 \leq t\{9.95, 11.93, 13.90\} \tag{47}$$

$$y_1 + 2y_2 \leq t\{7.98, 9.97, 11.96, 14.95\} \tag{48}$$

$$y_1, y_2 \geq 0, t > 0. \tag{49}$$

Finally, we convert the resulting MMP problem in to standard non-linear mixed integer multi-objective mathematical programming using Lagrange interpolating polynomial. Therefore, using (26–33), the non-linear mixed integer multi-objective mathematical programming is expressed by

$$\max : h_1 = (-3u_1^2 + u_1 + 2)y_1 + (1/2u_2^2 + 5/2u_2 + 3)y_2 - 5t \tag{50}$$

$$\max : h_2 = (-3u_3 - 3)y_1 + (2u_4 + 3)y_2 + 4t \tag{51}$$

subject to

$$(2v_1 + 1)y_2 + 4t = 1 \tag{52}$$

$$-y_1 + 4y_2 \leq t(0.52w_1^2 + 1.46w_1 + 9.95) \tag{53}$$

$$y_1 + 2y_2 \leq t(2.16w_2^3 - 4.53w_2^2 + 2.47w_2 + 7.9) \tag{54}$$

$$0 \leq u_1 \leq 2, 0 \leq u_2 \leq 2, 0 \leq u_3 \leq 1 \tag{55}$$

$$0 \leq u_4 \leq 1, 0 \leq v_1 \leq 1, 0 \leq w_1 \leq 2 \tag{56}$$

$$0 \leq w_2 \leq 3, y_1, y_2 \geq 0, t > 0 \tag{57}$$

$$u_1, u_2, u_3, u_4, v_1, w_1, w_2 \in \mathbb{Z}. \tag{58}$$

To find the compromise solutions of the above non-linear mixed multi-objective programming, we use $\epsilon-$constrained method. Solving each objective function with the given constraint by rejecting the other objective function we get the ideal solution of the two objective functions. After constructing the payoff matrix using the ideal solutions, we get the best upper and worst lower bounds of the two objective functions which are given by $4.184168 \leq h_1 \leq 15.18$ and $-23.645 \leq h_2 \leq 2.932149$.

Therefore, the bounds of $\epsilon_q, q = 1, 2$ are the points in the range of the objective function $h_q(x)$, i.e., $4.184168 \leq \epsilon_1 \leq 15.18$ and $-23.645 \leq \epsilon_2 \leq 2.932149$. Now, using $\epsilon_q, q = 1, 2$, we define two different single objective programming problems by taking one objective function as constraint as follows:

i: The first single objective h_1 programming problem is given by considering the second objective function h_2 as constraint which is given below:

$$\max : h_1 = (-3u_1^2 + u_1 + 2)y_1 + (1/2u_2^2 + 5/2u_2 + 3)y_2 - 5t \tag{59}$$

subject to

$$(-3u_3 - 3)y_1 + (2u_4 + 3)y_2 + 4t \geq \epsilon_2 \tag{60}$$

$$(2v_1 + 1)y_2 + 4t = 1 \tag{61}$$

$$-y_1 + 4y_2 \leq t(0.52w_1^2 + 1.46w_1 + 9.95) \tag{62}$$

$$y_1 + 2y_2 \leq t(2.16w_2^3 - 4.53w_2^2 + 2.47w_2 + 7.9) \tag{63}$$

Table 2 Efficient solution and values of Z_1 and Z_2 for different values of ϵ_2

ϵ_2	x_1	x_2	Z_1	Z_2
−23.645	32.8600	0.0000	15.18	−23.645
−20.00000	30.5535	0.450197	14.21417	−20.00000
−15.00000	30.199992	1.329999	12.88931	−15.00000
−5.00000	23.53749	4.66125	10.23957	−5.00000
0.000000	0.35000	3.825001	4.338658	0.00000
2.000000	3.20555	4.538887	5.480807	2.00000
2.932149	0.0000	3.737501	4.184168	2.932149

Table 3 Efficient solution and values of Z_1 and Z_2 for different values of ϵ_1

ϵ_1	x_1	x_2	Z_1	Z_2
4.184168	7.73750	3.73750	6.1841695	2.932149
6.0000	4.683335	4.908334	6.0000	1.626754
8.0000	11.81000	6.689999	8.00000	0.1889616
10.0000	22.499988	5.179998	10.0000	−4.095861
12.0000	28.620001	2.119999	12.0000	−11.64379
14.0000	31.68000	0.590000	14.0000	−19.19172
15.1800	32.86000	0.00000	15.1800	−23.6450

$$0 \le u_1 \le 2, 0 \le u_2 \le 2, 0 \le u_3 \le 1 \tag{64}$$

$$0 \le u_4 \le 1, 0 \le v_1 \le 1, 0 \le w_1 \le 2 \tag{65}$$

$$0 \le w_2 \le 3, y_1, y_2 \ge 0, t > 0 \tag{66}$$

$$u_1, u_2, u_3, u_4, v_1, w_1, w_2 \in \mathbb{Z}. \tag{67}$$

Similarly, the second single objective programming problem is formulated by considering h_2 as an objective function and h_1 as a constraint in the above equation. Solving (59–67) by taking different values of ϵ_2 in the interval $[-23.645, 2.932149]$ by using LINGO software and applying the transformation $x_i = y_i/t, i = 1, 2$, the solution of the original programming problem is given by Table 2. Similarly, replacing h_1 by h_2 and solving (59–67) by taking different values of ϵ_1 in the interval $[4.184168, 15.18]$ by using LINGO software and applying the transformation $x_i = y_i/t, i = 1, 2$, the solution of the original programming problem is given by Table 3.

5 Conclusion

In this paper, MFP problem is considered in which the parameters in the objective functions are multi-choice and the right-hand side parameters are multi-choice random variables following Cauchy distribution. Multi-choice parameters are replaced by Lagrange interpolating polynomial and random variables are handled by chance-constrained programming method. Compromise solutions for numerical example are obtained by using different values of ϵ using ϵ-constrained method. The upper and lower bounds of ϵ_1 and ϵ_2 are obtained by solving single objective non-linear mixed integer programming problems by LINGO software. The problem can be further solved by stochastic simulation-based genetic algorithm.

References

1. Biswal, M., Acharya, S.: Transformation of a multi-choice linear programming problem. Appl. Math. Comput. **210**(1), 182–188 (2009). https://doi.org/10.1016/j.amc.2008.12.080
2. Biswal, M., Acharya, S.: Solving multi-choice linear programming problems by interpolating polynomials. Math. Comput. Modell. **54**(5–6), 1405–1412 (2011). https://doi.org/10.1016/j.mcm.2011.04.009
3. Chakraborty, M., Gupta, S.: Fuzzy mathematical programming for multi objective linear fractional programming problem. Fuzzy Sets Syst. **125**(3), 335–342 (2002). https://doi.org/10.1016/S0165-0114(01)00060-4
4. Chang, C.T.: Multi-choice goal programming. Omega **35**(4), 389–396 (2007). https://doi.org/10.1016/j.omega.2005.07.009
5. Chang, C.T.: Revised multi-choice goal programming. Appl. Math. Modell. **32**(12), 2587–2595 (2008). https://doi.org/10.1016/j.apm.2007.09.008
6. Charnes, A., Cooper, W.W.: Programming with linear fractional functionals. Naval Res. Logist. Quart. **9**(3–4), 181–186 (1962). https://doi.org/10.1002/nav.3800090303
7. Charnes, A., Cooper, W.W.: Deterministic equivalents for optimizing and satisficing under chance constraints. Operat. Res. **11**(1), 18–39 (1963). https://doi.org/10.1287/opre.11.1.18
8. Dutta, D., Rao, J., Tiwari, R.: A restricted class of multiobjective linear fractional programming problems. Eur. J. Oper. Res. **68**(3), 352–355 (1993). https://doi.org/10.1016/0377-2217(93)90191-O
9. Kornbluth, J.S., Steuer, R.E.: Multiple objective linear fractional programming. Manag. Sci. **27**(9), 1024–1039 (1981). https://doi.org/10.1287/mnsc.27.9.1024
10. Luhandjula, M.K.: Fuzzy approaches for multiple objective linear fractional optimization. Fuzzy Sets Syst. **13**(1), 11–23 (1984). https://doi.org/10.1016/0165-0114(84)90023-X
11. Nykowski, I., Zołkiewski, Z.: A compromise procedure for the multiple objective linear fractional programming problem. Eur. J. Oper. Res. **19**(1), 91–97 (1985). https://doi.org/10.1016/0377-2217(85)90312-1
12. Pradhan, A., Biswal, M.P.: Linear programming problems with some multi-choice fuzzy parameters. Yugosl. J. Oper. Res. **28**(2), 249–264 (2018). https://doi.org/10.2298/YJOR160520004P
13. Pradhan, A., Biswal, M.: Multi-choice probabilistic linear programming problem. Opsearch **54**(1), 122–142 (2017). https://doi.org/10.1007/s12597-016-0272-7

Approximate Solution of Hypersingular Integral Equation by Using Differential Transform Method

Subhabrata Mondal and B. N. Mandal

Abstract In this article, we present differential transform method for solving a first kind hypersingular integral equation in which the kernel is containing a strong singular part and a regular part. Also some simple examples with exact solutions are considered to establish the validity and applicability of the proposed mechanism. Numerical outcomes reveal that our proposed technique works well and has good accuracy.

Keywords Hypersingular integral equation · Hadamard finite part · Differential transform method

1 Introduction

Here we have considered the first kind hypersingular integral equation (HSIE) of the form given by

$$\int_{-1}^{1}\left[\frac{R(\tau,s)}{(\tau-s)^2}+S(\tau,s)\right]u(\tau)d\tau=h(s)\ ,\quad -1\le s\le 1\ , \tag{1}$$

with $u(\pm 1)=0$, where $R(\tau,s)$ and $S(\tau,s)$ are regular and as well square integrable functions of τ and s, and $R(s,s)\neq 0$. In the left side of (1), the integral involving $\frac{1}{(\tau-s)^2}$ is defined as

S. Mondal (✉)
Department of Applied Mathematics, University of Calcutta, 92, A.P.C. Road, Kolkata 700009, India
e-mail: subhamath644@gmail.com

B. N. Mandal
Physics and Applied Mathematics Unit, Indian Statistical Institute, 203, B.T. Road, Kolkata 700108, India

D. Giri et al. (eds.), *Proceedings of the Fifth International Conference on Mathematics and Computing*, Advances in Intelligent Systems and Computing 1170, https://doi.org/10.1007/978-981-15-5411-7_13

$$\int_{-1}^{1} \frac{u(\tau)}{(\tau-s)^2}d\tau = \lim_{\epsilon\to 0^+}\left[\int_{-1}^{s-\epsilon} \frac{u(\tau)}{(\tau-s)^2}d\tau + \int_{s+\epsilon}^{1} \frac{u(\tau)}{(\tau-s)^2}d\tau - \frac{u(s+\epsilon)+u(s-\epsilon)}{\epsilon}\right],$$
$$-1 \le s \le 1, \tag{2}$$

and is figured out in the form of Hadamard finite part of order 2.

When $R(\tau, s) = 1$, then (1) can be written in the form

$$\int_{-1}^{1} u(\tau)\left[\frac{1}{(\tau-s)^2} + S(\tau, s)\right]d\tau = h(s)\ , \quad -1 \le s \le 1, \tag{3}$$

with $u(\pm 1) = 0$, which arises naturally in various types of problem in mathematical physics, engineering, etc. Here we have used a new technique named differential transform method to obtain the approximate solution of Eq. (1).

2 Differential Transform Method

The differential transform of a function $h(s)$ is denoted by $H(r)$ defined by Odibat [1], Zhou [2]

$$H(r) = \frac{1}{r!}\left[\frac{d^r h(s)}{ds^r}\right]_{s=s_0} \tag{4}$$

and the differential inverse transform of $H(r)$ is defined as

$$h(s) = \sum_{r=0}^{\infty} H(r)(s-s_0)^r. \tag{5}$$

Result 1. If $h(s) = s^p$, then $H(r) = \delta(r-p)$, where

$$\delta(r-p) = \begin{cases} 1 & , r = p\ , \\ 0 & , r \neq p\ . \end{cases}$$

Result 2. If $h(s) = ag(s)$, then $H(r) = aG(r)$, where a is a constant and $G(r)$ is the differential transform of $g(s)$.

Result 3. If $h(s) = p(s)q(s)$, then $H(r) = \sum_{l=0}^{r} P(l)Q(r-l)$, where $P(r)$ and $Q(r)$ are the differential transforms of $p(s)$ and $q(s)$, respectively.

Results 1, 2, and 3 can be deduced from Eq. (4) assuming $s_0 = 0$.

3 Method of Solution

The unspecified function $u(s)$ of the hypersingular integral Eq. (1) satisfying the end conditions $u(\pm 1) = 0$ can be written in the form

$$u(s) = \sqrt{1-s^2}v(s) \ , \quad -1 \leq s \leq 1 \tag{6}$$

where $v(s)$ is a well-mannered unknown function of s in $[-1, 1]$. For the approximate solution of (1), $v(s)$ can be approximated using Maclaurin series expansion in $-1 < s < 1$ as

$$v(s) = \sum_{r=0}^{N} V(r)s^r, \qquad N \to \infty \tag{7}$$

where $V(r)$'s are the differential transform of $v(s)$.

Now using (6) and (7) in (1), we obtain

$$\sum_{r=0}^{N} V(r)\left[\int_{-1}^{1} \frac{\sqrt{1-\tau^2}K(\tau,s)\tau^r}{(\tau-s)^2} d\tau + \int_{-1}^{1} \sqrt{1-\tau^2}S(\tau,s)\tau^r d\tau\right] = h(s), \tag{8}$$

$$-1 \leq s \leq 1.$$

Both the functions $R(\tau, s)$ and $S(\tau, s)$ can be approximated as (for fixed s; cf. Chakrabarti and Vander Bergh [3])

$$R(\tau,s) \approx \sum_{c=0}^{m} R_c(s)\tau^c \ , \quad S(\tau,s) \approx \sum_{d=0}^{w} S_d(s)\tau^d \tag{9}$$

where $R_c(s)$ and $S_d(s)$ are the known expressions. Then (8) gives

$$\begin{aligned} &\textstyle\sum_{r=0}^{N} V(r) \sum_{c=0}^{m} R_c(s)\left[\int_{-1}^{1} \frac{\sqrt{1-\tau^2}\tau^{c+r}}{(\tau-s)^2} d\tau\right] + \\ &\textstyle\sum_{r=0}^{N} V(r) \sum_{d=0}^{w} L_d(s)\left[\int_{-1}^{1} \sqrt{1-\tau^2}\tau^{d+r} d\tau\right] = h(s) \quad , \end{aligned} \tag{10}$$

which can be rewritten as

$$\begin{aligned} &\textstyle\sum_{r=0}^{N} V(r) \sum_{c=0}^{m} R_c(s)\left[\frac{d}{dx}\int_{-1}^{1} \frac{\sqrt{1-\tau^2}\tau^{c+r}}{(\tau-s)} d\tau\right] + \\ &\textstyle\sum_{r=0}^{N} V(r) \sum_{d=0}^{w} L_d(s)\left[\int_{-1}^{1} \frac{\sqrt{1-\tau^2}\tau^{d+r}}{\sqrt{1-\tau^2}} d\tau\right] = h(s) \quad . \end{aligned} \tag{11}$$

Now by using the following results:

$$\int_{-1}^{1} \frac{\sqrt{1-\tau^2}\tau^j}{(\tau-s)} d\tau = -\pi s^{j+1} + \sum_{i=0}^{j-1} \frac{1+(-1)^i}{4} \frac{\Gamma(\frac{1}{2})\Gamma(\frac{i+1}{2})}{\Gamma(\frac{i+4}{2})} s^{j-i-1} ,$$

$$j = 1, 2, 3, ...$$

and

$$f(p) = \int_{-1}^{1} \frac{\tau^p}{\sqrt{1-\tau^2}} d\tau = \begin{cases} 0 & , \ p = 1, 3, 5, ... \\ \pi & , \ p = 0 \\ \frac{1.3.5....(p-1)}{2^{\frac{p}{2}}(\frac{p}{2})!} & , \ p = 2, 4, 6, ... \end{cases}$$

in Eq. (11), we obtain

$$\begin{aligned} &\sum_{r=0}^{N}\sum_{c=0}^{m} V(r)R_c(s)\left[-\pi(c+r+1)s^{c+r}\right] + \sum_{r=0}^{N}\sum_{c=0}^{m} V(r)R_c(s) \\ &\sum_{i=0}^{c+r-1}\left[\frac{1+(-1)^i}{4}\frac{\Gamma(\frac{1}{2})\Gamma(\frac{i+1}{2})}{\Gamma(\frac{i+4}{2})}(c+r-i-1)s^{c+r-i-2}\right] + \sum_{r=0}^{N}\sum_{d=0}^{w} V(r) \\ &S_d(s)\left[f(d+r) - f(d+r+2)\right] = h(s) \ . \end{aligned} \tag{12}$$

Now by using the differential transform method and its properties, we obtain

$$\begin{aligned} &\sum_{r=0}^{N}\sum_{c=0}^{m} V(r)[-\pi(c+r+1)]\sum_{l=0}^{\bar{k}} \mathbf{R}_c(l)\delta(\bar{k}-l-c-r) + \sum_{r=0}^{N}\sum_{c=0}^{m} V(r) \\ &\sum_{i=0}^{c+r-1} \frac{1+(-1)^i}{4}\frac{\Gamma(\frac{1}{2})\Gamma(\frac{i+1}{2})}{\Gamma(\frac{i+4}{2})}(c+r-i-1)\sum_{l=0}^{\bar{k}} \mathbf{R}_c(l)\delta(\bar{k}-l-c-r+i+2)+ \\ &\sum_{r=0}^{N}\sum_{d=0}^{w} V(r)\left[f(d+r) - f(d+r+2)\right]\mathbf{S}_d(\bar{k}) = \ H(\bar{k}) \ , \ \bar{k} = 1, 2, 3, ..., N, \end{aligned} \tag{13}$$

which is a system of linear equations in $V(r)$ (where $\mathbf{R}_c(\bar{k})$ and $\mathbf{S}_d(\bar{k})$ are the differential transforms of $R_c(s)$ and $S_d(s)$, respectively).

Therefore, the approximate solution of Eq. (1) is of the form (after $V(r)$'s are obtained from (13))

$$u(s) = \sqrt{1-s^2}\sum_{r=0}^{N} V(r)s^r. \tag{14}$$

4 Illustrative Examples

In order to demonstrate the performance of our technique, we present in this section, numerical results of some simple examples of hypersingular integral equations of first kind, solved by the our method of this study.

Example 1 If we take $R(\tau, s) \equiv 1, \ S(\tau, s) \equiv 0$, then Eq. (1) reduces to a simple hypersingular integral equation whose exact solution was obtained by Martin [4].

Table 1 Comparison with exact solution

s	0	±0.2	±0.4	±0.6	±0.8
$u(s)$[present method]	−0.3183	−0.3119	−0.2917	−0.2546	−0.1920
$u(s)$[exact values]	−0.3183	−0.3119	−0.2917	−0.2546	−0.1920

Here we will use the above method to obtain the solution for the function $h(s) = 1$. For this case $u(s)$ has the exact solution $u(s) = -\frac{1}{\pi}(1-s^2)^{\frac{1}{2}}$.

In this case, $R_c(s)$ and $S_d(s)$ in (9) are given by

$$R_0(s) = 1,\ R_c(s) = 0\ (c > 0),\ \text{and}\ S_d(s) = 0\ (d \geq 0).$$

Hence, relation (13) generates

$$-\pi \sum_{r=0}^{N} V(r)(r+1)\delta(\bar{k}-r) + \sum_{r=1}^{N}\sum_{i=0}^{r-1} V(r)\frac{1+(-1)^i}{4}\frac{\Gamma(\frac{1}{2})\Gamma(\frac{i+1}{2})}{\Gamma(\frac{i+4}{2})}$$

$$(r-i-1)\delta(d-r+i+2) = \delta(\bar{k}) \quad , \quad \bar{k} = 0, 1, 2, ..., N.$$

Choosing $N = 2$, $V(r)(r = 0, 1, 2)$ are obtained here and thus by using (14), approximate values of $u(s)$ is obtained. In Table 1, these values of $u(s)$ at $s = 0, \pm 0.2, \pm 0.4, \pm 0.6, \pm 0.8$ are given along with the exact values. It is noticed that both the values coincide.

Example 2 Now for the equation

$$\int_{-1}^{1}\left[\frac{1}{(\tau-s)^2} + (\tau+s)\right]u(\tau)d\tau = 1 + 2\tau\ ,\ \ -1 \leq \tau \leq 1, \tag{15}$$

with $u(\pm 1) = 0$. Its approximate solution was earlier obtained by Mandal and Bera [5] using a different method.

In view of (9), clearly we see that

$$R_0(s) = 1,\ \ R_c(s) = 0\ \ (c \geq 1)$$

and

$$S_0(s) = s,\ \ S_1(s) = 1,\ \ S_d(s) = 0\ \ (d \geq 2).$$

Table 2 Comparison with Mandal and Bera [5]

s	0	0.2	0.4	0.6	0.8
$u(s)$[present method]	−0.3697	−0.4427	−0.4894	−0.4929	−0.4189
$u(s)$[Mandal and Bera (2006)]	−0.3697	−0.4427	−0.4894	−0.4929	−0.4189

Therefore, from (13), we obtain

$$-\pi\sum_{r=0}^{N}V(r)(r+1)\delta(\bar{k}-r)+\sum_{r=1}^{N}\sum_{i=0}^{r-1}V(r)\frac{1+(-1)^i}{4}\frac{\Gamma(\frac{1}{2})\Gamma(\frac{i+1}{2})}{\Gamma(\frac{i+4}{2})}$$

$$(r-i-1)\delta(\bar{k}-r+i+2)+\sum_{r=0}^{N}V(r)[f(r)-f(r+2)]\delta(\bar{k}-1)+\sum_{r=0}^{N}V(r)$$

$$[f(r+1)-f(r+3)]\delta(\bar{k})=\delta(\bar{k})+2\delta(\bar{k}-1)\quad,\ \bar{k}=0,1,2,...,N,$$

choosing $N=3$, $V(r)$ $[r=0,1,2,3]$ are obtained here and by using (14), approximate values of $u(s)$ are obtained as given by Table 2.

Acknowledgements The first author thanks CSIR for the award of a research fellowship (CSIR award letter no. 09/028(0971)/2016-EMR-I, Dated: 19/10/2016).

References

1. Odibat, Z.M.: Differential transform method for solving Volterra integral equation with separable Kernels. Math. Comput. Modelling **48**, 1144–1149 (2008)
2. Zhou, J.K.: Differential transformation and its applications for electrical circuits. Huazhong University Press, Wuhan, China (1986) (in Chinese)
3. Chakrabarti, A., Vanden, G.: Approximate solution of singular integral equations. Appl. Math. Lett. **17**, 553–559 (2004)
4. Martin, P.A.: Exact solution of a system of a simple hypersingular integral equation. J. Integral. Equ. Appl. **4**, 197–204 (1992)
5. Mandal, B.N., Bera, G.H.: Approximate solution for a class of hypersingular integral equations. Appl. Math. Lett. **19**, 1286–1290 (2006)

Topological and Algebraic Properties of Triple n-normed Spaces

Tanweer Jalal and Ishfaq Ahmad Malik

Abstract In this paper, we introduce new triple n-normed sequence spaces and establish some of their topological and algebraic properties.

Keywords Triple difference sequence spaces · Ideal convergence · Musielak–Orlicz function · Ideal · n-normed space

1 Introduction

A triple sequence real (complex) $x = (x_{ijk})$ is a function $x : \mathbb{N} \times \mathbb{N} \times \mathbb{N} \to \mathbb{R}(\mathbb{C})$. The symbols $\mathbb{N}, \mathbb{R},$ and $\mathbb{C}$ have their usual meanings. By the symbol ω''' we mean the class of all complex triple sequences (x_{ijk}), $i, j, k \in \mathbb{N}$ which is a linear space under the coordinate-wise addition and scalar multiplication.

Sahiner et al. [24] initially introduced and investigated statistical convergence of various types of triple sequences. The concept of statistical convergent double sequence was given by Mursaleen and Edely [21]. The generalization of statistical convergence is ideal convergence which was given by Kostyrko et al. [17]. Later on this concept was extended to double and triple sequences and in this direction various interesting results were obtained by several authors [11–13, 23, 27].

The nonlinear generalization of a normed linear space is 2-normed space introduced by Gähler [10] and then extended it to n-normed spaces by Misiak [20]. The further development in this context can be found in [1, 6, 14–16, 22, 26].

T. Jalal (✉)
Department of Mathematics, National Institute of Technology, Srinagar-190006, Hazratbal, Jammu and Kashmir, India
e-mail: tjalal@nitsri.net

I. A. Malik
Department of Mathematics, National Institute of Technology, Srinagar-190006, Hazratbal, Jammu and Kashmir, India
e-mail: ishfaq_2phd15@nitsri.net

D. Giri et al. (eds.), *Proceedings of the Fifth International Conference on Mathematics and Computing*, Advances in Intelligent Systems and Computing 1170, https://doi.org/10.1007/978-981-15-5411-7_14

In this paper, we define the spaces $W^3[M, p\|\cdot, \ldots, \cdot\|]^I$, $W_0^3[M, p\|\cdot, \ldots, \cdot\|]^I$, and $W_\infty^3[M, p\|\cdot, \ldots, \cdot\|]^I$ and establish some of their topological and algebraic properties.

2 Definitions and Preliminaries

Definition 1 Let $X \neq \phi$, a class $I \subset 2^X$ (power set of X) is said to be an ideal in X if the following conditions hold good:

(i) I is additive, that is, if $A, B \in I$ then $A \cup B \in I$;
(ii) I is hereditary, that is, if $A \in I$, and $B \subset A$ then $B \in I$.

I is called non-trivial ideal if $X \notin I$.

Definition 2 [24, 25] A triple sequence (x_{ijk}) is said to be convergent to L in Pringsheim's sense if for every $\epsilon > 0$, there exists $N \in \mathbb{N}$ such that

$$|x_{ijk} - L| < \epsilon \qquad \text{whenever} \quad i, j, k \geq N$$

and written as $\lim_{i,j,k \to \infty} x_{ijk} = L$.

Example Define the sequence (x_{ijk}) by

$$x_{ijk} = \begin{cases} i + j \text{ for all } i = j \text{ and } k = 1 \\ \frac{1}{i^2 jk} \qquad\qquad \text{e.w} \end{cases}$$

Then $x_{ijk} \to 0$ in Pringsheim's sense but is unbounded.

Definition 3 A triple sequence (x_{ijk}) is said to be I-convergent to a number L if for every $\epsilon > 0$,

$$\left\{(i, j, k) \in \mathbb{N} \times \mathbb{N} \times \mathbb{N} : |x_{ijk} - L| \geq \epsilon\right\} \in I.$$

In this case, we write $I - \lim x_{ijk} = L$.

Definition 4 A triple sequence (x_{ijk}) is said to be I-null if $L = 0$. In this case, we write $I - \lim x_{ijk} = 0$.

Definition 5 [24, 25] A triple sequence (x_{ijk}) is said to be Cauchy sequence if for every $\epsilon > 0$, there exists $N \in \mathbb{N}$ such that

$$|x_{ijk} - x_{lmn}| < \epsilon \quad \text{whenever } i \geq l \geq N, j \geq m \geq N, k \geq n \geq N.$$

Definition 6 A triple sequence (x_{ijk}) is said to be I—Cauchy sequence if for every $\epsilon > 0$, there exists $N \in \mathbb{N}$ such that

$$\{(i, j, k) \in \mathbb{N} \times \mathbb{N} \times \mathbb{N} : |x_{ijk} - x_{lmn}| \geq \epsilon\} \in I$$

whenever $i \geq l \geq N, j \geq m \geq N, k \geq n \geq N.$

Definition 7 [24, 25] A triple sequence (x_{ijk}) is said to be bounded if there exists $M > 0$, such that $|x_{ijk}| < M$ for all $i, j, k \in \mathbb{N}$.

Definition 8 A triple sequence (x_{ijk}) is said to be $I-$bounded if there exists $M > 0$, such that $\{(i, j, k) \in \mathbb{N} \times \mathbb{N} \times \mathbb{N} : |x_{ijk}| \geq M\} \in I$ for all $i, j, k \in \mathbb{N}$.

Definition 9 A triple sequence space E is said to be solid if $(\alpha_{ijk}x_{ijk}) \in E$ whenever $(x_{ijk}) \in E$ and for all sequences (α_{ijk}) of scalars with $|\alpha_{ijk}| \leq 1$, for all $i, j, k \in \mathbb{N}$.

Definition 10 ***n*-normed Space**
Let $n \in \mathbb{N}$ and X be a linear space over the field $\mathbb{K}$ of reals (complex) of dimension d, where $d \geq n$. A real-valued function $\| \cdot, ..., \cdot \|$ on X^n satisfying the following four conditions:

(1) $\|x_1, x_2, \ldots, x_n\| = 0$ if and only if $x_1, x_2, \ldots, x_n$ are linearly dependent in X;
(2) $\|x_1, x_2, \ldots, x_n\|$ is invariant under permutation;
(3) $\|\alpha x_1, x_2, \ldots, x_n\| = |\alpha| \|x_1, x_2, \ldots, x_n\|$ for any $\alpha \in \mathbb{R}$; and
(4) $\|x_1 + x_1^{'}, x_2, \ldots, x_n\| \leq \|x_1, x_2, \ldots, x_n\| + \|x_1^{'}, x_2, \ldots, x_n\|$

is called an n-norm on X and $(X, \| \cdot, ..., \cdot \|)$ is called an n-normed space over the field $\mathbb{K}$.

Definition 11 Orlicz function and Musielak–Orlicz function
An Orlicz function is a function $\mathcal{M} : [0, \infty) \to [0, \infty)$ which is continuous, non-decreasing, and convex with $\mathcal{M}(0) = 0$, $\mathcal{M}(x) > 0$, for $x > 0$ and $\mathcal{M}(x) \to \infty$ as $x \to \infty$. If convexity of Orlicz function $\mathcal{M}$ is replaced by $\mathcal{M}(x + y) \leq \mathcal{M}(x) + \mathcal{M}(y)$, then this function is called modulus function.
Lindenstrauss and Tzafriri [19] used the idea of Orlicz function to construct Orlicz sequence space:

$$\ell_{\mathcal{M}} = \left\{ x \in \omega : \sum_{k=1}^{\infty} \mathcal{M}\left(\frac{|x_k|}{\rho}\right) < \infty, \quad \text{for some } \rho > 0 \right\}.$$

The space $\ell_{\mathcal{M}}$ with the norm

$$\|x\| = \inf \left\{ \rho > 0 : \sum_{k=1}^{\infty} \mathcal{M}\left(\frac{|x_k|}{\rho}\right) \leq 1 \right\}$$

becomes a Banach space which is called an Orlicz sequence space. A sequence $M = (\mathcal{M}_{ijk})$ of Orlicz function is called a Musielak–Orlicz function. A sequence $N = (\mathcal{N}_{ijk})$ defined by

$$\mathcal{N}_{ijk}(v) = \sup\{|v|u - (\mathcal{M}_{ijk})(u) : u \geq 0\} \ , \quad i, j, k = 1, 2, 3, \cdots$$

is called the complementary function of a Musielak–Orlicz function M. For a given Musielak–Orlicz function M, the Musielak–Orlicz sequence space t_M is defined as follows:

$$t_M = \left\{ x \in w''' : I_M \left(|x_{ijk}|\right)^{\frac{1}{i+j+k}} \to 0 \ \text{ as } \ i, j, k \to \infty \right\},$$

where I_M is a convex modulus function defined by

$$I_M(x) = \sum_{i,j,k=1}^{\infty} \mathcal{M}_{ijk} \left(|x_{ijk}|\right)^{\frac{1}{i+j+k}} , \qquad x = \left(x_{ijk}\right) \in t_M.$$

Definition 12 Paranormed Space
Let X be a linear metric space. A function $p : X \to \mathbb{R}$ is called paranorm, if

(1) $p(x) \geq 0$, for all $x \in X$;
(2) $p(-x) = p(x)$, for all $x \in X$;
(3) $p(x + y) \leq p(x) + p(y)$, for all $x, y \in X$; and
(4) If (λ_n) is a sequence of scalars with $\lambda_n \to \lambda$ as $n \to \infty$ and (x_n) is a sequence of vectors with $p(x_n - x) \to 0$ as $n \to \infty$, then $p(\lambda_n x_n - \lambda x) \to 0$ as $n \to \infty$.

Let $p = (p_{ijk})$ be any bounded sequence of positive numbers, I an admissible ideal, and $(X, \|\cdot, \ldots, \cdot\|)$ an n-normed space. Then for the Musielak–Orlicz function $M = (\mathcal{M}_{ijk})$, $z_1, \ldots, z_{n-1} \in X$, $K > 0$ and $L \in \mathbb{C}$, we define the following sequence spaces:

$$W^3[M, p, \|\cdot, \ldots, \cdot\|]^I = \Bigg\{ x = (x_{ijk}) \in \omega''' : \forall \epsilon > 0, \Bigg\{ (i, j, k) \in \mathbb{N} \times \mathbb{N} \times \mathbb{N} :$$
$$\frac{1}{lmn} \sum^{l,m,n} \left[M \left(\left\| \frac{x_{ijk} - L}{\rho}, z_1, \cdots, z_{n-1} \right\| \right) \right]^{p_{ijk}} \geq \epsilon \Bigg\} \in I, \text{ for some } \rho > 0 \Bigg\},$$

$$W_0^3[M, p, \|\cdot, \ldots, \cdot\|]^I = \Bigg\{ x = (x_{ijk}) \in \omega''' : \forall \epsilon > 0, \Bigg\{ (i, j, k) \in \mathbb{N} \times \mathbb{N} \times \mathbb{N} :$$
$$\frac{1}{lmn} \sum^{l,m,n} \left[M \left(\left\| \frac{x_{ijk}}{\rho}, z_1, \cdots, z_{n-1} \right\| \right) \right]^{p_{ijk}} \geq \epsilon \Bigg\} \in I, \text{ for some } \rho > 0 \Bigg\},$$

$$W_\infty^3[M, p, \|\cdot, \ldots, \cdot\|]^I = \Bigg\{ x = (x_{ijk}) \in \omega''' : \Bigg\{ (i, j, k) \in \mathbb{N} \times \mathbb{N} \times \mathbb{N} :$$
$$\sup_{i,j,k} \frac{1}{lmn} \sum^{l,m,n} \left[M \left(\left\| \frac{x_{ijk}}{\rho}, z_1, \cdots, z_{n-1} \right\| \right) \right]^{p_{ijk}} \geq K \Bigg\} \in I, \text{for } K > 0 \text{ and for some } \rho > 0 \Bigg\}.$$

Some Special Cases:

(i) If $p = (p_{ijk}) = 1$ then

$$W^3[M, \|\cdot, \ldots, \cdot\|]^I = \left\{x = (x_{ijk}) \in \omega''' : \forall \epsilon > 0, \left\{(i, j, k) \in \mathbb{N} \times \mathbb{N} \times \mathbb{N} : \frac{1}{lmn} \sum^{l,m,n} \left[M\left(\left\|\frac{x_{ijk} - L}{\rho}, z_1, \cdots, z_{n-1}\right\|\right)\right] \geq \epsilon\right\} \in I, \text{ for some } \rho > 0\right\},$$

$$W_0^3[M, \|\cdot, \ldots, \cdot\|]^I = \left\{x = (x_{ijk}) \in \omega''' : \forall \epsilon > 0, \left\{(i, j, k) \in \mathbb{N} \times \mathbb{N} \times \mathbb{N} : \frac{1}{lmn} \sum^{l,m,n} \left[M\left(\left\|\frac{x_{ijk}}{\rho}, z_1, \cdots, z_{n-1}\right\|\right)\right] \geq \epsilon\right\} \in I, \text{ for some } \rho > 0\right\},$$

$$W_\infty^3[M, \|\cdot, \ldots, \cdot\|]^I = \left\{x = (x_{ijk}) \in \omega''' (: \left\{(i, j, k) \in \mathbb{N} \times \mathbb{N} \times \mathbb{N} : \sup_{i,j,k} \frac{1}{lmn} \sum^{l,m,n} \left[M\left(\left\|\frac{x_{ijk}}{\rho}, z_1, \cdots, z_{n-1}\right\|\right)\right] \geq K\right\} \in I, \text{ for } K > 0 \text{ and for some } \rho > 0\right\}.$$

(ii) If $M(x) = x$ then

$$W^3[p, \|\cdot, \ldots, \cdot\|]^I = \left\{x = (x_{ijk}) \in \omega''' : \forall \epsilon > 0, \left\{(i, j, k) \in \mathbb{N} \times \mathbb{N} \times \mathbb{N} : \frac{1}{lmn} \sum^{l,m,n} \left[\left\|x_{ijk} - L, z_1, \cdots, z_{n-1}\right\|\right]^{p_{ijk}} \geq \epsilon\right\} \in I, \text{ for some } \rho > 0\right\},$$

$$W_0^3[p, \|\cdot, \ldots, \cdot\|]^I = \left\{x = (x_{ijk}) \in \omega''' : \forall \epsilon > 0, \left\{(i, j, k) \in \mathbb{N} \times \mathbb{N} \times \mathbb{N} : \frac{1}{lmn} \sum^{l,m,n} \left[\left\|x_{ijk}, z_1, \cdots, z_{n-1}\right\|\right]^{p_{ijk}} \geq \epsilon\right\} \in I, \text{ for some } \rho > 0\right\},$$

$$W_\infty^3[p, \|\cdot, \ldots, \cdot\|]^I = \left\{x = (x_{ijk}) \in \omega''' : \left\{(i, j, k) \in \mathbb{N} \times \mathbb{N} \times \mathbb{N} : \sup_{i,j,k} \frac{1}{lmn} \sum^{l,m,n} \left[\left\|x_{ijk}, z_1, \cdots, z_{n-1}\right\|\right]^{p_{ijk}} \geq K\right\} \in I, \text{ for } K > 0 \text{ and for some } \rho > 0\right\}.$$

The following inequality is very useful , if $p = (p_{ijk})$ be a triple sequence of positive real numbers with $0 < p_{ijk} \leq \sup_{i,j,k} p_{ijk} = H$, and $D = \max\{1, 2^{H-1}\}$, then for sequences $(a_{ijk}), (b_{ijk})$

$$|a_{ijk} + b_{ijk}|^{p_{ijk}} \leq D\left(|a_{ijk}|^{p_{ijk}} + |b_{ijk}|^{p_{ijk}}\right). \tag{1}$$

3 Topological and Algebraic Properties of the New Sequence Spaces

Theorem 1 *Let M be Musielak–Orlicz function then the triple sequence spaces $W^3[M, p, \|\cdot, \ldots, \cdot\|]^I$, $W_0^3[M, p, \|\cdot, \ldots, \cdot\|]^I$ and $W_\infty^3[M, p, \|\cdot, \ldots, \cdot\|]^I$ are linear.*

Proof We prove the result for the sequence space $W^3[M, p, \|\cdot, \ldots, \cdot\|]^I$.
Let $x = (x_{ijk}), y = (y_{ijk}) \in W^3[M, p, \|\cdot, \ldots, \cdot\|]^I$ and $\alpha, \beta \in \mathbb{C}$, then for $z_1, z_2, \ldots, z_{n-1} \in X$ and $L_1, L_2 \in \mathbb{C}$

$$\left\{(i,j,k)\in\mathbb{N}\times\mathbb{N}\times\mathbb{N}:\frac{1}{lmn}\sum^{l,m,n}\left[M\left(\left\|\frac{x_{ijk}-L_1}{\rho_1},z_1,\cdots,z_{n-1}\right\|\right)\right]^{p_{ijk}}\geq\epsilon\right\}\in I,\ \rho_1>0, \tag{2}$$

and

$$\left\{(i,j,k)\in\mathbb{N}\times\mathbb{N}\times\mathbb{N}:\frac{1}{lmn}\sum^{l,m,n}\left[M\left(\left\|\frac{y_{ijk}-L_2}{\rho_2},z_1,\cdots,z_{n-1}\right\|\right)\right]^{p_{ijk}}\geq\epsilon\right\}\in I,\ \rho_2>0. \tag{3}$$

As $\|\cdot,\ldots,\cdot\|$ is an n-norm and M is an Musielak–Orlicz function, we have

$$\begin{aligned}
&\frac{1}{lmn}\sum^{l,m,n}\left[M\left(\left\|\frac{(\alpha x_{ijk}+\beta y_{ijk})-(\alpha L_1+\beta L_2)}{|\alpha|\rho_1+|\beta|\rho_2},z_1,\cdots,z_{n-1}\right\|\right)\right]^{p_{ijk}}\\
&=\frac{1}{lmn}\sum^{l,m,n}\left[M\left(\left\|\frac{\alpha(x_{ijk}-L_1)+\beta(y_{ijk}-L_2)}{|\alpha|\rho_1+|\beta|\rho_2},z_1,\cdots,z_{n-1}\right\|\right)\right]^{p_{ijk}}\\
&\leq D\frac{1}{lmn}\sum^{l,m,n}\left[\frac{|\alpha|}{|\alpha|\rho_1+|\beta|\rho_2}M\left(\left\|\frac{x_{ijk}-L_1}{\rho_1},z_1,\cdots,z_{n-1}\right\|\right)\right]^{p_{ijk}}\\
&+D\frac{1}{lmn}\sum^{l,m,n}\left[\frac{|\beta|}{|\alpha|\rho_1+|\beta|\rho_2}M\left(\left\|\frac{y_{ijk}-L_2}{\rho_2},z_1,\cdots,z_{n-1}\right\|\right)\right]^{p_{ijk}} \qquad \text{using (1)}\\
&\leq D\mathsf{A}\frac{1}{lmn}\left[\sum^{l,m,n}\left[M\left(\left\|\frac{x_{ijk}-L_1}{\rho_1},z_1,\cdots,z_{n-1}\right\|\right)\right]^{p_{ijk}}+\sum^{l,m,n}\left[M\left(\left\|\frac{y_{ijk}-L_2}{\rho_2},z_1,\cdots,z_{n-1}\right\|\right)\right]^{p_{ijk}}\right],
\end{aligned}$$

where

$$\mathsf{A}=\max\left\{1,\left(\frac{|\alpha|}{|\alpha|\rho_1+|\beta|\rho_2}\right)^H,\left(\frac{|\beta|}{|\alpha|\rho_1+|\beta|\rho_2}\right)^H\right\}.$$

From the above inequality, we get

$$\begin{aligned}
&\left\{(i,j,k)\in\mathbb{N}\times\mathbb{N}\times\mathbb{N}:\frac{1}{lmn}\sum^{l,m,n}\left[M\left(\left\|\frac{(\alpha x_{ijk}+\beta y_{ijk})-(\alpha L_1+\beta L_2)}{|\alpha|\rho_1+|\beta|\rho_2},z_1,\cdots,z_{n-1}\right\|\right)\right]^{p_{ijk}}\geq\epsilon\right\}\\
&\subseteq\left\{(i,j,k)\in\mathbb{N}\times\mathbb{N}\times\mathbb{N}:D\mathsf{A}\lim_{i,j,k\to\infty}\left[M\left(\left\|\frac{x_{ijk}-L_1}{\rho_1},z_1,\cdots,z_{n-1}\right\|\right)\right]^{p_{ijk}}\geq\frac{\epsilon}{2}\right\}\\
&\bigcup\left\{(i,j,k)\in\mathbb{N}\times\mathbb{N}\times\mathbb{N}:D\mathsf{A}\lim_{i,j,k\to\infty}\left[M\left(\left\|\frac{y_{ijk}-L_2}{\rho_2},z_1,\cdots,z_{n-1}\right\|\right)\right]^{p_{ijk}}\geq\frac{\epsilon}{2}\right\}.
\end{aligned}$$

Thus, using (2), (3), we get

$$\left\{(i,j,k)\in\mathbb{N}\times\mathbb{N}\times\mathbb{N}:\frac{1}{lmn}\sum^{l,m,n}\left[M\left(\left\|\frac{(\alpha x_{ijk}+\beta y_{ijk})-(\alpha L_1+\beta L_2)}{|\alpha|\rho_1+|\beta|\rho_2},z_1,\cdots,z_{n-1}\right\|\right)\right]^{p_{ijk}}\geq\epsilon\right\}\in I.$$

Therefore, $(\alpha x+\beta y)\in W^3[M,p,\|\cdot,\ldots,\cdot\|]^I$, which proves its linearity. In the same way, linearity of other spaces can be proved.

Theorem 2 *For fixed* $(i,j,k)\in\mathbb{N}\times\mathbb{N}\times\mathbb{N}$, $W^3_\infty[M,p,\|\cdot,\ldots,\cdot\|]^I$ *is a paranormed space with respect to paranorm given by*

$$\mathsf{g}(x) = \inf\left\{\rho^{\frac{p_{lmn}}{H}} : \sup_{i,j,k} \frac{1}{lmn} \sum^{l,m,n} \left[M \left\| \frac{x_{ijk}}{\rho}, z_1, \ldots, z_{n-1} \right\|\right]^{p_{ijk}} \leq 1\right\},$$

for all $z_1, \ldots, z_{n-1} \in X$.

Proof We prove that all the conditions of Definition 12 are satisfied.
(i). $x = 0$ implies $\|0, z_1, \ldots, z_{n-1}\| = 0$, since the set containing the zero element is linearly dependent. Also $M(0) = 0$ gives $\mathsf{g}(0) = 0$.
(ii). It is obvious that $\mathsf{g}(x) = \mathsf{g}(-x)$.
(iii). Let $x = (x_{ijk})$, $y = (y_{ijk}) \in W^3_\infty[M, p, \|\cdot, \ldots, \cdot\|]^I$.
Then there exists $\rho_1, \rho_2 > 0$ such that

$$\sup_{i,j,k} \frac{1}{lmn} \sum^{l,m,n} \left[M\left(\left\| \frac{x_{ijk}}{\rho_1}, z_1, \ldots, z_{n-1} \right\|\right)\right]^{p_{ijk}} \leq 1$$

and

$$\sup_{i,j,k} \frac{1}{lmn} \sum^{l,m,n} \left[M\left(\left\| \frac{y_{ijk}}{\rho_2}, z_1, \ldots, z_{n-1} \right\|\right)\right]^{p_{ijk}} \leq 1$$

for each $z_1, \ldots, z_{n-1} \in X$.
Now using the properties of Musielak–Orlicz function, we have

$$\begin{aligned}
&\sup_{i,j,k} \frac{1}{lmn} \sum^{l,m,n} \left[M\left(\left\| \frac{x_{ijk}+y_{ijk}}{\rho_1+\rho_2}, z_1, \ldots, z_{n-1} \right\|\right)\right]^{p_{ijk}} \\
&\quad \leq \frac{\rho_1}{\rho_1+\rho_2} \sup_{i,j,k} \frac{1}{lmn} \sum^{l,m,n} \left[M\left(\left\| \frac{x_{ijk}}{\rho_1}, z_1, \ldots, z_{n-1} \right\|\right)\right]^{p_{ijk}} \\
&\quad + \frac{\rho_1}{\rho_1+\rho_2} \sup_{i,j,k} \frac{1}{lmn} \sum^{l,m,n} \left[M\left(\left\| \frac{y_{ijk}}{\rho_2}, z_1, \ldots, z_{n-1} \right\|\right)\right]^{p_{ijk}}.
\end{aligned}$$

Thus, $\sup_{i,j,k} \frac{1}{lmn} \sum^{l,m,n} \left[M\left(\left\| \frac{x_{ijk}+y_{ijk}}{\rho_1+\rho_2}, z_1, \ldots, z_{n-1} \right\|\right)\right]^{p_{ijk}} \leq 1.$
So for all $z_1, \ldots, z_{n-1} \in X$, we have

$$\begin{aligned}
\mathsf{g}(x+y) &= \inf\left\{\rho^{\frac{p_{lmn}}{H}} : \sup_{i,j,k} \frac{1}{lmn} \sum^{l,m,n} \left[M \left\| \frac{x_{ijk}+y_{ijk}}{\rho_1+\rho_2}, z_1, \ldots, z_{n-1} \right\|\right]^{p_{ijk}} \leq 1\right\} \\
&\leq \inf\left\{\rho^{\frac{p_{lmn}}{H}} : \sup_{i,j,k} \frac{1}{lmn} \sum^{l,m,n} \left[M \left\| \frac{x_{ijk}}{\rho_1}, z_1, \ldots, z_{n-1} \right\|\right]^{p_{ijk}} \leq 1\right\} \\
&\quad + \inf\left\{\rho^{\frac{p_{lmn}}{H}} : \sup_{i,j,k} \frac{1}{lmn} \sum^{l,m,n} \left[M \left\| \frac{y_{ijk}}{\rho_2}, z_1, \ldots, z_{n-1} \right\|\right]^{p_{ijk}} \leq 1\right\} \\
&= \mathsf{g}(x) + \mathsf{g}(y).
\end{aligned}$$

We now show that if $\lambda^m \to \lambda$ and $\mathsf{g}(x^m - x) \to 0$ as $m \to \infty$, then $\mathsf{g}(\lambda^m x^m - \lambda x) \to 0$, as $m \to \infty$.
For all $z_1, \ldots, z_{n-1} \in X$ let $x = x_{ijk} \in W_\infty^3[M, p, \|\cdot, \ldots, \cdot\|]^I$ and $\rho_{m_1}, \rho_{m_2} > 0$ be such that

$$\sup_{i,j,k} \frac{1}{lmn} \sum^{l,m,n} \left[M \left(\left\| \frac{x_{ijk}^m}{\rho_{m_1}}, z_1, \ldots, z_{n-1} \right\| \right) \right]^{p_{ijk}} \leq 1$$

and

$$\sup_{i,j,k} \frac{1}{lmn} \sum^{l,m,n} \left[M \left(\left\| \frac{x_{ijk}^m - x_{ijk}}{\rho_{m_2}}, z_1, \ldots, z_{n-1} \right\| \right) \right]^{p_{ijk}} \leq 1.$$

Then, we have

$$\begin{aligned}
&\sup_{i,j,k} \frac{1}{lmn} \sum^{l,m,n} \left[M \left(\left\| \frac{\lambda^m x_{ijk}^m - \lambda x_{ijk}}{\rho_{m_1}|\lambda^m - \lambda| + \rho_{m_2}|\lambda|}, z_1, \ldots, z_{n-1} \right\| \right) \right]^{p_{ijk}} \\
&\leq \sup_{i,j,k} \frac{1}{lmn} \sum^{l,m,n} \left[M \left(\left\| \frac{\lambda^m x_{ijk}^m - \lambda x_{ijk}^m}{\rho_{m_1}|\lambda^m - \lambda| + \rho_{m_2}|\lambda|}, z_1, \ldots, z_{n-1} \right\| \right.\right. \\
&\qquad \left.\left. + \left\| \frac{\lambda x_{ijk}^m - \lambda x_{ijk}}{\rho_{m_1}|\lambda^m - \lambda| + \rho_{m_2}|\lambda|}, z_1, \ldots, z_{n-1} \right\| \right) \right]^{p_{ijk}} \\
&\leq \frac{\rho_{m_1}|\lambda^m - \lambda|}{\rho_{m_1}|\lambda^m - \lambda| + \rho_{m_2}|\lambda|} \sup_{i,j,k} \frac{1}{lmn} \sum^{l,m,n} \left[M \left(\left\| \frac{x_{ijk}^m}{\rho_{m_1}}, z_1, \ldots, z_{n-1} \right\| \right) \right]^{p_{ijk}} \\
&\qquad + \frac{\rho_{m_2}|\lambda|}{\rho_{m_1}|\lambda^m - \lambda| + \rho_{m_2}|\lambda|} \sup_{i,j,k} \frac{1}{lmn} \sum^{l,m,n} \left[M \left(\left\| \frac{x_{ijk}^m - x_{ijk}}{\rho_{m_2}}, z_1, \ldots, z_{n-1} \right\| \right) \right]^{p_{ijk}}.
\end{aligned}$$

Thus,

$$\sup_{i,j,k} \frac{1}{lmn} \sum^{l,m,n} \left[M \left(\left\| \frac{\lambda^m x_{ijk}^m - \lambda x_{ijk}}{\rho_{m_1}|\lambda^m - \lambda| + \rho_{m_2}|\lambda|}, z_1, \ldots, z_{n-1} \right\| \right) \right]^{p_{ijk}} \leq 1$$

Hence,

$$\begin{aligned}
\mathsf{g}(\lambda^m x^m - \lambda x) = \inf \Bigg\{ &\left(\rho_{m_1}|\lambda^m - \lambda| + \rho_{m_2}|\lambda|\right)^{\frac{p_{lmn}}{H}} : \\
&\sup_{i,j,k} \frac{1}{lmn} \sum^{l,m,n} \left[M \left(\left\| \frac{\lambda^m x_{ijk}^m - \lambda x_{ijk}}{\rho_{m_1}|\lambda^m - \lambda| + \rho_{m_2}|\lambda|}, z_1, \ldots, z_{n-1} \right\| \right) \right]^{p_{ijk}} \leq 1 \Bigg\} \\
\leq \left(|\lambda^m - \lambda|\right)^{\frac{p_{lmn}}{H}} \inf &\left\{ \rho_{m_1}^{\frac{p_{lmn}}{H}} : \sup_{i,j,k} \frac{1}{lmn} \sum^{l,m,n} \left[M \left(\left\| \frac{x_{ijk}^m}{\rho_{m_1}}, z_1, \ldots, z_{n-1} \right\| \right) \right]^{p_{ijk}} \leq 1 \right\} \\
+ \left(|\lambda|\right)^{\frac{p_{lmn}}{H}} \inf &\left\{ \rho_{m_2}^{\frac{p_{lmn}}{H}} : \sup_{i,j,k} \frac{1}{lmn} \sum^{l,m,n} \left[M \left(\left\| \frac{x_{ijk}^m - x_{ijk}}{\rho_{m_2}}, z_1, \ldots, z_{n-1} \right\| \right) \right]^{p_{ijk}} \leq 1 \right\}.
\end{aligned}$$

Since $\lambda^m \to \lambda$ and $\mathsf{g}(x^m - x) \to 0$ as $m \to \infty$. Hence, $\mathsf{g}(\lambda^m x^m - \lambda x) \to 0$ as $m \to \infty$, which completes the proof.

Theorem 3 *The triple sequences* $W_0^3[M, p, \|\cdot, \ldots, \cdot\|]^I$ *and* $W_\infty^3[M, p, \|\cdot, \ldots, \cdot\|]^I$ *are solid.*

Proof We prove the result for $W_0^3[M, p, \|\cdot, \ldots, \cdot\|]^I$.
Consider $x = (x_{ijk}) \in W_0^3[M, p, \|\cdot, \ldots, \cdot\|]^I$, and a sequence of scalar (α_{ijk}) such that $|\alpha_{ijk}| \leq 1$ for all $i, j, k \in \mathbb{N}$.
Then, we have

$$\left\{(i, j, k) \in \mathbb{N} \times \mathbb{N} \times \mathbb{N} : \frac{1}{lmn}\sum^{l,m,n}\left[M\left(\left\|\frac{(\alpha_{ijk})(x_{ijk})}{\rho}, z_1, \ldots, z_{n-1}\right\|\right)\right]^{p_{ijk}}\right\}$$
$$\subset \left\{(i, j, k) \in \mathbb{N} \times \mathbb{N} \times \mathbb{N} : \frac{\mathsf{K}}{lmn}\sum^{l,m,n}\left[M\left(\left\|\frac{x_{ijk}}{\rho}, z_1, \ldots, z_{n-1}\right\|\right)\right]^{p_{ijk}} \geq \epsilon\right\} \in I,$$

where $\mathsf{K} = \max\{1, |\alpha_{ijk}|^H\}$. Hence, $\alpha x = (\alpha_{ijk}x_{ijk}) \in W_0^3[M, p, \|\cdot, \ldots, \cdot\|]^I$ for all scalars (α_{ijk}) with $|\alpha_{ijk}| \leq 1$ with $i, j, k \in \mathbb{N}$, whenever $x = (x_{ijk}) \in W_0^3[M, p, \|\cdot, \ldots, \cdot\|]^I$.
Hence $W_0^3[M, p, \|\cdot, \ldots, \cdot\|]^I$ is solid.
In the same way, we can show that $W_\infty^3[M, p, \|\cdot, \ldots, \cdot\|]^I$ is solid.

Theorem 4 *Let* M_1 *and* M_2 *be Musielek–Orlicz functions, then we have*

$$W_\infty^3[M_1, p, \|\cdot, \ldots, \cdot\|]^I \cap W_\infty^3[M_2, p, \|\cdot, \ldots, \cdot\|]^I \subseteq W_\infty^3[M_1 + M_2, p, \|\cdot, \ldots, \cdot\|]^I.$$

Proof Let $x = (x_{ijk}) \in W_\infty^3[M_1, p, \|\cdot, \ldots, \cdot\|]^I \cap W_\infty^3[M_2, p, \|\cdot, \ldots, \cdot\|]^I$. Then there exists finite positive numbers K_1, K_2 such that

$$\left\{(i, j, k) \in \mathbb{N} \times \mathbb{N} \times \mathbb{N} : \sup_{i,j,k} \frac{1}{lmn}\sum^{l,m,n}\left[M_1\left(\left\|\frac{x_{ijk}}{\rho_1}, z_1, \cdots, z_{n-1}\right\|\right)\right]^{p_{ijk}} \geq K_1\right\} \in I, \quad \text{for some } \rho_1 > 0$$

and

$$\left\{(i, j, k) \in \mathbb{N} \times \mathbb{N} \times \mathbb{N} : \sup_{i,j,k} \frac{1}{lmn}\sum^{l,m,n}\left[M_1\left(\left\|\frac{x_{ijk}}{\rho_2}, z_1, \cdots, z_{n-1}\right\|\right)\right]^{p_{ijk}} \geq K_2\right\} \in I, \quad \text{for some } \rho_2 > 0.$$

Let $\rho = \max\{\rho_1, \rho_2\}$. The result follows from the inequality

$$\begin{aligned}
&\sum^{l,m,n}\left[(M_1 + M_2)\left(\left\|\frac{x_{ijk}}{\rho}, z_1, \cdots, z_{n-1}\right\|\right)\right]^{p_{ijk}} \\
&= \sum^{l,m,n}\left[M_1\left(\left\|\frac{x_{ijk}}{\rho}, z_1, \cdots, z_{n-1}\right\|\right) + M_2\left(\left\|\frac{x_{ijk}}{\rho}, z_1, \cdots, z_{n-1}\right\|\right)\right]^{p_{ijk}} \\
&\leq \sum^{l,m,n}\left[M_1\left(\left\|\frac{x_{ijk}}{\rho_1}, z_1, \cdots, z_{n-1}\right\|\right)\right]^{p_{ijk}} + \sum^{l,m,n}\left[M_2\left(\left\|\frac{x_{ijk}}{\rho_2}, z_1, \cdots, z_{n-1}\right\|\right)\right]^{p_{ijk}}.
\end{aligned}$$

Thus, $x = (x_{ijk}) \in W_\infty^3[M_1 + M_2, p, \|\cdot, \ldots, \cdot\|]^I$, which completes the proof.

References

1. Alotaibi, A., Mursaleen M., Sharma, S.K.: Double sequence spaces over $n-$normed spaces defined by a sequence of orlicz functions. J. Ineq. Appl. **2014**, 216 (2014)
2. Debnath, S., Sarma, B., Das, B.C.: Some generalized triple sequence spaces of real numbers. J. Non. Anal. Opt. **6**, 71–79 (2015)
3. Debnath, S., Subramanian, N.: Generalized rough lacunary statistical triple difference sequence spaces in probability of fractional order defined by musielak-orlicz function. Bol. Soc. Paran. Mat. (3s) **37**(1), 55–62 (2019)
4. Esi, A.: On some triple almost lacunary sequence spaces defined by orlicz functions. Res. Rev.: Discret. Math. Struct. **1**, 16–25 (2014)
5. Esi, A., Catalbas, M.N.: Almost convergence of triple sequences G. J. Math. Anal. **2**, 6–10 (2014)
6. Esi, A., Savas, E.: On lacunary statically convergent triple sequences in probabilistic normed space. Appl. Math. Inf. Sci. **9**, 2529–2534 (2015)
7. Fast, H.: Surla convergence statistique. Colloq. Math. **2**, 241–244 (1951)
8. Fridy, J.A.: On statistical convergence. Analysis **5**, 301–313 (1985)
9. Fridy, J.A.: Statistical limit points. Proc. Am. Math. Soc. **11**, 1187–1192 (1993)
10. Gähler, S.: Linear 2-normietre Rume. Math. Nachr. **28**, 1–43 (1965)
11. Jalal, T.: Some new $I-$convergent sequence spaces defined by using a sequence of modulus functions in n-normed spaces. Int. J. Math. Arch. **5**(9), 202–209 (2014)
12. Jalal, T.: Some new $I-$lacunary generalized difference sequence spaces defined in n-normed spaces. Springer Proc. Math. Sat. **171**, 249–258 (2016)
13. Jalal, T.: Some new lacunary sequence spaces of invariant means defined by musielak-Orlicz functions on n-normed space. Int. J. P. Appl. Math. **119**(7), 1–11 (2018)
14. Jalal, T., Malik, I.A.: Some new triple sequence spaces over $n-$normed space. Proyecciones J. Math. **37**(3), 547–564 (2018)
15. Jalal, T., Malik, I.A.: I-convergent triple sequence spaces over $n-$normed space. Asia Pac. J. Math. **5**(2), 233–242 (2018)
16. Jalal, T., Malik, I.A.: I-convergence triple difference sequence spaces over $n-$normed space. Tbil. Math. J. **11**(4), 93–102 (2018)
17. Kostyrko, P., Salat, T., Wilczynski, W.: $I-$convergence. Real Anal. Exch. **26**(2), 669–686 (2000)
18. Kumar, V.: On $I-$convergence of double sequences. Math. Commun. **12**, 171–181 (2007)
19. Lindenstrauss, J., Tzafriri, L.: On orlicz sequence spaces. Israel J. Math. **10**, 379–390 (1971)
20. Misiak, A.: $n-$Inner product spaces. Math. Nachr. **140**, 299–319 (1989)
21. Mursaleen, M., Edely, O.H.H.: Statistical convergence of double sequences. J. Math. Anal. Appl. **288**, 223–231 (2003)
22. Mursaleen, M., Raj, K., Sharma, S.K.: Some spaces of differences and lacunary statistical convergence in $n-$normed space defined by sequence of orlicz functions. Miskolc Math. Notes **16**(1), 283–304 (2015)
23. Salat, T., Tripathy , B.C., Ziman, M.: On some properties of $I-$convergence. Tatra Mountain Mathematical Publications, pp. 669–686 (2000)
24. Sahiner, A., Gurdal, M., Duden, F.K.: Triple Sequences and their statistical convergence. Selcuk J. Appl. Math. **8**, 49–55 (2007)
25. Sahiner, A., Tripathy, B.C.: Some I related properties of triple sequences. Selcuk J. Appl. Math. **9**, 9–18 (2008)
26. Sharma, S.K., Esi, A.: Some $I-$convergent sequence spaces defined by using sequence of moduli and n-normed space. J. Egyp. Math. Soc. **21**(2), 103–107 (2013)
27. Tripathy, B.C.: Statistically convergent double sequence. Tamkang. J. Math. **34**(3), 231–237 (2003)

On λ-Ideal Statistical Convergence of Weighted h for Multiple Sequences

Ekrem Savaş

Abstract The goal of this paper is to extend the recently introduced summability method of [21] and present the new notion, namely, $\mathcal{I}_\lambda$-double statistical convergence of weight h via ideal.

Keywords Ideal · Filter · $\mathcal{I}$-double statistical convergence of order α · $\mathcal{I}_\lambda$-double statistical convergence of weight h

Mathematics Subject Classification (2010) 40A05 · 40D25

1 Introduction

Statistical convergence is a generalization of the usual notion of convergence. The concept of statistical convergence was introduced by Steinhause [23] and Fast [5] and later reintroduced by Schoenberg [22] independently. Statistical convergence turned out to be one of the most active areas of research in summability theory after the work of Fridy [6] and Šalát [13]. The more general idea of λ-statistical convergence was introduced by Mursaleen in [8]. This idea from single to double sequences of scalars was extended by Mursaleen and Edely [9]. Quite recently, a generalization of statistical convergence, namely, ideal convergence has been introduced and studied by Kostyrko et al. [7]. Several papers deal with this problem, see [4, 14–16, 19, 20]. Recently, in [4, 19], we also used ideals to introduce the concepts of $\mathcal{I}$-statistical convergence, $\mathcal{I}$-lacunary statistical convergence, and $\mathcal{I}_\lambda$-statistical convergence and observe their properties.

E. Savaş (✉)
Department of Mathematics, Uşak University, Uşak, Turkey
e-mail: ekremsavas@yahoo.com

D. Giri et al. (eds.), *Proceedings of the Fifth International Conference on Mathematics and Computing*, Advances in Intelligent Systems and Computing 1170, https://doi.org/10.1007/978-981-15-5411-7_15

In [2, 3], the notions of statistical convergence of order α- and λ-statistical convergence of order α were introduced and studied. Quite recently Savas [21] introduced the concepts of $\mathcal{I}$-double statistical convergence and $\mathcal{I}_\lambda$-double statistical convergence.

The goal of this paper is to introduce a new and further general notion, namely, $\mathcal{I}_\lambda-$double statistical convergence of weight h.

Throughout by a sequence $y = (y_{ij})$, we shall mean a double sequences of real numbers.

2 Preliminaries

For easy understanding of the material incorporated in this paper, we began with some known definitions and notions will be needed in the sequel.

Definition 2.1 (see [7]) A family $\mathcal{I} \subset 2^{\mathbb{N}}$ is said to be an ideal of $\mathbb{N}$ if the following conditions hold:
(a) $R, S \in \mathcal{I}$ implies $R \cup S \in \mathcal{I}$,
(b) $R \in \mathcal{I},\ S \subset R$ implies $S \in \mathcal{I}$.

Definition 2.2 (see [7]) A nonempty family $F \subset 2^{\mathbb{N}}$ is said to be a filter of $\mathbb{N}$ if the following conditions hold:
(a) $\phi \notin F$,
(b) $R, S \in F$ implies $R \cap S \in F, and$
(c) $R \in F,\ R \subset S$ implies $S \in F$.

If $\mathcal{I}$ is a proper ideal of $\mathbb{N}$ (i.e., $\mathbb{N} \notin \mathcal{I}$), then the family of sets $F(\mathcal{I}) = \{M \subset \mathbb{N} : \exists\ R \in \mathcal{I} : M = \mathbb{N} \setminus R\}$ is a filter of $\mathbb{N}$. It is called the filter associated with the ideal.

Definition 2.3 A proper ideal $\mathcal{I}$ is said to be admissible if $\{n\} \in \mathcal{I}$ for each $n \in \mathbb{N}$.

Throughout $\mathcal{I}$ will stand for a proper admissible ideal of $\mathbb{N}$.

Definition 2.4 (See [7]) Let $\mathcal{I} \subset 2^{\mathbb{N}}$ be a proper admissible ideal in $\mathbb{N}$. The sequence (y_i) is said to be $\mathcal{I}$-convergent to $\beta \in \mathbb{R}$ if for each $\gamma > 0$ the set $A(\gamma) = \{i \in \mathbb{N} : |y_i - \beta| \geq \gamma\} \in \mathcal{I}$.

Let $y = (y_{ij})$ be a double sequence, if for each $\gamma > 0$ there exists $N \in \mathbb{N}$ such that $\left|y_{ij} - \beta\right| < \gamma$ whenever $i, j \geq N$, then we say that $y = (y_{ij})$ is convergent in Pringsheim's sense or $P-$convergent to β and β is called Pringsheim limit (denoted by $P - limy = \beta$), (see, [10]). Further investigations in this direction and other applications can be found in the earlier works [11, 12, 17, 18] and many other authors.

Let $E \subseteq \mathbb{N} \times \mathbb{N}$ be a two-dimensional set of positive integers and let $E_{u,v}$ be the numbers of (i, j) in E such that $i \leq u$ and $j \leq v$. Then the lower asymptotic density of E is defined as

$$P - \liminf_{u,v} \frac{E_{u,v}}{uv} = \delta_2(E).$$

In the case when the sequence $(\frac{E_{u,v}}{uv})_{u,v=1,1}^{\infty,\infty}$ has a limit then we say that E has a natural density and is defined as

$$P - \lim_{u,v} \frac{E_{u,v}}{uv} = \delta_2(E).$$

Mursaleen and Edely have presented the main definition as follows.

Definition 2.5 ([9]) Let $y = (y_{ij})$ be a double sequence and y is double statistically convergent to β provided that for each $\gamma > 0$

$$P - \lim_{u,v} \frac{1}{uv}\{\text{number of } (i, j) : i < u \text{ and } j < v, |y_{ij} - \beta| \geq \gamma\} = 0.$$

In this case, $s^2 - \lim_{i,j} y_{ij} = \beta$ and we denote the set of all statistical convergent double sequences by s^2.

Definition 2.6 Let $y = (y_{ij})$ be a double sequence of real numbers, if for each $\gamma > 0$,

$$R(\gamma) = \left\{(i, j) \in \mathbb{N} \times \mathbb{N} : \left|y_{ij} - \beta\right| \geq \gamma\right\} \in \mathcal{I},$$

then we say that $y = (y_{ij})$ is convergent to the number β with respect to the ideal $\mathcal{I}$. In this case, $\mathcal{I} - \lim_{ij} y_{ij} = \beta$.

Note that if $\mathcal{I}$ is the ideal $\mathcal{I}_0$ then $\mathcal{I}$- convergence coincides with the convergence in Pringsheim's sense and if we take $\mathcal{I}_d = \{R \subset \mathbb{N} \times \mathbb{N} : \delta_2(R) = 0\}$ then $\mathcal{I}_d-$convergence becomes statistical convergence for double sequences.

Let $h : \mathbb{N} \to [0, \infty)$ be a function with $\lim_{n\to\infty} h(n) = \infty$. The upper density of weight h was defined in [1] by the formula

$$\overline{d}_h(R) = \limsup_{n\to\infty} \frac{R(1, v)}{h(v)}$$

for $R \subset \mathbb{N}$. Then the family

$$\mathcal{I}_h = \{R \subset \mathbb{N} : \overline{d}_h(R) = 0\}$$

sets an ideal. It has been observed in [1] that $\mathbb{N} \in \mathcal{I}_h$ iff $\frac{v}{h(v)} \to 0$. So we additionally assume that $v/h(v) \nrightarrow 0$ as $v \to \infty$ so that $\mathbb{N} \notin \mathcal{I}_h$ and $\mathcal{I}_h$ is a proper admissible ideal of $\mathbb{N}$. The set of all such weight functions h satisfying the above properties will be denoted by H.

3 Main Results

We have

Definition 3.1 Let $y = \{y_{ij}\}$ be a double sequence, if for each $\gamma > 0$ and $\rho > 0$

$$\{(u, v) \in \mathbb{N} \times \mathbb{N} : \frac{1}{h(uv)} |\{i \leq u \text{ and } j \leq v : |x_{ij} - \beta| \geq \gamma\}| \geq \rho\} \in \mathcal{I}.$$

Then we say that $y = \{y_{ij}\}$ is $\mathcal{I}$-statistically convergent of weight h to β or $s_2(\mathcal{I})^h$-convergent to β. In this case, we write $y_{ij} \to \beta(s_2(\mathcal{I})^h)$. The class of all $\mathcal{I}$-double statistically convergent sequences of weight h will be denoted by simply $s_2(\mathcal{I})^h$.

Remark 3.2 If we consider $\mathcal{I} = \mathcal{I}_{fin} = \{R \subseteq \mathbf{N} : R \text{ is a finite subset }\}$, $s_2(\mathcal{I})^h$-convergence coincides with double statistical convergence of weight h. Further if we take $h(uv) = (uv)^\alpha$, it reduces to $\mathcal{I}-$statistical convergence of order α.

Let $\lambda = (\lambda_u)$ and $\mu = (\mu_v)$ be two non-decreasing sequences of positive real numbers both of which tend to ∞ as u and v approach ∞, respectively. Also let $\lambda_{u+1} \leq \lambda_u + 1$, $\lambda_1 = 1$, and $\mu_{v+1} \leq \mu_v + 1$, $\mu_1 = 1$. The collection of such sequence $(\bar{\lambda}_{u,v})$ will be denoted by Δ.

We write the generalized double de la Vallee-Poussin mean by

$$t_{u,v}(x) = \frac{1}{h(\lambda_u \mu_v)} \sum_{i \in I_u, j \in J_v} y_{ij},$$

where $I_u = [u - \lambda_u + 1, u]$ and $J_v = [v - \mu_v + 1, v]$.

A sequence $y = (y_{ij})$ is said to be $[V^2, \bar{\lambda}]^h(\mathcal{I})$-summable to a number $\beta \in X$, if

$$\mathcal{I} - lim_{u,v} t_{u,v}(y) \to \beta,$$

i.e., for any $\gamma > 0$,

$$\left\{(u, v) \in \mathbb{N} \times \mathbb{N} : \left|t_{u,v}(y) - \beta\right| \geq \gamma\right\} \in \mathcal{I}.$$

Throughout this paper we shall denote $\lambda_u \mu_v$ by $\bar{\lambda}_{u,v}$ and $(i \in I_u, j \in I_v)$ by $(i, j) \in I_{u,v}$.

If $\mathcal{I} = \mathcal{I}_{fin}$, $([V, \bar{\lambda}]^h(\mathcal{I})-$ summability becomes $[V, \bar{\lambda}]^g$ summability.

Definition 3.3 Let $y = (y_{ij})$ be a double sequence, if for every $\gamma > 0$ and $\rho > 0$,

$$\left\{(u, v) \in N \times N : \frac{1}{h(\bar{\lambda}_{uv})} | \left\{(i, j) \in I_{uv} : |y_{ij} - \beta| \geq \gamma\right\} | \geq \rho\right\} \in \mathcal{I}.$$

Then we say that $y = (y_{ij})$ is $\mathcal{I}_\lambda-$double statistically convergent of weight h. In this case, $s^h_{\bar{\lambda}}(\mathcal{I}) - \lim y = \beta$ or $y_{ij} \to \beta$ $(s^h_{\bar{\lambda}}(\mathcal{I}))$.

Remark 3.4 If we consider $\mathcal{I} = \mathcal{I}_{fin}, s^h_{\bar{\lambda}}(\mathcal{I})$-convergence coincides with $\lambda-$ double statistical convergence of weight h. If we consider $\bar{\lambda}_{uv} = uv$, Definition 3.1 reduces to Definition 2.1.

We shall denote by $s_{\bar{\lambda}}(\mathcal{I})^h$ and $\left[V, \bar{\lambda}\right]^h(\mathcal{I})$ the set of all $\mathcal{I}_\lambda$-double statistically convergent of weight h and $\left[V, \bar{\lambda}\right](\mathcal{I})$-convergent of weight h sequences, respectively.

Theorem 3.5 *Let $h_1, h_2 \in H$ be such that there exist $K > 0$ and $(i_0, j_0) \in \mathbb{N} \times \mathbb{N}$ such that $\frac{h_1(\bar{\lambda}_{uv})}{h_2(\bar{\lambda}_{uv})} \leq K$ for all $u \geq i_0$ and $v \geq j_0$. Then $s_{\bar{\lambda}}(\mathcal{I})^{h_1} \subset s_{\bar{\lambda}}(\mathcal{I})^{h_2}$.*

Proof For any $\gamma > 0$, we write

$$\frac{|\{(i, j) \in I_{uv} : |y_{ij} - \beta| \geq \gamma\}|}{h_2(\bar{\lambda}_{uv})} \leq \frac{|\{(i, j) \in I_{uv} : |y_{ij} - \beta| \geq \gamma\}|}{h_1(\bar{\lambda}_{uv})},$$

and so for any $\rho > 0$,

$$\{(u, v) \in \mathbb{N} \times \mathbb{N} : \frac{|\{(i, j) \in I_{uv} : |y_{ij} - \beta| \geq \gamma\}|}{h_2(\bar{\lambda}_{uv})} \geq \rho\}$$

$$\subset \{(u, v) \in \mathbb{N} \times \mathbb{N} : \frac{|\{(i, j) \in I_{uv} : |y_{ij} - \beta| \geq \gamma\}|}{h_1(\bar{\lambda}_{uv})} \geq \rho\}.$$

Finally, we get that $s_{\bar{\lambda}}(\mathcal{I})^{h_1} \subset s_{\bar{\lambda}}(\mathcal{I})^{h_2}$.

Proof Similarly, we can prove the following.

Theorem 3.6 *Let $h_1, h_2 \in H$ be such that there exist $K > 0$ and $(i_0, j_0) \in \mathbb{N} \times \mathbb{N}$ such that $\frac{h_1(\bar{\lambda}_{uv})}{h_2(\bar{\lambda}_{uv})} \leq K$ for all $u \geq i_0$ and $v \geq j_0$. Then*
(i) $s_2(\mathcal{I})^{h_1} \subset s_2(\mathcal{I})^{h_2}$.
(ii) $\left[V, \bar{\lambda}\right]^{h_1}(\mathcal{I}) \subset \left[V, \bar{\lambda}\right]^{h_2}(\mathcal{I})$.

Theorem 3.7 *Let* $\bar{\lambda} = (\bar{\lambda}_{uv}) \in \Delta$. *Then* $y_{ij} \to \beta\left[V, \bar{\lambda}\right]^h(\mathcal{I}) \Rightarrow y_{ij} \to \beta\left(s_{\bar{\lambda}}(I)^h\right)$.

Proof Let $\gamma > 0$ and $y_{ij} \to \beta\left[V, \bar{\lambda}\right]^h(\mathcal{I})$. We have

$$\sum_{(ij)\in I_{uv}} |y_{ij} - \beta| \geq \sum_{(ij)\in I_{uv} \& |y_{ij}-\beta|>\gamma} |y_{ij} - \beta| \geq \gamma. \left|\left\{(ij) \in I_{uv} : |y_{ij} - \beta| \geq \gamma\right\}\right|.$$

Hence for $\rho > 0$,

$$\frac{1}{h(\bar{\lambda}_{uv})}\left|\left\{(i, j) \in I_{uv} : |y_{ij} - \beta| \geq \gamma\right\}\right| \geq \rho \Rightarrow \frac{1}{h(\bar{\lambda}_{uv})}\sum_{(ij)\in I_{uv}} |y_{ij} - \beta| \geq \gamma\rho,$$

i.e.,

$$\left\{(u, v) \in \mathbb{N} \times \mathbb{N} : \frac{1}{h(\bar{\lambda}_{uv})}\left|\left\{(i, j) \in I_{uv} : |y_{ij} - \beta| \geq \gamma\right\}\right| \geq \rho\right\}$$

$$\subset \left\{(u, v) \in \mathbb{N} \times \mathbb{N} : \frac{1}{h(\bar{\lambda}_{uv})}\left\{\sum_{(ij)\in I_{uv}} |y_{ij} - \beta| \geq \gamma\right\} \geq \gamma\rho\right\}.$$

Since $y_{ij} \to \beta\left[V, \bar{\lambda}\right]^h(\mathcal{I})$, so the set on the right-hand side belongs to $\mathcal{I}$ and so it follows that $y_{ij} \to \beta\left(s_{\bar{\lambda}}(\mathcal{I})\right)^h$.

Theorem 3.8 $s_2(\mathcal{I})^h \subset s_{\bar{\lambda}}(\mathcal{I})^h$ *if*

$$\liminf_{uv} \frac{h(\bar{\lambda}_{uv})}{h(uv)} > 0. \tag{1}$$

Proof For given $\gamma > 0$,

$$\begin{aligned}\frac{1}{h(uv)}\left|\left\{i \leq u \text{ and } j \leq v : |y_{ij} - \beta| \geq \gamma\right\}\right| &\geq \frac{1}{h(uv)}|\{(i, j) \in I_{uv} : |x_{uv} - \beta| \geq \gamma\}| \\ &\geq \frac{h(\bar{\lambda}_{uv})}{h(uv)}\frac{1}{h(\bar{\lambda}_{uv})}\left|\left\{(i, j) \in I_{uv} : |y_{ij} - \beta| \geq \gamma\right\}\right|.\end{aligned}$$

If $\liminf\limits_{uv\to\infty} \frac{h(\bar{\lambda}_{uv})}{h(uv)} = a$ then from definition $\left\{(u, v) \in N \times N : \frac{h(\bar{\lambda}_{uv})}{h(uv)} < \frac{a}{2}\right\}$is finite. For $\rho > 0$,

$$\left\{(u,v)\in\mathbb{N}\times\mathbb{N}:\frac{1}{h(\bar{\lambda}_{uv})}\left|\{(i,j)\in I_{uv}:|y_{ij}-\beta|\geq\gamma\}\right|\geq\rho\right\}$$
$$\subset\left\{(u,v)\in\mathbb{N}\times\mathbb{N}:\frac{1}{g(uv)}\left|\{(k,l)\in I_{uv}:|y_{ij}-L|\geq\gamma\}\right|\geq\frac{a}{2}\rho\right\}\cup$$
$$\left\{(u,v)\in\mathbb{N}\times\mathbb{N}:\frac{g(\bar{\lambda}_{uv})}{h(u,v)}<\frac{a}{2}\right\}.$$

Since $\mathcal{I}$ is admissible, the set on the right-hand side belongs to $\mathcal{I}$. □

Finally, we complete this paper by presenting the following theorem.

Theorem 3.9 *Let $h_1, h_2 \in H$ and let $\bar{\lambda}=(\bar{\lambda}_{uv})$, $\bar{\mu}=(\bar{\mu}_{uv})$ be two sequences in Λ such that $\bar{\lambda}_{uv}\leq\bar{\mu}_{uv}$ for all $(u,v)\in\mathbb{N}\times\mathbb{N}$. If*

$$\lim_{(u,v)\to\infty}\inf\frac{h_1(\bar{\lambda}_{uv})}{h_2(\bar{\mu}_{uv})}>0, \tag{2}$$

then $S_{\bar{\mu}}(\mathcal{I})^{h_2}\subset S_{\bar{\lambda}}(\mathcal{I})^{h_1}$.

Proof Assume that $\bar{\lambda}_{uv}\leq\bar{\mu}_{uv}$ for all $(u,v)\in\mathbb{N}\times\mathbb{N}$ and let (2) be satisfied. For $\gamma>0$, we obtain

$$\left\{(i,j)\in\bar{J}_{uv}:\left|y_{ij}-\beta\right|\geq\gamma\right\}\supseteq\left\{(i,j)\in I_{uv}:\left|y_{ij}-\beta\right|\geq\gamma\right\},$$

where $\bar{J}_u=[u-\mu_u+1,u]$ and we denote $(i\in\bar{J}_v, j\in\bar{J}_v)$ by $(i,j)\in\bar{J}_{u,v}$.
Therefore, we may write

$$\frac{1}{h_2(\bar{\mu}_{uv})}\left|\left\{(i,j)\in\bar{J}_{uv}:\left|y_{ij}-\beta\right|\geq\gamma\right\}\right|\geq\frac{h_1(\bar{\lambda}_{uv})}{h_2(\bar{\mu}_{uv})}\frac{1}{h_1(\bar{\lambda}_{uv})}\left|\left\{(i,j)\in I_{uv}:\left|y_{ij}-\beta\right|\geq\gamma\right\}\right|$$

and so for all $(u,v)\in\mathbb{N}\times\mathbb{N}$ we have

$$\left\{(u,v)\in\mathbb{N}\times\mathbb{N}:\frac{1}{h_1(\bar{\lambda}_{uv})}\left|\left\{(i,j)\in I_{uv}:\left|y_{ij}-\beta\right|\geq\gamma\right\}\right|\geq\rho\right\}$$

$$\subseteq\left\{(u,v)\in\mathbb{N}\times\mathbb{N}:\frac{1}{h_2(\bar{\mu}_{uv})}\left|\left\{(i,j)\in J_{uv}:\left|y_{ij}-\beta\right|\geq\gamma\right\}\right|\geq\rho\frac{h_1(\bar{\lambda}_{uv})}{h_2(\bar{\mu}_{uv})}\right\}\in\mathcal{I}.$$

Therefore $S_{\bar{\mu}}(\mathcal{I})^{h_2}\subset S_{\bar{\lambda}}(\mathcal{I})^{h_1}$. □

References

1. Balcerzak, M., Das, P., Filipczak, M., Swaczyna, J.: Generalized kinds of density and the associated ideals. Acta Math. Hungar. **147**(1), 97–115 (2015)
2. Colak, R.: Statistical convergence of order α. Modern Methods in Analysis and Its Applications, pp. 121–129. Anamaya Publication, New Delhi, (2010)
3. Colak, R., Bektas, C.A.: λ-statistical convergence of order α. Acta Math. Sci. **31B**(3), 953–959 (2011)
4. Das, P., Savaş, E., Ghosal, S.K.: On generalizations of certain summability methods using ideals. Appl. Math. Lett. **24**, 1509–1514 (2011)
5. Fast, H.: Sur la convergence statistique. Colloq. Math. **2**, 241–244 (1951)
6. Fridy, J.A.: On statistical convergence. Analysis **5**, 301–313 (1985)
7. Kostyrko, P., Šalát, T., Wilczynki, W.: I-convergence. R. Anal. Exch. **26**(2), 669–685 (2000/2001)
8. Mursaleen, M.: λ-statistical convergence. Math. Slovaca **50**, 111–115 (2000)
9. Mursaleen, M., Edely, O.H.: Statistical convergence of double sequences. J. Math. Anal. Appl. **288**, 223–231 (2003)
10. Pringsheim, A.: Zur theorie der zweifach unendlichen Zahlenfolgen. Math. Ann. **53**, 289–321 (1900)
11. Patterson, R.F.: Equivalence of methods for the summation of double sequences. Appl. Math. Lett. **24**(2), 107–110 (2011)
12. Patterson, R.F.: Rates of convergence for double sequences. Southeast Asian Bull. Math. **26**(3), 469–478 (2003)
13. Šalát, T.: On Statistically convergent sequences of real numbers. Math. Slovaca **30**, 139–150 (1980)
14. Savaş, E.: Δ^m -strongly summable sequences spaces in 2-normed spaces defined by ideal convergence and an Orlicz function. App. Math. Comp. **217**, 271–276 (2010)
15. Savaş, E., A-sequence spaces in 2-normed space defined by ideal convergence and an Orlicz function. Abst. Appl. Analy. **2011**, Article ID 741382 (2011)
16. Savaş, E.: On some new sequence spaces in 2-normed spaces using ideal convergence and an Orlicz function. J. Ineq. Appl., Article Number: 482392 (2010). https://doi.org/10.1155/2010/482392
17. Savaş, E.: On $\overline{\lambda}$-statistically convergent double sequences of fuzzy numbers. J. Inequal. Appl. Art. ID 147827, 6 p. (2008)
18. Savaş, E.: $\overline{\lambda}$-double sequence spaces of fuzzy real numbers defined by Orlicz function. Math. Commun. **14**(2), 287–297 (2009)
19. Savaş, E., Das, P.: A generalized statistical convergence via ideals. Appl. Math. Lett. **24**, 826–830 (2011)
20. Savaş, E., Das, P., Dutta, S.: A note on strong matrix summability via ideals. Appl. Math Lett. **25**(4), 733–738 (2012)
21. Savaş, E.: On generalized double statistical convergence via ideals. In: The Fifth Saudi Science Conference, 16–18 Apr 2012
22. Schoenberg, I.J.: The integrability of certain functions and related summability methods. Amer. Math. Mon. **66**, 361–375 (1959)
23. Steinhaus, H.: Sur la convergence ordinaire et la convergence asymptotique. Colloquium Math. **27**, 73–74 (1951)

A-Fractal Rational Functions and Their Positivity Aspects

S. K. Katiyar and A. K. B. Chand

Abstract Coalescence hidden variable fractal interpolation function (CHFIF) proves more versatile than classical interpolant and fractal interpolation function (FIF). Using rational functions and CHFIF, a general construction of **A**-fractal rational functions is introduced for the first time in the literature. This construction of **A**-fractal rational function also allows us to insert shape parameters for positivity-preserving univariate interpolation. The convergence analysis of the proposed scheme is established. With suitably chosen numerical examples and graphs, the effectiveness of the positivity-preserving interpolation scheme is illustrated.

Keywords Iterated function system · Fractals · Fractal interpolation functions · Hidden variable fractal interpolation function · Convergence · Shape preservation · Positivity

1 Introduction and Motivation

Barnsley [1] framed fractal interpolation function (FIF) with the help of iterated function system (IFS) which can model a large class of self-affine or self-similar or self-referential objects. FIF captures the irregularity of data very effectively in comparison with the classical interpolants and gives a successful instrument for modeling data sampled from real-world signals. To make corresponding FIF smooth, one can choose scaling factors suitably [2]. Barnsley et al. [3] have introduced univariate hid-

A. K. B. Chand is grateful to Project No. MTR/2017/000574 of Department of Science and Technology, India.

S. K. Katiyar (✉)
Department of Mathematics, SRM Institute of Science and Technology, Chennai 603203, India
e-mail: sbhkatiyar@gmail.com

A. K. B. Chand
Department of Mathematics, Indian Institute of Technology Madras, Chennai 600036, India
e-mail: chand@iitm.ac.in

D. Giri et al. (eds.), *Proceedings of the Fifth International Conference on Mathematics and Computing*, Advances in Intelligent Systems and Computing 1170, https://doi.org/10.1007/978-981-15-5411-7_16

den variable FIF, which are more irregular, flexible, appealing, and diverse than a FIF. In some practical applications, the simulating curves exhibit partly self-referential and partly non-self-referential nature. Later, Chand and Kapoor [4, 5] introduced the concept of coalescence hidden variable FIF (CHFIF). By using suitable IFS, Chand et al. [6, 13] obtained a fractal perturbation $\mathbf{g}[\mathbf{A}]$ for a given continuous function $\mathbf{g} : I \to \mathbb{R}^2$ and developed positivity- and monotonicity-preserving $\mathscr{C}^1$-cubic HFIFs.

The traditional nonrecursive shape-preserving interpolation schemes available in the literature ([9–11, 21–23] and reference quoted therein) produce an interpolant with a specified degree of smoothness. But the derivative function of these shape-preserving interpolants is infinitely differentiable except possibly at finite number of points, viz., the knots. That's why, these schemes are inadequate. Recently, by using polynomial and rational interpolants, our group has developed fractal interpolation function with shape preservation/approximation (see, for instance, [7, 8, 12–18, 20, 24]).

Our search for a method of shape preservation with rational cubic spline with hidden variable came up empty-handed, and this article is undertaken in this direction. We try to present the rational cubic interpolation in a framework of the CHFIF. The classical $\mathscr{C}^1$-rational cubic interpolant [23] and its fractal generalization, namely, the $\mathscr{C}^1$-rational cubic FIF [24] are special cases of the proposed scheme. One can construct infinite number of self-referential or non-self-referential rational spline FIFs with suitable choices of hidden variables, scaling parameters involved in the construction which is the major advantage of this method. An upper bound for the uniform distance between an original smooth function and this scheme is studied. With a suitable choice of the shape parameters and the scaling parameters, the interpolant can be made positive according to the behavior of the positive data. This attribute of shape-preserving scheme may find potential applications in various nonlinear phenomena.

The remainder of the paper is organized as follows. In Sect. 2, we review briefly the notion of CHFIFs. In Sect. 3, we construct A-fractal rational functions with u_n, and v_n known as tension (shape) parameters. Section 4 focuses on constraining the shape parameters and scaling parameters involved in the scheme so as to preserve positivity. In Sect. 5, we find the convergence analysis by finding an upper bound for the interpolation error. Section 6 contains some numerical examples which illustrate the proposed interpolation scheme and the effects of hidden variables, and the scaling parameters involved in the construction. We end in Sect. 7 with some final comments.

2 Preliminaries

In this section, we provide some theoretical background like the notions of CHFIF. The interested reader is addressed to [1, 3, 6, 19] for more details about IFS, FIF. For $r \in \mathbb{N}$, let $\mathbb{N}_r$ denote the subset $\{1, 2, \ldots, r\}$ of $\mathbb{N}$, set of natural numbers.

2.1 Hidden Variable Fractal Interpolation Function

Consider a generalized set of data $\tilde{\Delta} = \{(t_n, x_n, d_n, z_n, d_n^*) : n \in \mathbb{N}_N\}$ of $\Delta = \{(t_n, x_n, d_n) : n \in \mathbb{N}_N\}$. Here $\{z_n : n \in \mathbb{N}_N\}$ are real parameters. Let $\mathbb{R}^2$ be endowed with the Manhattan metric $d_M\big((t_1, x_1), (t_2, x_2)\big) = |t_1 - t_2| + |x_1 - x_2|$, which is induced by the l^1-norm. For $n \in \mathbb{N}_{N-1}$, let the contraction homeomorphisms $L_n : I = [x_1, x_N] \to I_n = [t_n, x_{n+1}] \subset I$ be given as

$$L_n(t) = a_n t + b_n, \; n \in \mathbb{N}_{N-1}. \tag{1}$$

The functions $H_n : I \times \mathbb{R}^2 \to \mathbb{R}^2$ are expressed as

$$H_n(t, \mathbf{x}) = H_n(t, x, z) = \big(H_n^1(t, x, z), H_n^2(t, z)\big)^t := A_n(t, x)^t + \big(p_n(t), q_n(t)\big)^t, \tag{2}$$

where t denotes the transpose, A_n are upper triangular matrices $\begin{bmatrix} \alpha_n & \beta_n \\ 0 & \gamma_n \end{bmatrix}$, and p_n, q_n are suitable real-valued continuous functions so that the following conditions are satisfied for all $n \in \mathbb{N}_{N-1}$:
(i) $d_M(H_n(t, x, z), H_n(t^\diamond, x, z)) \le k_1|t - t^\diamond|$ for some constant $k_1 > 0$;
(ii) $d_M\big(H_n(t, x, z), H_n(t, x^\diamond, z^\diamond)\big) \le k_2\, d_M\big((x, z)(x^\diamond, z^\diamond)\big)$ for $0 \le k_2 < 1$;
and (iii) $H_n(t_1, x_1, z_1) = (x_n, z_n)$ and $H_n(t_N, x_N, z_N) = (x_{n+1}, z_{n+1})$.
Define $w_n : I \times \mathbb{R}^2 \to I \times \mathbb{R}^2$ by $w_n(t, x, z) = \big(L_n(t), H_n(t, x, z)\big)$. In the IFS $\{I \times \mathbb{R}^2; w_n : n \in \mathbb{N}_{N-1}\}$, we consider the special case $p_n(t) = g_1 \circ L_n(t) - \alpha_n b_1(t) - \beta_n b_2(t)$, $q_n(t) = g_2 \circ L_n(t) - \gamma_n b_2(t)$, where $\mathbf{b} = (b_1, b_2) \in \mathrm{Lip}(I, \mathbb{R}^2)$ satisfies $\mathbf{b}(t_1) = \mathbf{g}(t_1)$ and $\mathbf{b}(t_N) = \mathbf{g}(t_N)$. The variables α_n, β_n, and γ_n are chosen such that $\|A_n\|_1 < 1$ for all $n \in \mathbb{N}_{N-1}$. It follows from the conditions on L_n and H_n that w_n are contraction maps with respect to the metric d_M^* defined on $I \times \mathbb{R}^2$ by $d_M^*\big((t, x, z), (t^*, x^*, z^*)\big) = |t - t^*| + \theta d_M\big((x, z), (x^*, z^*)\big)$, where $\theta = \frac{1-a}{2c_1}$, and $a = \max\{\frac{h_n}{x_N - x_1} : n \in \mathbb{N}_{N-1}\}$. Consequently, the generalized IFS $\{I \times \mathbb{R}^2; w_n : n \in \mathbb{N}_{N-1}\}$ admits an attractor $G \subseteq I \times \mathbb{R}^2$. In this case, the IFS provides an attractor that is the graph of a continuous function denoted here as $\mathbf{g}[\mathbf{A}] = (g_1[\mathbf{A}], g_2[\mathbf{A}])$. This function $\mathbf{g}[\mathbf{A}]$ is called as **A**-fractal function of **f** with respect to the partition Δ and the function **b**. We stipulate that $\mathbf{g}[\mathbf{A}]$ satisfies (details can be found in [6])

$$\mathbf{g}[\mathbf{A}](t) = \mathbf{g}(t) + A_n(\mathbf{g}[\mathbf{A}] - \mathbf{b})\big(L_n^{-1}(t)\big), t \in I_n, \; n \in \mathbb{N}_{N-1}. \tag{3}$$

Note that the fractal function $\mathbf{g}[\mathbf{A}] = \big(g_1[A], g_2[A]\big)$ can be written componentwise as

$$\begin{aligned} g_1[\mathbf{A}]\big(L_n(t)\big) &= \alpha_n g_1[\mathbf{A}](t) + \beta_n g_2[\mathbf{A}](t) + g_1\big(L_n(t)\big) - \alpha_n b_1(t) - \beta_n b_2(t), \\ g_2[\mathbf{A}]\big(L_n(t)\big) &= \gamma_n g_2[\mathbf{A}](t) + g_2\big(L_n(t)\big) - \gamma_n b_2(t). \end{aligned} \tag{4}$$

3 Construction of A-Fractal Rational Function with Shape Parameters

Consider the interpolation dataset $\Delta = \{(t_n, x_n, d_n) : n \in \mathbb{N}_N\}$ generated by a smooth real-valued function Ψ (i.e., $\Psi(t_n) = x_n$ and $\Psi^{(1)}(t_n) = d_n$ for $n \in \mathbb{N}_N$), where $t_1 < t_2 < \cdots < t_N$. Extend this data to $\tilde{\Delta} = \{(t_n, x_n, d_n, z_n, d_n^*) : n \in \mathbb{N}_N\}$ with real parameters z_n, d_n^*, where d_n^* denote the approximate derivative values from the data $\{(t_n, z_n) : n \in \mathbb{N}_N\}$ at the grid point t_n. Let g_1 and g_2 are the traditional $\mathscr{C}^1$-rational cubic splines [23] corresponding to the datasets Δ and $\overline{\Delta} = \{(t_n, z_n, d_n^*) : n \in \mathbb{N}_N\}$, respectively. With $\theta = \frac{t-t_1}{t_N-t_1}$, and $h_n = t_{n+1} - t_n$,

$$g_1\big(L_n(t)\big) = \frac{A_{1n}(1-\theta)^3+A_{2n}\theta(1-\theta)^2+A_{3n}\theta^2(1-\theta)+A_{4n}\theta^3}{v_n(1-\theta)^2+2u_nv_n\theta(1-\theta)+u_n\theta^2}, \tag{5}$$

$$g_2\big(L_n(t)\big) = \frac{A^*_{1n}(1-\theta)^3+A^*_{2n}\theta(1-\theta)^2+A^*_{3n}\theta^2(1-\theta)+A^*_{4n}\theta^3}{v_n(1-\theta)^2+2u_nv_n\theta(1-\theta)+u_n\theta^2}, \tag{6}$$

where $A_{1n} = v_nx_n$, $A_{2n} = (2u_nv_n + v_n)x_n + v_nh_nd_n$, $A_{3n} = (2u_nv_n + u_n)x_{n+1} - u_nh_nd_{n+1}$, $A_{4n} = u_nx_{n+1}$, $A^*_{1n} = v_nz_n$, $A^*_{2n} = (2u_nv_n + v_n)z_n + v_nh_nd_n^*$, $A^*_{3n} = (2u_nv_n + u_n)z_{n+1} - u_nh_nd^*_{n+1}$, $A^*_{4n} = u_nz_{n+1}$. To ensure a strictly positive denominator, we always select shape parameters $v_n > 0$ and $u_n > 0$. To develop the **A**-fractal rational cubic spline corresponding to **g**, we choose scaling parameters α_n, β_n, and γ_n such that $\|A_n\|_1 < 1$ for all $n \in \mathbb{N}_{N-1}$ and select a family $B = \{\mathbf{b_n} = (b_{n1}, b_{n2}) \in \mathscr{C}^1(I, \mathbb{R}^2) : n \in \mathbb{N}_N\}$ satisfying the conditions (E): $b_{n1}(t_1) = g_1(t_1) = x_1$, $b_{n1}(t_N) = g_1(t_N) = x_N$, $b^{(1)}_{n1}(t_1) = g^{(1)}_1(t_1) = d_1$, $b^{(1)}_{n1}(x_N) = g^{(1)}_1(t_N) = d_N$, and $b_{n2}(t_1) = g_2(t_1) = z_1$, $b_{n2}(t_N) = g_2(t_N) = z_N$, $b^{(1)}_{n2}(t_1) = g^{(1)}_2(t_1) = d^*_1$, $b^{(1)}_{n2}(t_N) = g^{(1)}_2(t_N) = d^*_N$. There are variety of choices for B. A natural choice of $\mathbf{b_n} = (b_{n1}, b_{n2})$ is the one in which b_{n1} and b_{n2} are the same form as in (5) (with knots at t_1 and t_N) corresponding to g_1 and g_2, respectively, i.e.,

$$b_{n1} = \frac{B_{1n}(1-\theta)^3+B_{2n}\theta(1-\theta)^2+B_{3n}\theta^2(1-\theta)+B_{4n}\theta^3}{v_n(1-\theta)^2+2u_nv_n\theta(1-\theta)+u_n\theta^2}, \tag{7}$$

$$b_{n2} = \frac{B^*_{1n}(1-\theta)^3+B^*_{2n}\theta(1-\theta)^2+B^*_{3n}\theta^2(1-\theta)+B^*_{4n}\theta^3}{v_n(1-\theta)^2+2u_nv_n\theta(1-\theta)+u_n\theta^2}, \tag{8}$$

where the coefficients B_{in}, B^*_{in} for $i = 1, 2, 3, 4$ are determined through conditions (E). With simple manipulations, we get $B_{1n} = v_nx_1$, $B_{2n} = (2u_nv_n + v_n)x_1 + v_nd_1(t_N - t_1)$,

$B_{3n} = (2u_n v_n + u_n)x_N - u_n d_N(t_N - t_1)$, $B_{4n} = u_n x_N$, $B^*_{1n} = v_n z_1$, $B^*_{2n}=(2u_n v_n + v_n)z_1 + v_n d^*_1(t_N - t_1)$, $B^*_{3n} = (2u_n v_n + u_n)z_N - u_n d^*_N(t_N - t_1)$, $B^*_{4n} = u_n z_N$. Now from (4), we have

$$g_1[\mathbf{A}]\big(L_n(t)\big) = \alpha_n g_1[\mathbf{A}](t) + \beta_n g_2[\mathbf{A}](t) + \frac{R_n(t)}{S_n(t)}, \tag{9}$$

$$g_2[\mathbf{A}]\big(L_n(t)\big) = \gamma_n g_2[\mathbf{A}](t) + \frac{R^*_n(t)}{S_n(t)}, \tag{10}$$

where $R_n(t) = v_n(x_n - \alpha_n x_1 - \beta_n z_1)(1-\theta)^3 + (2u_n v_n + v_n)(x_n - \alpha_n x_1 - \beta_n z_1) + v_n[h_n d_n - \alpha_n d_1(t_N - t_1) - \beta_n d^*_1(t_N - t_1)]\theta(1-\theta)^2 + (2u_n v_n + u_n)(x_{n+1} - \alpha_n x_N - \beta_n z_N) - u_n[h_n d_{n+1} - \alpha_n d_N(t_N - t_1) - \beta_n d^*_N(t_N - t_1)]\theta^2(1-\theta) + u_n(x_{n+1} - \alpha_n x_N - \beta_n z_N)\theta^3$, $R^*_n(t) = v_n(z_n - \gamma_n z_1)(1-\theta)^3 + (2u_n v_n + v_n)(z_n - \gamma_n z_1) + v_n [h_n d^*_n - \gamma_n d^*_1(t_N - t_1)]\theta(1-\theta)^2 + (2u_n v_n + u_n)(z_{n+1} - \gamma_n z_N) - u_n[h_n d^*_{n+1} - \gamma_n d^*_N(t_N - t_1)]\theta^2(1-\theta) + u_n(z_{n+1} - \gamma_n z_N)\theta^3$, and $S_n(t) = v_n(1-\theta)^2 + 2u_n v_n \theta(1-\theta) + u_n\theta^2$. The function $g_1[\mathbf{A}] : I \to \mathbb{R}$ satisfies $g_1[\mathbf{A}](t_n) = x_n$ and $g_1[\mathbf{A}]^{(1)}(t_n) = d_n$ is the $\mathscr{C}^1$-RCSCHFIF corresponding to Δ.

Remark 1 If we select $z_n = x_n$ and $d^*_n = d_n$ for all $n \in \mathbb{N}_N$, and $\alpha_n + \beta_n = \gamma_n$ for all $n \in \mathbb{N}_{N-1}$, then $g_1[\mathbf{A}]$ coincides with $g_2[\mathbf{A}]$, representing a self-referential $\mathscr{C}^1$-rational cubic spline FIF [24]. $g_1[\mathbf{A}]$ is, in general, non-self-referential for other choices of the hidden variables and parameters. When $\alpha_n = \beta_n = 0$ for all $n \in \mathbb{N}_{N-1}$ in (9) then this scheme recovers the classical $\mathscr{C}^1$-rational cubic spline [23].

Remark 2 This scheme coincides with the cubic Hermite CHFIF (CHCHFIF) when $u_n = v_n = 1 \;\forall\, n \in \mathbb{N}_{N-1}$. If $u_n = v_n = 1$ and $\beta_n = 0 \;\forall\, n \in \mathbb{N}_{N-1}$, then we get $\mathscr{C}^1$-cubic Hermite FIF. When $\alpha_n = 0$, $\beta_n = 0$, and $u_n = v_n = 1$, for all $n \in \mathbb{N}_{N-1}$, this scheme recovers the classical piecewise cubic interpolant.

4 Positivity of A-Fractal Rational Function

In general, this scheme developed in the previous section does not preserve positivity. Thus, it is compulsory to choose suitable values to the scaling factors and shape parameters so that it generates a positivity-preserving interpolant. The sufficient condition for the positivity of developed scheme is given in the following theorem.

Theorem 1 *Let* $g_1[\mathbf{A}]$ *be a* $\mathscr{C}^1$*-RCSCHFIF obtained via vector-valued FIF* $\mathbf{g}[\mathbf{A}] = \big(g_1[\mathbf{A}], g_2[\mathbf{A}]\big)$ *that corresponds to the data* $\tilde{\Delta}$ *with* $x_n > 0, z_n > 0 \;\forall\, n \in \mathbb{N}_N$. *If for each* $n \in \mathbb{N}_{N-1}$,
(i) the scaling parameters α_n, γ_n, *and* β_n *such that* $0 \le \gamma_n \le \min\left\{a_n, \frac{z_n}{z_1}, \frac{z_{n+1}}{z_N}\right\}$, $0 \le \alpha_n + \beta_n \le \min\left\{\frac{x_n}{l_1}, \frac{x_{n+1}}{l_2}\right\}$, *where* $l_1 = \max\{x_1, z_1\}$ *and* $l_2 = \max\{x_N, z_N\}$.
(ii) the shape parameters u_n *and* v_n *satisfy*
$u_n > \left\{\frac{-h_n d_n - x_n + \alpha_n[x_1 + d_1(t_N - t_1)] + \beta_n[z_1 + d^*_1(t_N - t_1)]}{2(x_n - \alpha_n x_1 - \beta_n z_1)}\right\}$,
$v_n > \left\{\frac{h_n d_{n+1} - x_{n+1} - \alpha_n[d_N(t_N - t_1) - x_N] - \beta_n[d^*_N(t_N - t_1) - z_N]}{2(x_{n+1} - \alpha_n x_N - \beta_n z_N)}\right\}$. *Then* $g_1[\mathbf{A}]$ *is positive on* I.

Proof We wish to construct a $\mathscr{C}^1$-RCSCHFIF $g_1[\mathbf{A}]$ (9) which is positive on the whole interval $I = [t_1, t_N]$. To achieve this aim, first we take $g_2[\mathbf{A}](t) > 0$ for all $t \in [t_1, t_N]$ because $g_1[\mathbf{A}]$ is defined recursively through an implicit function $g_1[\mathbf{A}]\big(L_n(t)\big) = \alpha_n g_1[\mathbf{A}](t) + \beta_n g_2[\mathbf{A}](t) + \frac{R_n(t)}{S_n(t)}$ and the sufficient conditions for positivity of $g_2[\mathbf{A}] > 0$ will be derived at the end using the self-referential conditions from $g_1[\mathbf{A}]$. Hence, for $g_1[\mathbf{A}](t) > 0 \,\forall\, t \in [t_1, t_N]$, we show that $\alpha_n g_1[\mathbf{A}](.) + \beta_n g_2[\mathbf{A}](.) + \frac{R_n(.)}{S_n(.)} > 0$ whenever $g_1[\mathbf{A}](.) > 0$, $g_2[\mathbf{A}](.) > 0$. To get strictly positive denominator $S_n(.)$ in the rational function $\frac{R_n(.)}{S_n(.)}$, $n \in \mathbb{N}_{N-1}$, we choose $u_n > 0, v_n > 0$. We assume $\alpha_n \geq 0$, $\beta_n \geq 0$ for all $n \in \mathbb{N}_{N-1}$ to get positivity. With these initial assumptions on the shape parameters and the scaling parameters, it follows from (9) that positivity of $g_1[\mathbf{A}]\big(L_n(t_j)\big)$ depends on positivity of $g_2[\mathbf{A}]$ and $\frac{R_n(.)}{S_n(.)}|_{t=t_j}$. Since the RCSCHFIF is generated iteratively from function values at knot points, then $g_1[\mathbf{A}]\big(L_n(t_j)\big) > 0$ for all $n \in \mathbb{N}_{N-1}$ and for all knot points t_j, which will ensure positivity of $g_1[\mathbf{A}]$ on I. Based on the foregoing assumptions on the shape parameters and the parameters involved in the IFS, the positivity of $g_1[\mathbf{A}]$ depends on the positivity of cubic polynomial $R_n(.)$ (see (9)). Hence, we require

$$R_n(\theta) = C_{1n}(1-\theta)^3 + C_{2n}\theta(1-\theta)^2 + C_{3n}\theta^2(1-\theta) + C_{4n}\theta^3 > 0. \tag{11}$$

If we put $\theta = \frac{\nu}{\nu+1}$ in (11) then we get
$R_n(\nu) = C_{4n}\nu^3 + C_{3n}\nu^2 + C_{2n}\nu + C_{1n} > 0$ for all $\nu > 0$.

The coefficients $C_{1n} \geq 0, C_{2n} \geq 0, C_{3n} \geq 0$, and $C_{4n} \geq 0$ only if the above expression is positive [22].

$$C_{1n} > 0 \Leftrightarrow v_n(x_n - \alpha_n x_1 - \beta_n z_1) > 0 \Leftrightarrow \alpha_n x_1 + \beta_n z_1 < x_n.$$

Similarly

$$C_{4n} > 0 \Leftrightarrow u_n(x_{n+1} - \alpha_n x_N - \beta_n z_N) > 0 \Leftrightarrow \alpha_n x_N + \beta_n z_N < x_{n+1}.$$

Take $l_1 = \max\{x_1, z_1\}$ and $l_2 = \max\{x_N, z_N\}$. Hence, we constrain the scaling parameters according to $0 \leq \alpha_n + \beta_n \leq \min\{\frac{x_n}{l_1}, \frac{x_{n+1}}{l_2}\}$. Now,

$$\begin{aligned} C_{2n} > 0 \Leftrightarrow & (2u_n v_n + v_n)(x_n - \alpha_n x_1 - \beta_n z_1) + v_n[h_n d_n - \alpha_n d_1(t_N - t_1) - \beta_n d_1^* \times \\ & \qquad (t_N - t_1)] > 0 \\ \Leftrightarrow u_n > & \frac{-h_n d_n - x_n + \alpha_n[x_1 + d_1(t_N - t_1)] + \beta_n[z_1 + d_1^*(t_N - t_1)]}{2(x_n - \alpha_n x_1 - \beta_n z_1)}. \end{aligned}$$

A similar analysis gives that
$C_{3n} > 0 \Leftrightarrow v_n > \left\{\frac{h_n d_{n+1} - x_{n+1} - \alpha_n[d_N(t_N - t_1) - x_N] - \beta_n[d_N^*(t_N - t_1) - z_N]}{2(x_{n+1} - \alpha_n x_N - \beta_n z_N)}\right\}$.
In the construction of $\mathscr{C}^1$-RCCHFIF, we assume $\|A_n\|_1 < a_n$ for all $n \in \mathbb{N}_{N-1}$.

Remark 3 Now, if we set $\alpha_n = \gamma_n, \beta_n = 0 \,\forall\, n \in \mathbb{N}_{N-1}$ and $t_n = z_n, d_n = d_n^* \,\forall\, n \in \mathbb{N}_N$ in the aforementioned theorem, the sufficient condition for the positivity of

RCCHFIF $g_1[\mathbf{A}]$ will give positivity condition for FIF $g_2[\mathbf{A}]$ by restricting γ_n, u_n, v_n as follows:
$0 \leq \gamma_n \leq \min\left\{a_n, \frac{z_n}{z_1}, \frac{z_{n+1}}{z_N}\right\}$, and
$u_n > \left\{\frac{-h_n d_n^* - z_n + \gamma_n[z_1 + d_1^*(x_N - x_1)]}{2(z_n - \gamma_n z_1)}\right\}, v_n > \left\{\frac{h_n d_{n+1}^* - z_{n+1} - \gamma_n[d_N^*(x_N - x_1) - z_N]}{2(z_{n+1} - \gamma_n z_N)}\right\}$, which coincide with the condition in [24].

Remark 4 If we set $\alpha_n = 0$, $\beta_n = 0 \,\forall\, n \in \mathbb{N}_{N-1}$, the positivity condition of RCCHFIF $g_1[\mathbf{A}]$ on u_n, v_n reduces to $u_n > \dfrac{-h_n d_n - x_n}{2x_n}$, $v_n > \dfrac{h_n d_{n+1} - y_{n+1}}{2y_{n+1}}$, which provide sufficient condition for the positivity of the classical rational cubic spline [23]. This demonstrates that this scheme can be used to solve the positivity preservation problems as well.

5 Convergence Analysis of RCSCHFIF

In this section, the uniform error bound for a RCSCHFIF $g_1[\mathbf{A}]$ is obtained from the generalized data $\{(t_n, x_n, d_n, z_n, d_n^*) : n \in \mathbb{N}_N\}$. Now let us assume that the data $\{(t_n, x_n) : n \in \mathbb{N}_N\}$ satisfying $t_1 < t_2 < \cdots < t_N$, being interpolated and generated from a function $\Psi \in \mathscr{C}^2(I)$. We will use triangle inequality $\|\Psi - f_1[\mathbf{A}]\|_\infty \leq \|\Psi - C\|_\infty + \|C - f_1[\mathbf{A}]\|_\infty$ to get our desired result. To make our presentation simple, we introduce the following notations: $|\alpha|_\infty = \max\{|\alpha_n| : n \in \mathbb{N}_{N-1}\}$, $|\beta|_\infty = \max\{|\beta_n| : n \in \mathbb{N}_{N-1}\}$, $h = \max\{h_n : n \in \mathbb{N}_{N-1}\}$, and $|I| = x_N - x_1$. We need the following result in the subsequent work of this section.

Proposition 1 ([23]) *Let $\Psi(x) \in \mathscr{C}^2[x_1, x_N]$, where $x_n = \Psi(t_n)$ and $d_n = \Psi^{(1)}(t_n)$. Then $C(x)$ satisfies*

$$\|\Psi - C\|_\infty \leq ch^2 \|\Psi^{(2)}\|_\infty, \tag{12}$$

where c is the optimal error constant. For the detail, reader may refer [23].

Theorem 2 *Let $g_1[\mathbf{A}]$ be a RCSCHFIF obtained from the generalized data $\{(t_n, x_n, d_n, z_n, d_n^*) : n \in \mathbb{N}_N\}$ and C be its classical nonrecursive counterpart. Assume that there exist positive constants Z_0, Z_0^* verifying $|\frac{\partial p_n(\tau_n, \beta_n, t)}{\partial \alpha_n}| \leq Z_0$ and $|\frac{\partial p_n(\alpha_n, \eta_n, t)}{\partial \beta_n}| \leq Z_0^*$, for $|\tau_n| \in (0, a_n)$, $|\eta_n| \in (0, a_n)$, $t \in I_n$, $n \in \mathbb{N}_{N-1}$. Then,*

$$\|g_1[\mathbf{A}] - C\|_\infty \leq \frac{1}{1 - |\alpha|_\infty}\Big[|\alpha|_\infty(\|C\|_\infty + Z_0) + |\beta|_\infty(\|g_2\|_\infty + Z_0^*)\Big].$$

Proof Denote $\Upsilon = \Big(\bigotimes_{n=1}^{n=N-1}(-a_n, a_n), \bigotimes_{n=1}^{n=N-1}(-a_n, a_n)\Big)$, and let $(\alpha, \beta) = (\alpha_1, \alpha_2, \ldots, \alpha_{N-1}; \beta_1, \beta_2, \ldots, \beta_{N-1}) \in \Upsilon$. We shall note that, RCSCHFIF $g_1[\mathbf{A}]$ is the unique fixed point of $\mathscr{T}_{1(\alpha,\beta)}$ defined as $\mathscr{T}_{1(\alpha,\beta)} g_1[\mathbf{A}](L_n(t)) = \alpha_n g_1[\mathbf{A}](t) +$

$\beta_n g_2[\mathbf{A}](t) + p_n(\alpha_n, \beta_n, t)$ for a fixed set of hidden variables $\{(z_n, d_n^*) : n \in \mathbb{N}_N\}$, scaling parameters γ_n, and $(\alpha, \beta) \in \Upsilon$ with $\alpha_n \neq 0$, $\beta_m \neq 0$ for some n and m in $\mathbb{N}_{N-1}$. Here the subscript (α, β) is used to emphasize the dependence of $\mathcal{T}_1$ on (α, β). We take $p_n(x) = p_n(\alpha_n, \beta_n, x)$ to avoid the confusion. If $(\alpha^*, \beta^*) = (0, 0, \ldots, 0; 0, 0, \ldots, 0) \in \Upsilon$, and with the same hidden variables and scaling parameters γ_n as mentioned above, the classical rational cubic interpolant C can be obtained as the fixed point of $\mathcal{T}_{1(\alpha^*,\beta^*)}$. Hence, for $t \in I_n$, $n \in \mathbb{N}_{N-1}$,

$$\begin{aligned}|\mathcal{T}_{1(\alpha,\beta)} g_1[\mathbf{A}](t) - \mathcal{T}_{1(\alpha,\beta)} C(t)| &= \Big|[\alpha_n g_1[\mathbf{A}](L_n^{-1}(t)) + \beta_n g_2[\mathbf{A}](L_n^{-1}(t)) + p_n(\alpha_n, \beta_n, t)] \\ &\quad - [\alpha_n C(L_n^{-1}(t)) + \beta_n g_2[\mathbf{A}](L_n^{-1}(t)) + p_n(\alpha_n, \beta_n, t)]\Big|, \\ &= |\alpha_n| |g_1[\mathbf{A}](L_n^{-1}(t)) - C(L_n^{-1}(t))|, \\ &\leq |\alpha|_\infty \|g_1[\mathbf{A}] - C\|_\infty.\end{aligned}$$

Further, we obtain

$$\|\mathcal{T}_{1(\alpha,\beta)} g_1[\mathbf{A}] - \mathcal{T}_{1(\alpha,\beta)} C\|_\infty \leq |\alpha|_\infty \|g_1[\mathbf{A}] - C\|_\infty. \tag{13}$$

Applying the mean value theorem for functions of several variables and using the expression for the componentwise operator $\mathcal{T}_1$,

$$\begin{aligned}|\mathcal{T}_{1(\alpha,\beta)} C(t) - \mathcal{T}_{1(\alpha^*,\beta^*)} C(t)| &= \Big|\alpha_n C(L_n^{-1}(t)) + \beta_n g_2[\mathbf{A}](L_n^{-1}(t)) + p_n(\alpha_n, \beta_n, t) - \\ &\quad p_n(0, 0, t)\Big|, \\ &\leq |\alpha_n| \|C\|_\infty + |\beta_n| \|g_2[\mathbf{A}]\|_\infty + |\alpha_n| \left|\frac{\partial p_n(\tau_n, \beta_n, L_n^{-1}(t))}{\partial \alpha_n}\right| \\ &\quad + |\beta_n| \left|\frac{\partial p_n(\alpha_n, \eta_n, L_n^{-1}(t))}{\partial \beta_n}\right|, \\ &\leq |\alpha|_\infty (\|C\|_\infty + Z_0) + |\beta|_\infty (\|g_2[\mathbf{A}]\|_\infty + Z_0^*).\end{aligned}$$

We get,

$$\|\mathcal{T}_{1(\alpha,\beta)} C - \mathcal{T}_{1(\alpha^*,\beta^*)} C\|_\infty \leq |\alpha|_\infty (\|C\|_\infty + Z_0) + |\beta|_\infty (\|g_2[\mathbf{A}]\|_\infty + Z_0^*). \tag{14}$$

The triangle inequality,

$$\begin{aligned}\|g_1[\mathbf{A}] - C\|_\infty &= \|\mathcal{T}_{1(\alpha,\beta)} g_1[\mathbf{A}] - \mathcal{T}_{1(\alpha^*,\beta^*)} C\|_\infty \\ &\leq \|\mathcal{T}_{1(\alpha,\beta)} g_1[\mathbf{A}] - \mathcal{T}_{1(\alpha,\beta)} C\|_\infty + \|\mathcal{T}_{1(\alpha,\beta)} C - \mathcal{T}_{1(\alpha^*,\beta^*)} C\|_\infty,\end{aligned} \tag{15}$$

coupled with (13)–(14) implies that

$$\|g_1[\mathbf{A}] - C\|_\infty \leq \frac{1}{1 - |\alpha|_\infty} \Big[|\alpha|_\infty (\|C\|_\infty + Z_0) + |\beta|_\infty (\|g_2[\mathbf{A}]\|_\infty + Z_0^*)\Big].$$

The stated result in the following theorem directly follows by coupling the results in Proposition and Theorem in triangle inequality.

Theorem 3 *Let $g_1[\mathbf{A}]$ and C, respectively, be the RCSCHFIF and the classical rational cubic spline for the original function $\Psi \in \mathscr{C}^2(I)$ with respect to the interpolation data $\{(t_n, x_n) : n \in \mathbb{N}_N\}$, and let $g_1[\mathbf{A}]^{(1)}(t_n) = C^{(1)}(t_n) = d_n$. Assume the hypothesis of Theorem 2 is true. Then,*

$$\|\Psi - g_1[\mathbf{A}]\|_\infty \le ch^2\|\Psi^{(2)}\|_\infty + \frac{1}{1-|\alpha|_\infty}\Big[|\alpha|_\infty(\|C\|_\infty + Z_0) + |\beta|_\infty(\|g_2[\mathbf{A}]\|_\infty + Z_0^*)\Big]. \tag{16}$$

Convergence Result: From $|\alpha_n| < a_n = \frac{h_n}{t_N - t_1}$, it follows that $\frac{|\alpha|_\infty}{1-|\alpha|_\infty} \le \frac{h}{|I|-h}$. Similarly, from the assumptions on β_n, we get $\frac{|\beta|_\infty}{1-|\alpha|_\infty} \le \frac{h}{|I|-h}$. With these observations, the error bound for the RCSCHFIF $g_1[\mathbf{A}]$ in the previous theorem verifies $\|\Psi - g_1[\mathbf{A}]\|_\infty = O(h)$, and consequently the RCSCHFIF converges uniformly to the original function $\Psi(x) \in \mathscr{C}^2[x_1, x_N]$ when the mesh norm of the partition tends to zero. If the scaling factors satisfy $|\alpha_n| < a_n^2 = \frac{h_n^2}{(t_N - t_1)^2}$, then as above discussion, we can get $\|\Psi - f_1[\mathbf{A}]\|_\infty = O(h^2)$. Thus, with some mild conditions on shape parameters and scaling parameters, the convergence properties of a RCCHFIF are similar to that of its classical counterpart. We close this section with the following remark.

Remark 5 Let the rational cubic interpolation q_n associated with the FIF $g_2[\mathbf{A}]$ satisfy $|\frac{\partial q_n(\varrho_n, x)}{\partial \gamma_n}| \le Z_0^{**}$ for $|\varrho_n| \in (0, a_n)$, $x \in I_n$, $n \in \mathbb{N}_{N-1}$. Then, in particular with $\alpha_n = \gamma_n$, $\beta_n = 0$, for all $n \in \mathbb{N}_{N-1}$, implies the following error estimates for $g_2[\mathbf{A}]$: $\|g_2[\mathbf{A}] - C^*\|_\infty \le \frac{|\gamma|_\infty}{1-|\gamma|_\infty}(\|C^*\|_\infty + Z_0^{**})$, where $|\gamma|_\infty = \max\{|\gamma_n| : n \in \mathbb{N}_{N-1}\}$ and C^* is the classical rational cubic interpolant for the data $\{(t_n, z_n, d_n^*) : n \in \mathbb{N}_N\}$. The main results of upper bound for $g_2[\mathbf{A}]$ can be obtained by replacing the above estimates with rightmost expression in (16).

6 Numerical Illustration

First, we extend the positive data $\{(t_n, x_n, d_n) = (0, 2, -4.86), (1, 0.07, 1), (2, 4, 3.47), (3, 7, 2.54)\}$ to a generalized set of data $\{(t_n, x_n, d_n, z_n, d_n^*) = (0, 2, -4.86, 1, 3.5), (1, 0.07, 1, 3, 0.5), (2, 4, 3.47, 2, 1), (3, 7, 2.54, 5, 5)\}$. The d_n, d_n^*, $n \in \{1, 2, 3, 4\}$ is estimated using the arithmetic mean method [11]. A nonpositive non-self-referential $\mathscr{C}^1$-RCSCHFIF is displayed in Fig. 1a with randomly chosen scaling parameters and shape parameters as $\alpha = (-0.4, 0.2, 0.2)$, $\beta = (0.004, 0.016, 0.2)$, $\gamma = (0.4004, 0.216, 0.4)$, $u = (0.2, 0.5, 0.5)$, and $v = (0.5, 0.5, 0.5)$, respectively. This illustrates the need of a process that ensures positive RCSCHFIF for a prescribed set of positive data. Now, we choose scaling and shape parameters as $\alpha = (0.004, 0.02, 0.2)$, β=(0.004, 0.01, 0.1), γ=(0.3, 0.3, 0.3), $u = (0.7182, 1, 0.15)$, and $v = (21, 1, 2)$, as the prescription given in Theorem 1. We obtain positive non-self-referential RCSCHFIF in Fig. 1b.

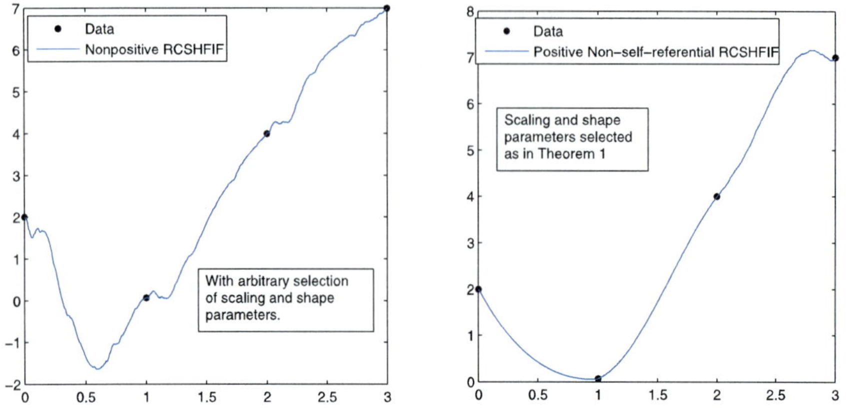

(a) Nonpositive RCSCHFIF (b) Positive RCSCHFIF

Fig. 1 RCSCHFIF

7 Closure

In the present work, a latest method is established for the construction of RCSCHFIF on a generalized dataset with the help of hidden variables. The method is $O(h)$ convergent. Meanwhile, the numerical examples indicate that our method can be also applied to generate nice feature-preserving space curves.

References

1. Barnsley, M.F.: Fractal functions and interpolation. Constr. Approx. **2**(1), 303–329 (1986)
2. Barnsley, M.F., Harrington, A.N.: The calculus of fractal interpolation functions. J. Approx. Theory **57**(1), 14–34 (1989)
3. Barnsley, M.F., Elton, J., Hardin, D., Massopust, P.: Hidden variable fractal interpolation functions. SIAM J. Math. Anal. **20**(5), 1218–1242 (1989)
4. Chand, A.K.B., Kapoor, G.P.: Spline coalescence hidden variable fractal interpolation function. J. Appl. Math. Article ID **36829**, 1–17 (2006)
5. Chand, A.K.B., Kapoor, G.P.: Smoothness analysis of coalescence hidden variable fractal interpolation functions. Int. J. Nonlinear Sci. **3**, 15–26 (2007)
6. Chand, A.K.B., Katiyar, S.K., Viswanathan, P.: Approximation using hidden variable fractal interpolation functions. J. Fractal Geom. **2**(1), 81–114 (2015)
7. Chand, A.K.B., Katiyar, S.K.: Quintic Hermite fractal interpolation in a strip: preserving copositivity. Springer Proc. Math. Stat. **143**, 463–475 (2015)
8. Chand, A.K.B., Navascus, M.A., Viswanathan, P., Katiyar, S.K.: Fractal trigonometric polynomials for restricted range approximation. Fractals **24**(2), 11 (2016)
9. Fritsch, F.N., Carlson, R.E.: Monotone piecewise cubic interpolations. SIAM J. Num. Anal. **17**(2), 238–246 (1980)

10. Gregory, J.A., Delbourgo, R.: Piecewise rational quadratic interpolation to monotonic data. IMA J. Numer. Anal. **2** (1982)
11. Gregory, J.A., Delbourgo, R.: Determination of derivative parameters for a monotonic rational quadratic interpolant. IMA J. Numer. Anal. **5**(1), 397–406 (1985)
12. Katiyar, S.K., Chand, A.K.B.: Toward a unified methodology for fractal extension of various shape preserving spline interpolants. Springer Proc. Math. Stat. **139**, 223–238 (2015)
13. Katiyar, S.K., Chand, A.K.B., Navascués, M.A.: Hidden variable **A**-fractal functions and their monotonicity aspects. Rev. R. Acad. Cienc. Zaragoza **71**, 7–30 (2016)
14. Katiyar, S.K., Reddy, K.M., Chand, A.K.B.: Constrained data visualization using rational bi-cubic fractal functions. Comp. Inform. Sci. Springer, 321–330 (2017)
15. Katiyar, S. K.: Shape preserving rational and coalescence fractal interpolation functions and approximation by variable scaling fractal functions. Ph.D. thesis, Indian Institute of Technology Madras, India (2017)
16. Katiyar, S.K., Chand, A.K.B., Saravana Kumar, G.: A new class of rational cubic spline fractal interpolation function and its constrained aspects. Appl. Math. Comp. **346**, 319–335 (2019)
17. Katiyar, S.K., Chand, A.K.B.: Shape preserving rational quartic fractal functions. Fractals. **27**(8) 1–15 (2019)
18. Katiyar, S.K., Chand, A.K.B.: A new class of monotone/convex rational fractal function. arXiv:1809.10682
19. Massopust, P.R.: Fractal Functions, Fractal Surfaces and Wavelets. Academic Press, Cambridge (1994)
20. Navascués, M.A., Viswanathan, P., Chand, A.K.B., Sebastián, M.V., Katiyar, S.K.: Fractal bases for Banach spaces of smooth functions. Bull. Aust. Math. Soc. **92**(3), 405–419 (2015)
21. Sarfraz, M., Hussain, M.Z.: Data visualization using rational spline interpolation. J. Comp. Appl. Math. **189**, 513–525 (2006)
22. Schimdt, J.W., Heß, W.: Positivity of cubic polynomial on intervals and positive spline interpolation. BIT Numer. Anal. **28**, 340–352 (1988)
23. Tian, M.: Monotonicity preserving piecewise rational cubic interpolation. Int. J. Math. Anal. **5**, 99–104 (2011)
24. Viswanathan, P., Chand, A.K.B.: A $\mathscr{C}^1$-Rational cubic fractal interpolation function: convergence and associated parameter identification problem. Acta Appl. Math. **136**, 19–41 (2014)

A Class of Sequence Spaces Defined by l-Fractional Difference Operator

Sanjay Kumar Mahto and P. D. Srivastava

Abstract The fractional order difference operator is generalized using l-Pochhammer symbol and l-fractional difference operator is defined. The l-fractional difference operator is further used to introduce a class of difference sequence spaces. Some properties and duals of the newly defined spaces are studied.

Keywords Sequence space · Normed space · Dual · Pochhammer symbol · Difference operator

1 Introduction

Fractional order difference operators defined on the set w of all sequences of complex numbers have recently got an attention to many researchers because of its applicability in the numerical analysis [3], statistical convergence [16], approximation theory [14], etc. Fractional order difference operators are basically generalizations of m-th-order difference operators, where m is a nonnegative integer. For a proper fraction a, Baliarsingh in [3] and Dutta and Baliarsingh in [8] introduced some fractional order difference operators Δ^a, $\Delta^{(a)}$, Δ^{-a}, and $\Delta^{(-a)}$. The fractional order difference operator $\Delta^{(a)} : w \to w$ is defined by

$$\Delta^{(a)}x = \left\{ \sum_{i=0}^{\infty} (-1)^i \frac{\Gamma(a+1)}{i!\Gamma(a-i+1)} x_{k-i} \right\}_k \tag{1}$$

S. K. Mahto (✉)
Department of Mathematics, R. N. A. R. College, Samastipur, Bihar 848101, India

Department of Mathematics, Indian Institute of Technology Kharagpur, Kharagpur 721302, India
e-mail: skmahto0777@gmail.com; kumarmahtosanjay@iitkgp.ac.in

P. D. Srivastava
Department of Mathematics, Indian Institute of Technology Bhilai, Chattisgarh 492015, India
e-mail: pds@maths.iitkgp.ac.in

D. Giri et al. (eds.), *Proceedings of the Fifth International Conference on Mathematics and Computing*, Advances in Intelligent Systems and Computing 1170, https://doi.org/10.1007/978-981-15-5411-7_17

for all $x = \{x_k\} \in w$. Assuming $x_{-s} = 0$ for all positive integers s, the fractional order difference operator (1) can be restated in terms of Pochhammer symbol as

$$\Delta^{(a)}x = \left\{ \sum_{i=0}^{k} \frac{(-a)_i}{i!} x_{k-i} \right\}_k, \tag{2}$$

where the Pochhammer symbol $(a)_k$ is defined by

$$(a)_k = \begin{cases} 1, & \text{when } k = 0 \\ a(a+1)(a+2)\dots(a+k-1), & \text{when } k \in \mathbb{N} \setminus \{0\} \end{cases}. \tag{3}$$

Difference operators are used to construct sequence spaces which are called difference sequence spaces. Kizmaz [17] was the first to develop the idea of difference sequence space. For the spaces $\mu = l_\infty$, c and c_0, he defined and studied some Banach spaces

$$\mu(\Delta) = \{x = \{x_k\} \in w : \Delta x \in \mu\}, \tag{4}$$

where $\Delta x = \{x_k - x_{k+1}\}$. Et and Colak [10] replaced the first-order difference operator by an m-th-order difference operator in $\mu(\Delta)$ and defined BK-spaces

$$\mu(\Delta^m) = \left\{x = \{x_k\} \in w : \Delta^m x \in \mu\right\}. \tag{5}$$

Later on, many authors such as Mursaleen [19], Et et al. [9], Aydin and Başar [1], Isik [13], Srivastava and Kumar [20], Maji and Srivastava [18], Bhardwaj and Bala [4], Djolović and Malkowsky [7], Chandra and Tripathy [5], Hazarika and Savas [12], Hazarika [11], etc. have generalized (4) and (5) and studied other difference sequence spaces. Recently, fractional difference operators have been used to define fractional difference sequence spaces such as in [2, 8, 15], etc.

One way to generalize the fractional order difference operator (2) is to apply l-Pochhammer symbol [6] instead of applying simply Pochhammer symbol. The l-Pochhammer symbol $(a)_{k,l}$, for a complex number a and a real number l, is introduced by Diaz and Pariguan in [6] as follows:

$$(a)_{k,l} = \begin{cases} 1, & \text{when } k = 0 \\ a(a+l)(a+2l)\dots(a+(k-1)l), & \text{when } k \in \mathbb{N} \setminus \{0\} \end{cases}. \tag{6}$$

In this paper, we introduce l-fractional difference operator $\Delta^{(a;l)}$ employing l-Pochhammer symbol. We further establish a class of sequence spaces $\mu(\Delta^{(a;l)}, v)$, where $\mu \in \{l_\infty, c_o, c\}$ which is defined by the use of l-fractional difference operator $\Delta^{(a;l)}$. We study some properties and determine α, β, and γ duals of the spaces $\mu(\Delta^{(a;l)}, v)$.

Throughout this paper, we use the following symbols: w, the set of all complex sequences; $\mathbb{C}$, the set of all complex numbers; $\mathbb{R}$, the set of all real numbers; $\mathbb{N}$, the set of all natural numbers; $\mathbb{N}_0 = \mathbb{N} \cup \{0\}$; l_∞, the space of all bounded sequences; c, the space of all convergent sequences; and c_0, the space of all null sequences.

2 l-Fractional Difference Operator

Now, we define l-fractional difference operator and study some properties of this operator. Results in this section are equivalent to the results in [2].

Definition 1 Let a be a real number. Then l-fractional difference operator $\Delta^{(a;l)} : w \to w$ is defined by

$$\Delta^{(a;l)} x = \left\{ \sum_{i=0}^{k} \frac{(-a)_{i,l}}{i!} x_{k-i} \right\}_k \tag{7}$$

for all $x = (x_k) \in w$. For convenience, we write the transformation of the kth term of a sequence $x = (x_k)$ by $\Delta^{(a;l)}$ as

$$\Delta^{(a;l)} x_k = \sum_{i=0}^{k} \frac{(-a)_{i,l}}{i!} x_{k-i}. \tag{8}$$

For some particular values of a and l, we have the following observations:

- For $l = 1$, the fractional difference operator $\Delta^{(a;l)}$ is reduced to the operator $\Delta^{(a)}$ (Eq. (2)).
- If $l = 1$ and $a = m$, a positive integer, then $\Delta^{(a;l)}$ is basically the m-th-order backward difference operator $\Delta^{(m)}$, where $\Delta^{(1)} x_k = x_k - x_{k-1}$ and $\Delta^{(m)} x_k = \Delta^{(m-1)}(\Delta^{(1)} x_k)$.
- $\Delta^{(a;a)} x_k = x_k - a x_{k-1}$.
- For $a = 2l$, we have $\Delta^{(a;l)} x_k = x_k - 2l x_{k-1} + l^2 x_{k-2}$.
- $\Delta^{(\frac{1}{2};\frac{1}{4})} x_k = x_k - \frac{1}{2} x_{k-1} + \frac{1}{16} x_{k-2}$.
- $\Delta^{(\frac{-1}{2};\frac{1}{4})} = x_k + \frac{1}{2} x_{k-1} + \frac{3}{16} x_{k-2} + \frac{3}{48} x_{k-3} + \frac{5}{256} x_{k-4} + \ldots + \frac{(\frac{-1}{2})(\frac{-1}{2}+\frac{1}{4})\ldots(\frac{-1}{2}+(k-1)\frac{1}{4})}{k!} x_0$.

Theorem 1 The l-fractional difference operator $\Delta^{(a;l)}$ is a linear operator.

Proof As the proof is straightforward, we omit it.

Theorem 2 *For real numbers a and b, the following results hold:*

1. $\Delta^{(a;l)}(\Delta^{(b;l)} x_k) = \Delta^{(a+b;l)} x_k = \Delta^{(b;l)}(\Delta^{(a;l)} x_k)$
2. $\Delta^{(a;l)}(\Delta^{(-a;l)} x_k) = x_k$.

Proof We give the proof of the first result only. The second result can be deduced from the first result. We consider

$$\Delta^{(a;l)}(\Delta^{(b;l)}x_k) = \Delta^{(a;l)} \sum_{i=0}^{k} \frac{(-b)_{i,l}}{i!} x_{k-i}. \tag{9}$$

As the l-fractional difference operator is linear,

$$\begin{aligned}\Delta^{(a;l)}(\Delta^{(b;l)}x_k) &= \sum_{i=0}^{k} \frac{(-b)_{i,l}}{i!} \Delta^{(a;l)} x_{k-i} \\ &= \sum_{i=0}^{k} \frac{(-b)_{i,l}}{i!} \sum_{j=0}^{k-i} \frac{(-a)_{j,l}}{j!} x_{k-i-j}.\end{aligned} \tag{10}$$

Taking the transformation $i + j = m$ or arranging the coefficients of x_{k-m} for $m = 0, 1, \ldots, k$, Eq. (10) is rewritten as

$$\Delta^{(a;l)}(\Delta^{(b;l)}x_k) = \sum_{m=0}^{k} \Big(\sum_{i=0}^{m} \frac{(-b)_{i,l}}{i!} \frac{(-a)_{(m-i),l}}{(m-i)!} \Big) x_{k-m}. \tag{11}$$

By mathematical induction on m, it is easy to verify that

$$\frac{(-(a+b))_{m,l}}{m!} = \Big(\sum_{i=0}^{m} \frac{(-b)_{i,l}}{i!} \frac{(-a)_{(m-i),l}}{(m-i)!} \Big). \tag{12}$$

Then Eqs. (11) and (12) imply that

$$\Delta^{(a;l)}(\Delta^{(b;l)}x_k) = \sum_{m=0}^{k} \frac{(-(a+b))_{m,l}}{m!} x_{k-m} = \Delta^{(a+b;l)} x_k.$$

Similarly, we can prove that

$$\Delta^{(b;l)}(\Delta^{(a;l)}x_k) = \Delta^{(a+b;l)} x_k.$$

This proves the theorem.

3 Sequence Spaces $\mu(\Delta^{(a;l)}, v)$ for $\mu = l_\infty, c_o$, and c

Now, we define a class of sequence spaces $\mu(\Delta^{(a;l)}, v)$, where $\mu \in \{l_\infty, c_o, c\}$ using l-fractional difference operator $\Delta^{(a;l)}$. Let V be the set of all real sequences $v = \{v_n\}$ such that $v_n \neq 0$ for all $n \in \mathbb{N}_0$. Then the class of sequence spaces $\mu(\Delta^{(a;l)}, v)$ is defined by

$$\mu(\Delta^{(a;l)}, v) = \left\{x = \{x_n\} \in w : \left\{\sum_{j=0}^{n} v_j \Delta^{(a;l)} x_j\right\}_n \in \mu\right\}.$$

Let y_n denotes the nth term of the sequence $\left\{\sum_{j=0}^{n} v_j \Delta^{(a;l)} x_j\right\}_n$, then

$$\begin{aligned} y_n &= \sum_{j=0}^{n} v_j \Delta^{(a;l)} x_j \\ &= \sum_{j=0}^{n} v_j \Big(\sum_{i=0}^{j} \frac{(-a)_{i,l}}{i!} x_{j-i}\Big). \end{aligned} \tag{13}$$

Taking the transformation $j - i = m$ or arranging the coefficients of x_m for $m = 0, 1, \ldots, n$, Eq. (13) can be rewritten as

$$y_n = \sum_{m=0}^{n} \Big(\sum_{i=0}^{n-m} \frac{(-a)_{i,l}}{i!} v_{i+m}\Big) x_m.$$

Now, we consider a matrix $C = (c_{nm})$ such that

$$c_{nm} = \begin{cases} \sum\limits_{i=0}^{n-m} \frac{(-a)_{i,l}}{i!} v_{i+m}, & \text{when } n \geq m \\ 0, & \text{when } n < m \end{cases}. \tag{14}$$

Then the sequence $\{y_n\}$ is C-transform of the sequence $\{x_n\}$.

Theorem 3 *The space $\mu(\Delta^{(a;l)}, v)$ for $\mu = l_\infty, c$ or c_0 is a normed linear space with respect to the norm $\|\cdot\|_{\mu(\Delta^{(a;l)}, v)}$ defined by*

$$\|x\|_{\mu(\Delta^{(a;l)}, v)} = \sup_n \left|\sum_{j=0}^{n} v_j \Delta^{(a;l)} x_j\right|.$$

Proof We observe that $\|\alpha x\|_{\mu(\Delta^{(a;l)}, v)} = |\alpha| \|x\|_{\mu(\Delta^{(a;l)}, v)}$ for all $\alpha \in \mathbb{C}$ and $\|x + y\|_{\mu(\Delta^{(a;l)}, v)} \leq \|x\|_{\mu(\Delta^{(a;l)}, v)} + \|y\|_{\mu(\Delta^{(a;l)}, v)}$ for all sequences $x = \{x_n\}$ and $y = \{y_n\}$ of $\mu(\Delta^{(a;l)}, v)$. Now suppose that $x = \{x_k\} \in \mu(\Delta^{(a;l)}, v)$ is such that $\|x\|_{\mu(\Delta^{(a;l)}, v)} = 0$.

Then

$$\sup_n |\sum_{j=0}^{n} v_j \Delta^{(a;l)} x_j| = 0.$$

This implies that

$$|\sum_{j=0}^{n} v_j \Delta^{(a;l)} x_j| = 0 \text{ for all } n \in \mathbb{N}_0.$$

This gives

$$\begin{aligned} v_0 \Delta^{(a;l)} x_0 &= 0 \\ v_1 \Delta^{(a;l)} x_1 &= 0 \\ &\vdots \end{aligned}$$

This shows that $x_n = 0$ for all $n \in \mathbb{N}_0$. That is, $x = (0, 0, \ldots)$. Thus $\|\cdot\|_{\mu(\Delta^{(a;l)}, v)}$ is a norm and $\mu(\Delta^{(a;l)}, v)$ is a normed linear space for $\mu = l_\infty, c_0$ or c.

Theorem 4 *The space $\mu(\Delta^{(a;l)}, v)$ for $\mu = l_\infty, c$ or c_0 is a complete normed linear space.*

Proof As proof of the theorem for the spaces $\mu(\Delta^{(a;l)}, v)$ for $\mu = l_\infty, c$, and c_0 run along similar lines, we just prove that the space $l_\infty(\Delta^{(a;l)}, v)$ is a complete normed linear space. For this, let $\{x^i\}_i = \{x^1 = \{x_k^1\}_{k=0}^\infty, x^2 = \{x_k^2\}_{k=0}^\infty, \ldots\}$ be a Cauchy sequence in $l_\infty(\Delta^{(a;l)}, v)$. Then, from definition of Cauchy sequence, there exists a natural number $n_0(\epsilon)$ corresponding to each real number $\epsilon > 0$ such that

$$\|x^i - x^m\|_{\mu(\Delta^{(a;l)}, v)} < \epsilon \tag{15}$$

for all $i, m \geq n_0(\epsilon)$. By definition of the norm $\|\cdot\|_{\mu(\Delta^{(a;l)}, v)}$, Eq. (15) gives

$$\sup_n \left| \sum_{j=0}^{n} v_j \Delta^{(a;l)} x_j^i - \sum_{j=0}^{n} v_j \Delta^{(a;l)} x_j^m \right| < \epsilon \tag{16}$$

for all $i, m \geq n_0(\epsilon)$. Inequality (16) can be rewritten with respect to C-transformation (see (14))as

$$\sup_n |(Cx^i)_n - (Cx^m)_n| < \epsilon$$

for all $i, m \geq n_0(\epsilon)$, where $(Cx)_n$ denotes the n-th term of the sequence $\{Cx\}$. This implies that for each nonnegative integer n

$$|(Cx^i)_n - (Cx^m)_n| < \epsilon \tag{17}$$

for all $i, m \geq n_0(\epsilon)$. Inequality (17) shows that for a fixed nonnegative integer n the sequence $\{(Cx^i)_n\}_i = \{(Cx^1)_n, (Cx^2)_n, \dots\}$ is a Cauchy sequence in the set of real numbers $\mathbb{R}$. As $\mathbb{R}$ is complete, the sequence $\{(Cx^i)_n\}_i$ converges to $(Cx)_n$ (say). Letting m tend to infinity in Inequality (17), we get

$$|(Cx^i)_n - (Cx)_n| < \epsilon \tag{18}$$

for all $i \geq n_0(\epsilon)$. Since Inequality (18) holds for each nonnegative integer n, we have

$$\sup_n |(Cx^i)_n - (Cx)_n| < \epsilon \tag{19}$$

for all $i \geq n_0(\epsilon)$, that is,

$$\|x^i - x\|_{\mu(\Delta^{(a;l)},v)} < \epsilon \tag{20}$$

for all $i \geq n_0(\epsilon)$. Now, it remains to show that $x = \{x_n\}$ belongs to $l_\infty(\Delta^{(a;l)}, v)$. Since the sequence $\{x^i\}$ belongs to $l_\infty(\Delta^{(a;l)}, v)$, there exists a real number M such that

$$\sup_n |(Cx^i)_n| \leq M. \tag{21}$$

With the help of Inequalities (19) and (21), we conclude that

$$\sup_n |(Cx)_n| = \sup_n |(Cx)_n - (Cx^i)_n| + \sup_n |(Cx^i)_n| \leq M + \epsilon. \tag{22}$$

It follows that $x = \{x_n\} \in l_\infty(\Delta^{(a;l)}, v)$. Thus, $l_\infty(\Delta^{(a;l)}, v)$ is a complete normed linear space.

Theorem 5 *The space $\mu(\Delta^{(a;l)}, v)$ (where $\mu = l_\infty, c_0$ or c) is linearly isomorphic to the space μ.*

Proof We need to show that there exists a mapping $T : \mu(\Delta^{(a;l)}, v) \to \mu$, which is linear and bijective. For our purpose, we suppose that T is defined by

$$Tx = \left\{ \sum_{j=0}^{n} v_j \Delta^{(a;l)} x_j \right\}_n = \{y_n\} \tag{23}$$

for all $x = \{x_n\} \in \mu(\Delta^{(a;l)}, v)$. We observe that the mapping T is linear and it satisfies the property "$Tx = \theta$ *implies* $x = \theta$," where $\theta = \{0, 0, \dots\}$. The property "$Tx = \theta$ *implies* $x = \theta$" shows that the mapping T is injective. Now it remains to prove that the mapping T is surjective. Therefore, we consider an element $\{y_n\} \in \mu$ and determine a sequence $\{x_n\}$ such that

$$x_n = \Delta^{(-a;l)} \left(\frac{y_n - y_{n-1}}{v_n} \right) \tag{24}$$

for all $n \in \mathbb{N}_0$. Then we see that

$$\sum_{j=0}^{n} v_j \Delta^{(a;l)} x_j = \sum_{j=0}^{n} v_j \Delta^{(a;l)} \Delta^{(-a;l)} \Big(\frac{y_j - y_{j-1}}{v_j} \Big). \tag{25}$$

By Theorem 2, Eq. (25) gives

$$\sum_{j=0}^{n} v_j \Delta^{(a;l)} x_j = \sum_{j=0}^{n} (y_j - y_{j-1}) = y_n. \tag{26}$$

This shows that $\left\{ \sum_{j=0}^{n} v_j \Delta^{(a;l)} x_j \right\}_n = \{y_n\} \in \mu$. Using Definition (1), we have $\{x_n\} \in \mu(\Delta^{(a;l)}, v)$. Thus, we have shown that for every sequence $\{y_n\} \in \mu$, there exists a sequence $\{x_n\} \in \mu(\Delta^{(a;l)}, v)$. That is, the mapping T is a surjective mapping. Hence, the space $\mu(\Delta^{(a;l)}, v)$ is linearly isomorphic to the space μ.

4 $\alpha-$, $\beta-$, and $\gamma-$ Duals of the Spaces $\mu(\Delta^{(a;l)}, v)$ for $\mu = l_\infty, c_o$, and c

Now, we determine $\alpha-$, $\beta-$, and $\gamma-$ duals of the spaces $\mu(\Delta^{(a;l)}, v)$ for $\mu = l_\infty, c_o$ and c. For two sequence spaces μ and λ, the multiplier space of μ and λ is defined by

$$S(\mu, \lambda) = \big\{ z = \{z_k\} \in w : zx = \{z_k x_k\} \in \lambda \text{ for all } x = \{x_k\} \in \mu \big\}.$$

In particular, $\alpha-$, $\beta-$, and $\gamma-$ duals of a space μ are defined by $\mu^\alpha = S(\mu, l_1)$, $\mu^\beta = S(\mu, cs)$, and $\mu^\gamma = S(\mu, bs)$, respectively, where l_1, cs, and bs are the space of absolutely summable sequences, the space of convergent series, and the space of bounded series, respectively. Let $B = (b_{nk})$ be an infinite matrix, then we have the following lemmas:

Lemma 1 ([21]) $B \in (l_\infty : l_1) = (c_0 : l_1) = (c : l_1)$ *if and only if*

$$\sup_{K \in F} \sum_n \Big| \sum_{k \in K} b_{nk} \Big| < \infty. \tag{27}$$

Lemma 2 ([21]) $B \in (l_\infty : c)$ *if and only if*

$$\lim_{n \to \infty} b_{nk} \text{ exists for all } k \textit{ and} \tag{28}$$

$$\lim_n \sum_k |b_{nk}| = \sum_k |\lim_n b_{nk}|. \tag{29}$$

Lemma 3 ([21]) *$B \in (c : c)$ if and only if (28),*

$$\sup_n \sum_k |b_{nk}| < \infty \text{ and} \tag{30}$$

$$\lim_n \sum_k b_{nk} \text{ exists.} \tag{31}$$

Lemma 4 ([21]) *$B \in (c_0 : c)$ if and only if (28) and (30) hold.*

Lemma 5 ([21]) *$B \in (l_\infty : l_\infty) = (c : l_\infty) = (c_0 : l_\infty)$ if and only if (30) holds.*

Now, we determine the duals of the spaces $\mu(\Delta^{(a;l)}, v)$ for $\mu = l_\infty, c_o$, and c through following theorem.

Theorem 6 *Let $D = (d_{nk})$ and $E = (e_{nk})$ are two matrices such that*

$$d_{nk} = \begin{cases} 0; & (k > n) \\ \frac{z_k}{v_k}; & (k = n) \quad \text{and} \\ z_n\left(\frac{(a)_{n-k,l}}{(n-k)!v_k} - \frac{(a)_{n-k-1,l}}{(n-k-1)!v_{k+1}}\right); & (k < n) \end{cases} \tag{32}$$

$$e_{nk} = \begin{cases} 0; & (k > n) \\ \frac{z_k}{v_k}; & (k = n) \\ \frac{1}{v_k}\sum_{i=0}^{n-k} \frac{(a)_{i,l}}{i!} z_{k+i} - \frac{1}{v_{k+1}} \sum_{i=0}^{n-k-1} \frac{(a)_{i,l}}{i!} z_{k+i+1}; & (k < n) \end{cases} \tag{33}$$

and consider the sets A_1, A_2, A_3, A_4, and A_5 which are defined as follows:

$$A_1 = \left\{z = (z_k) \in w : \sup_{K \in F} \sum_n \left| \sum_{k \in K} d_{nk} \right| < \infty \right\},$$

$$A_2 = \left\{z = (z_k) \in w : \lim_{n \to \infty} e_{nk} \text{ exists for all } k \right\},$$

$$A_3 = \left\{z = (z_k) \in w : \lim_n \sum_k |e_{nk}| = \sum_k |\lim_n e_{nk}| \right\},$$

$$A_4 = \left\{z = (z_k) \in w : \sup_n \sum_k |e_{nk}| < \infty \right\} \text{ and}$$

$$A_5 = \left\{z = (z_k) \in w : \lim_n \sum_k e_{nk} \text{ exists} \right\}.$$

Then,

$$\left\{l_\infty(\Delta^{(a;l)}, v)\right\}^\alpha = A_1,\ \left\{l_\infty(\Delta^{(a;l)}, v)\right\}^\beta = A_2 \cap A_3,\ \left\{l_\infty(\Delta^{(a;l)}, v)\right\}^\gamma = A_4$$

$$\left\{c_0(\Delta^{(a;l)}, v)\right\}^\alpha = A_1,\ \left\{c_0(\Delta^{(a;l)}, v)\right\}^\beta = A_2 \cap A_4,\ \left\{c_0(\Delta^{(a;l)}, v)\right\}^\gamma = A_4$$

$$\left\{c(\Delta^{(a;l)}, v)\right\}^\alpha = A_1,\ \left\{c(\Delta^{(a;l)}, v)\right\}^\beta = A_2 \cap A_4 \cap A_5 \text{ and } \left\{c(\Delta^{(a;l)}, v)\right\}^\gamma = A_4.$$

Proof We determine duals of the space $c_0(\Delta^{(a;l)}, v)$. Duals of other spaces $l_\infty(\Delta^{(a;l)}, v)$ and $c(\Delta^{(a;l)}, v)$ can be deduced on the similar lines as that of the $c_0(\Delta^{(a;l)}, v)$. By Eq. (24), we obtain that

$$z_n x_n = z_n \Delta^{(-a;l)}\Big(\frac{y_n - y_{n-1}}{v_n}\Big) = (Dy)_n \tag{34}$$

for all $\{x_n\} \in c_0(\Delta^{(a;l)}, v)$ and $\{y_n\} \in c_0$. Now the sequence $\{z_n\} \in \{c_0(\Delta^{(a;l)}, v)\}^\alpha$ iff $\{z_n x_n\} \in l_1$ for all $\{x_n\} \in c_0(\Delta^{(a;l)}, v)$. That is, $\{z_n\} \in \{c_0(\Delta^{(a;l)}, v)\}^\alpha$ iff $Dy \in l_1$ for all $\{y_n\} \in c_0$. In other words, $\{z_n\} \in \{c_0(\Delta^{(a;l)}, v)\}^\alpha$ iff $D \in (c_0 : l_1)$. Keeping this in view and using Lemma 1, we conclude that

$$\{c_0(\Delta^{(a;l)}, v)\}^\alpha = A_1. \tag{35}$$

To find β-dual, we define a sequence $\{t_n\}$ as follows:

$$t_n = \sum_{r=0}^{n} z_r x_r = (Ey)_n. \tag{36}$$

Then the sequence $\{z_n\} \in \{c_0(\Delta^{(a;l)}, v)\}^\beta$ iff $\{t_n\} \in c$ for all $\{x_n\} \in c_0(\Delta^{(a;l)}, v)$. That is, $\{z_n\} \in \{c_0(\Delta^{(a;l)}, v)\}^\beta$ iff $Ey \in c$ for all $\{y_n\} \in c_0$. In other words, $\{z_n\} \in \{c_0(\Delta^{(a;l)}, v)\}^\beta$ iff $E \in (c_0 : c)$. Keeping this in view and using Lemma 4, we conclude that

$$\{c_0(\Delta^{(a;l)}, v)\}^\beta = A_2 \cap A_4. \tag{37}$$

Similarly, using Lemma 5, we get

$$\{c_0(\Delta^{(a;l)}, v)\}^\gamma = A_4. \tag{38}$$

5 Conclusions

By introducing l-fractional difference operator, we have given a more general fractional order difference operator. Also, we have defined a class of difference sequence spaces with the help of l-fractional difference operator. Further, we have studied some topological properties and determined the duals of the spaces.

References

1. Aydın, C., Başar, F.: Some new difference sequence spaces. Appl. Math. Comput. **157**(3), 677–693 (2004)
2. Baliarsingh, P.: Some new difference sequence spaces of fractional order and their dual spaces. Appl. Math. Comput. **219**(18), 9737–9742 (2013)
3. Baliarsingh, P.: A unifying approach to the difference operators and their applications. Boletim da Sociedade Paranaense de Matemática **33**(1), 49–56 (2013)
4. Bhardwaj, V., Bala, I.: Generalized difference sequence space defined by $|\overline{N}, p_k|$ summability and an orlicz function in seminormed space. Math. Slovaca **60**(2), 257–264 (2010)
5. Chandra, P., Tripathy, B.: On generalised kothe-toeplitz duals of some sequence spaces. Indian J. Pure Appl. Math. **33**(8), 1301–1306 (2002)
6. Dıaz, R., Pariguan, E.: On hypergeometric functions and pochhammer k-symbol. Divulg. Matemtıcas **15**(2), 179–192 (2007)
7. Djolović, I., Malkowsky, E.: Characterizations of compact operators on some euler spaces of difference sequences of order m. Acta Math. Sci. **31**(4), 1465–1474 (2011)
8. Dutta, S., Baliarsingh, P.: A note on paranormed difference sequence spaces of fractional order and their matrix transformations. J. Egypt. Math. Soc. **22**(2), 249–253 (2014)
9. Et, M., Altinok, H., Altin, Y.: On some generalized sequence spaces. Appl. Math. Comput. **154**(1), 167–173 (2004)
10. Et, M., Çolak, R.: On some generalized difference sequence spaces. Soochow J. Math. **21**(4), 377–386 (1995)
11. Hazarika, B.: On difference ideal convergence of double sequences in random 2-normed spaces. Acta Math. Vietnam. **39**(3), 393–404 (2014)
12. Hazarika, B., Savas, E.: Some i-convergent lambda-summable difference sequence spaces of fuzzy real numbers defined by a sequence of orlicz functions. Math. Comput. Modell. **54**(11), 2986–2998 (2011)
13. Işik, M.: Generalized vector-valued sequence spaces defined by modulus functions. J. Inequalities Appl. **2010**(1), 457892 (2010)
14. Kadak, U.: Generalized weighted invariant mean based on fractional difference operator with applications to approximation theorems for functions of two variables. Results Math., pp. 1–22
15. Kadak, U.: Generalized lacunary statistical difference sequence spaces of fractional order. Int. J. Math. Math. Sci. **2015** (2015)
16. Kadak, U.: Weighted statistical convergence based on generalized difference operator involving (p, q)-gamma function and its applications to approximation theorems. J. Math. Anal. Appl. **448**(2), 1633–1650 (2017)
17. Kizmaz, H.: Certain sequence spaces. Can. Math. Bull. **24**(2), 169–176 (1981)
18. Maji, A., Srivastava, P.D.: On $b^{(m)}$-difference sequence spaces using generalized means and compact operators. Analysis **34**(3), 257–281 (2014)
19. Mursaleen: Generalized spaces of difference sequences. J. Math. Anal. Appl. **203**(3), 738–745 (1996)
20. Srivastava, P., Kumar, S.: Generalized vector-valued paranormed sequence space using modulus function. Appl. Math. Comput. **215**(12), 4110–4118 (2010)
21. Stieglitz, M., Tietz, H.: Matrixtransformationen von folgenräumen eine ergebnisübersicht. Math. Z. **154**(1), 1–16 (1977)

Zipper Rational Quadratic Fractal Interpolation Functions

Sangita Jha and A. K. B. Chand

Abstract In this article, we propose an interpolation method using a binary parameter called signature such that the graph of the interpolant is an attractor of a suitable zipper rational iterated function system. The presence of scaling factors and signature in the proposed zipper rational quadratic fractal interpolation functions (ZRQFIFs) gives the flexibility to produce a wide variety of interpolants. Using suitable conditions on the scale factor and shape parameter, we construct a $\mathcal{C}^1$-continuous ZRQFIF from a $\mathcal{C}^0$-continuous ZRQFIF. We also establish the uniform convergence of ZRQFIF to an original data-generating function. Further, we deduce suitable conditions on the IFS parameters and shape parameters to retain the positivity feature associated with a prescribed data by the proposed interpolant.

Keywords Zipper · Fractal interpolation · Rational quadratic FIF · Convergence · Positivity

1 Introduction

Iterated function systems (IFSs) have been at the heart of fractal geometry almost from its origin. The first systematic work on IFS was started by Hutchinson in [12]. IFSs form the foundation for constructing and characterizing self-referential fractals. Barnsley [3] introduced the concept of fractal interpolation function (FIF) using IFS. Also, one can construct fractals using flexible methods of IFS known as zipper

A. K. B. Chand is thankful for the project: MTR/2017/000574—MATRICS from the Science and Engineering Research Board (SERB), Government of India.

S. Jha (✉)
Department of Mathematics, NIT Rourkela, Rourkela, Odisha 769008, India
e-mail: jhasa@nitrkl.ac.in

A. K. B. Chand (✉)
Department of Mathematics, IIT Madras, Chennai 600036, India
e-mail: chand@iitm.ac.in

D. Giri et al. (eds.), *Proceedings of the Fifth International Conference on Mathematics and Computing*, Advances in Intelligent Systems and Computing 1170, https://doi.org/10.1007/978-981-15-5411-7_18

[19, 20]. The attractor of the IFS is the graph of a FIF which interpolates a given set of data points. Fractal functions include some familiar non-differentiable functions such as the Weierstrass function, space-filling curves and Takagi curves. Fractal interpolation function (FIF) constitutes an advance in the technique of approximation, since all the traditional functions of real-world data interpolation can be generalized through fractal techniques (see, for instance [6, 16]). In the theory of FIFs, many researchers have generalized the notion of FIF in different ways [4, 13–15]. Barnsley and Harrington studied the calculus of FIFs and initiated the concept of fractal splines. Fractal cubic splines have been developed in [6], where the FIFs are constructed using polynomial IFSs. The advantage of a polynomial FIF is that one can construct an infinite number of interpolants depending on the choice of the scale vector under the given boundary conditions. Chand and his group [7, 9, 10] have found that for preserving shapes of the original data, rational FIFs perform better than polynomial FIFs.

The construction of fractals using the notion of zipper was studied in detail by Aseev [2]. Tetenov [20] has studied multi-zipper construction, where he gave a detailed description of self-similar Jordan arcs in $\mathbb{R}^n$. Recently, Chand et al. [8] developed affine zipper fractal interpolation functions. The cubic zipper FIFs are studied in detail in [17]. The class of zipper FIFs is larger than the class of traditional FIFs. Since rational functions with shape parameters are an ideal tool for preserving the shape of the original data set, we construct zipper rational quadratic fractal interpolation function (ZRQFIF). In particular, we construct zipper rational quadratic FIF, where the interpolant consists of rational functions of the form $\frac{p_i(x)}{q_i(x)}$, where $p_i(x)$ are quadratic polynomials and $q_i(x)$ are preassigned quadratic polynomials that consist of one free shape parameter. Due to the presence of scaling vector and signature in the FIF, we obtain a wide variety of interpolants. If we choose all the scaling vectors and signatures as 0, then the ZRQFIF reduces to the classical quadratic rational function [11]. The proposed interpolant has zeroth-order continuity. The order of continuity is elevated up to first order by suitable selection of the shape parameters in each sub-interval. The proposed zipper fractal interpolation scheme is suitable to retain the positivity feature of any arbitrary complicated data set. Also, the proposed interpolant can be used to the Hermite data generating from a function whose derivative has irregularity in a dense subset of the domain of interpolation. The interpolants proposed here can be applied in different areas, namely, computer-aided geometric design, image compression, data visualization, and modern data analysis.

The paper is developed as follows. In Sect. 2, preliminaries of the zipper are collected and a univariate interpolation scheme is reviewed. In Sect. 3, we introduce the zipper rational quadratic spline FIFs with one family of shape parameters. Numerical examples are given to demonstrate the results. Then we establish the uniform convergence of ZRQFIF to an original function in Sect. 4. Suitable values of IFS parameters and shape parameters are deduced so that ZRQFIF generates a positive zipper fractal interpolant for a given positive data set in Sect. 5.

2 Fractal Interpolation Function Using Zipper

In this section, we review the construction of zipper fractal interpolation function from [8]. First we recall the definition of zipper.

Definition 1 Let w_i, $i = 1, 2, \ldots, n$ be non-surjective maps on a complete metric space X. Then the system $\mathcal{I} = \{X; w_i : i = 1, 2, \ldots, n\}$ is called a zipper [2] with vertices $(v_0, v_1, \ldots, v_n)$ and signature $\boldsymbol{\varepsilon} = (\varepsilon_1, \varepsilon_2, \ldots, \varepsilon_n) \in \{0, 1\}^n$ if for any $i = 1, 2, \ldots, n$,

$$w_i(v_0) = v_{i-1+\varepsilon_i}, \quad w_i(v_n) = v_{i-\varepsilon_i}.$$

The points v_0 and v_n are called the initial and the final point of the zipper. Any compact set $G \subset X$ satisfying the self-referential equation

$$G = \cup_{i=1}^{n} w_i(G)$$

is called the attractor or self-referential set corresponding to the zipper $\mathcal{I}$. Let $n > 2$ be a natural number and $\{(x_i, y_i) \in \mathbb{R}^2 : i = 0, 1, 2, \ldots, n\}$ be a given data set with $x_0 < x_1 < \cdots < x_n$. Let us denote the first n natural numbers by $\mathbb{N}_n$. Consider $I := [x_0, x_n]$ and $I_i := [x_{i-1}, x_i]$ for $i \in \mathbb{N}_n$. Let $L_i : I \to I_i$ be the affine map of the form $L_i(x) = a_i x + b_i$ and satisfy

$$L_i(x_0) = x_{i-1+\varepsilon_i}, \; L_i(x_n) = x_{i-\varepsilon_i}, \; i \in \mathbb{N}_n. \tag{1}$$

It is worth to note that I_i is contracted version of either I or flipped version of I. The value of a_i and b_i can be obtained from (1) as

$$a_i = \frac{x_{i-\varepsilon_i} - x_{i-1+\varepsilon_i}}{x_n - x_0}, \quad b_i = \frac{x_n x_{i-1+\varepsilon_i} - x_0 x_{i-\varepsilon_i}}{x_n - x_0}.$$

Let K be a compact subset of $\mathbb{R}$ containing the points y_i, $i = 0, 1, 2, \ldots, n$. Assume that $F_i : I \times K \to K$ is a function of the form

$$F_i(x, y) = \alpha_i y + q_i(x),$$

where α_i are scale factors and satisfy $|\alpha|_\infty = \max\{|\alpha_i| : i \in \mathbb{N}_n\} < 1$ and $q_i : I \to \mathbb{R}$ are continuous functions satisfying the "join-up" conditions:

$$F_i(x_0, y_0) = y_{i-1+\varepsilon_i}, \quad F_i(x_n, y_n) = y_{i-\varepsilon_i}, \quad i \in \mathbb{N}_n. \tag{2}$$

Here F_i contracted either the graph of a function or its flipped version from $[x_0, x_n]$ to $[x_{i-1}, x_i]$. The system $\mathcal{I} = \{I \times K; w_i(x, y) : i \in \mathbb{N}_n\}$, where $w_i(x, y) := (L_i(x), F_i(x, y))$ is a zipper with vertices $\{(x_i, y_i), i \in \mathbb{N}_n\}$ and signature $\boldsymbol{\varepsilon} \in \{0, 1\}^n$. Using Theorem 1 of [3], one can show that the zipper $\mathcal{I}$ has a unique attractor G and it is the graph of a continuous function that interpolates the data $\{(x_i, y_i) : i =$

$0, 1, 2, \ldots, n\}$. The corresponding interpolating function is called a zipper fractal interpolation function. To show explicitly the dependence on the signature $\boldsymbol{\varepsilon}$ and the scaling parameter $\alpha = (\alpha_1, \alpha_2 \ldots, \alpha_n) \in (-1, 1)^n$, we denote the zipper FIF by $f_{\boldsymbol{\varepsilon}}^{\alpha}$. The following results can be found from Ref. [8].

Theorem 1 *For the IFS $\mathcal{I} = \{I \times K; w_i : i = 1, 2, \ldots, n\}$, the following holds.*
(i) There exists a unique compact set $G^* \subset I \times K$ such that $G^* = \cup_{i=1}^{n} w_i(G^*)$.
(ii) G^* is the graph of a continuous function $f_{\boldsymbol{\varepsilon}}^{\alpha} : I \to \mathbb{R}$ which interpolates the data $\{(x_i, y_i) : i = 0, 1, 2, \ldots, n\}$, that is, $G^* = \{(x, f_{\boldsymbol{\varepsilon}}^{\alpha}(x)) : x \in I\}$ and $f_{\boldsymbol{\varepsilon}}^{\alpha}(x_i) = y_i$ for $i = 0, 1, 2, \ldots, n$.

3 Zipper Rational Quadratic FIF with Shape Parameters

In Sect. 3.1, we construct $\mathcal{C}^0$-continuous ZRQFIF containing one shape parameters in each sub-interval I_i using the IFS mentioned in Sect. 2. If we choose the derivative values at grid points suitably, then one can obtain $\mathcal{C}^1$-ZRQFIF in Sect. 3.2.

3.1 $\mathcal{C}^0$-Zipper Rational Quadratic FIF

Here we consider zipper rational IFS involving rational functions $r_{i,\varepsilon}(x) = \frac{A_{i,\varepsilon}(1-\theta)^2+B_{i,\varepsilon}\theta(1-\theta)+C_{i,\varepsilon}\theta^2}{1+(u_i-2)\theta(1-\theta)}$, $i \in \mathbb{N}_n$, $\theta = \frac{x-x_0}{x_n-x_0}$. The shape parameter u_i is selected such that it satisfies $u_i > -2$. Also, the condition $u_i > -2$ asserts strict positive denominator in the rational expression $r_{i,\varepsilon}$. For the construction of ZRQFIF we proceed as follows:

Let $\mathcal{G} = \{f \in \mathcal{C}(I) : f(x_0) = y_0, f(x_n) = y_n\}$. Then $\mathcal{G}$ is a complete metric space induced by the uniform norm on $\mathcal{G}$. Let $\alpha \in (-1, 1)^n$ and $\varepsilon \in \{0, 1\}^n$. Define the Read-Bajraktarević operator $T_{\varepsilon}^{\alpha} : \mathcal{G} \mapsto \mathcal{G}$ as

$$
\begin{aligned}
T_{\varepsilon}^{\alpha}(L_i(x)) &= F_i(x, f(x)) \\
&= \alpha_i f(x) + \frac{A_{i,\varepsilon}(1-\theta)^2 + B_{i,\varepsilon}\theta(1-\theta) + C_{i,\varepsilon}\theta^2}{1+(u_i-2)\theta(1-\theta)}, x \in I, i \in \mathbb{N}_n.
\end{aligned}
$$

It is easy to check that T_{ε}^{α} is a contraction on $\mathcal{G}$. The fixed point R_{ε}^{α} of T_{ε}^{α} satisfies the following functional equation:

$$
R_{\varepsilon}^{\alpha}(L_i(x)) = \alpha_i R_{\varepsilon}^{\alpha}(x) + \frac{A_{i,\varepsilon}(1-\theta)^2 + B_{i,\varepsilon}\theta(1-\theta) + C_{i,\varepsilon}\theta^2}{1+(u_i-2)\theta(1-\theta)}. \tag{3}
$$

The unknown coefficient in the expression of $r_{i,\varepsilon}$ can be determined using conditions (2), which is equivalent to determine from the relations $R_{\varepsilon}^{\alpha}(x_{i-1}) = y_{i-1+\varepsilon_i}$ and $R_{\varepsilon}^{\alpha}(x_i) = y_{i-\varepsilon_i}$. Again, to determine all the coefficients, we need to impose additional condition $R_\varepsilon^\alpha(x_i^+) = d_i$. It is known from [5] that for a simply differentiable FIF, we need to select the scale factor as $|\alpha_i| < a_i, i \in \mathbb{N}_n$. Now substituting $x = x_0$ in (3), we obtain

$$R_\varepsilon^\alpha(L_i(x_0)) = \alpha_i R_\varepsilon^\alpha(x_0) + A_{i,\varepsilon},$$
$$\Rightarrow R_{\varepsilon_i}^\alpha(x_{i-1+\varepsilon_i}) = \alpha_i y_0 + A_{i,\varepsilon},$$
$$\Rightarrow A_{i,\varepsilon_i} = y_{i-1+\varepsilon_i} - \alpha_i y_0.$$

Similarly, substituting $x = x_n$ in (3), we obtain

$$C_{i,\varepsilon} = y_{i-\varepsilon_i} - \alpha_i y_n.$$

Again, from (3), we have

$$a_i (R_\varepsilon^\alpha)^{(1)}(L_i(x)^+) = \alpha_i R_\varepsilon^\alpha(x^+) + r_{i,\varepsilon}^{(1)}(x).$$

Substituting $x = x_0$ in the above expression and using (3), we obtain

$$B_{i,\varepsilon} = u_i y_{i-1+\varepsilon_i} + h_i d_{i-1+\varepsilon_i} - \alpha_i[u_i y_0 + (x_n - x_0)d_0],$$

where $h_i = x_i - x_{i-1}$. Thus, with these values of $A_{i,\varepsilon}$, $B_{i,\varepsilon}$ and $C_{i,\varepsilon}$, we get the desired ZRQFIF in the form

$$R_\varepsilon^\alpha(L_i(x)) = \alpha_i R_\varepsilon^\alpha(x) + \frac{p_{i,\varepsilon}^*(x)}{q_{i,\varepsilon}^*(x)}, \tag{4}$$

$$p_{i,\varepsilon}^*(x) \equiv p_{i,\varepsilon}(\theta) = \{y_{i-1-\varepsilon_i} - \alpha_i y_0\}(1-\theta)^2 + \{u_i y_{i-1+\varepsilon_i} + h_i d_{i-1+\varepsilon_i} - \alpha_i[u_i y_0 + (x_n - x_0)d_0]\}\theta(1-\theta) + \{y_{i-\varepsilon_i} - \alpha_i y_n\}\theta^2,$$

$$q_{i,\varepsilon}^*(x) \equiv q_{i,\varepsilon}(\theta) = 1 + (u_i - 2)\theta(1-\theta).$$

To use the above FIF, we need to compute the derivative values $d_i, i \in \mathbb{N}_n$. If d_i are not given, then we can compute it by the arithmetic mean method (see [9]).

Remark 1 (i) If $\varepsilon_i = 0, i \in \mathbb{N}_n$ then ZRQFIF reduces to quadratic rational fractal interpolant [9].
(ii) If $\alpha_i = 0 = \epsilon_i, i \in \mathbb{N}_n$, then ZRQFIF reduces to the classical quadratic rational spline interpolant [11].
(iii) If all $\alpha_i = 0 = \epsilon_i, i \in \mathbb{N}_n$ and $u_i = 2$, then ZRQFIF reduces to classical quadratic spline interpolant [1].

3.2 $\mathcal{C}^1$-Zipper Rational Quadratic FIF

The construction given in Sect. 3 generates $\mathcal{C}^0$-continuous ZRQFIF. The quadratic polynomial $p_{i,\varepsilon}$ does not provide sufficient degree of freedom to make $(R_\varepsilon^\alpha)^{(1)}$ to be continuous by claiming $R_\varepsilon^\alpha(x_{i-1}) = d_{i-1}$ and $R_\varepsilon^\alpha(x_i) = d_i, i \in \mathbb{N}_n$. The following theorem provides the existence of $\mathcal{C}^k$-continuous zipper rational FIF. The proof of this can be found in Ref. [17].

Theorem 2 *Let* $\{(x_i, y_i), i = 0, 1, \dots, n\}$ *be a given interpolation data with increasing abscissae. For each* $i \in \mathbb{N}_n$, *let* L_i *be the affine map satisfying* (1). *For the polynomials* $p_{i,\varepsilon}^*$ *and* $q_{i,\varepsilon}^*$ *of suitable degrees, where* $q_{i,\varepsilon}^*(x) \neq 0 \,\forall\, x \in I$, *let* $r_{i,\varepsilon}(x) = \frac{p_{i,\varepsilon}^*(x)}{q_{i,\varepsilon}^*(x)}$ *and define* $F_i(x, y) = \alpha_i y + r_{i,\varepsilon}(x)$. *Suppose for some* $k \geq 0$, $|\alpha_i| < a_i^k$, $r_{i,\varepsilon} \in \mathcal{C}^k[x_0, x_n]$. *Let* $F_{i,k}(x, y) = \frac{\alpha_i y + r_{i,\varepsilon}^{(k)}(x)}{a_i^k}$ *and*

$$y_{0,k} = \begin{cases} \frac{r_{1,\varepsilon}^{(k)}(x_0)}{a_1^k - \alpha_1}, & \text{if } \varepsilon_1 = 0 \\ \frac{\alpha_1 y_{n,k} + r_{1,\varepsilon}^{(k)}(x_n)}{a_1^k}, & \text{if } \varepsilon_1 = 1 \end{cases}, \quad y_{n,k} = \begin{cases} \frac{r_{n,\varepsilon}^{(k)}(x_n)}{a_n^k - \alpha_n}, & \text{if } \varepsilon_n = 0 \\ \frac{\alpha_n y_{0,k} + r_{n,\varepsilon}^{(k)}(x_n)}{a_n^k}, & \text{if } \varepsilon_n = 1. \end{cases}$$

For $i \in \{2, 3, \dots, n\}$, *if*

$$F_{i-1,k}(x_n, y_{n,k}) = F_{i,k}(x_0, y_{0,k}), \varepsilon_{i-1} = 0, \varepsilon_i = 0;$$

$$F_{i-1,k}(x_n, y_{n,k}) = F_{i,k}(x_n, y_{n,k}), \varepsilon_{i-1} = 0, \varepsilon_i = 0;$$

$$F_{i-1,k}(x_0, y_{0,k}) = F_{i,k}(x_0, y_{0,k}), \varepsilon_{i-1} = 1, \varepsilon_i = 0;$$

$$F_{i-1,k}(x_0, y_{0,k}) = F_{i,k}(x_n, y_{n,k}), \varepsilon_{i-1} = 1, \varepsilon_i = 1,$$

then the rational IFS $\{I \times \mathbb{R}; L_i(x), F_i(x, y) : i \in \mathbb{N}_n\}$ *determines a zipper rational FIF* $R_\varepsilon^\alpha \in \mathcal{C}^k(I)$.

The following theorem gives the condition for R_ε^α to be $\mathcal{C}^1$-continuous.

Theorem 3 *Let* $\{(x_i, y_i) : i = 0, 1, \dots, n\}$ *be a given interpolation data and* R_ε^α *be the zipper rational quadratic FIF determined from* (4). *Then* R_ε^α *preserves* $\mathcal{C}^1$-*smoothness under the following conditions on the scaling and shape parameters:*

Choose α_i *such that*

(i) $|\alpha_i| < a_i, \; \alpha_i \neq \frac{y_{i-\varepsilon_i} - y_{i-1+\varepsilon_i}}{y_n - y_0}, \; \forall\, i \in \mathbb{N}_n$,

(ii) $u_i = \frac{h_i(d_{i-\varepsilon_i} + d_{i-1+\varepsilon_i}) - \alpha_i(x_n - x_0)(d_0 + d_n)}{y_{i-\varepsilon_i} - y_{i-1+\varepsilon_i} - \alpha_i(y_n - y_0)}$ *subject to* $u_i > -2 \,\forall\, i \in \mathbb{N}_n$.

Proof From Theorem 2, we know that for the construction of $\mathcal{C}^1$-smoothness zipper rational FIF, we need to impose the condition $|\alpha_i| < a_i$ for all $i \in \mathbb{N}_n$. We also need the following conditions on the maps F_i.

$$F_i(x_0, y_0) = y_{i-1+\varepsilon_i}, \; F_i(x_n, y_n) = y_{i-\varepsilon_i}, \; F_{i,1}(x_0, d_0) = d_{i-1+\varepsilon_i}, \; F_{i,1}(x_n, d_n) = d_{i-\varepsilon_i}. \tag{5}$$

The zipper rational quadratic FIF given in (4) satisfy the above constraints (5). From the conditions $F_{i,1}(x_n, d_n) = d_{i-\varepsilon_i}$, we obtain

$$a_i d_{i-\varepsilon_i} = \alpha_i d_n + \frac{(y_{i-\varepsilon_i} - \alpha_i y_n)u_i - u_i y_{i-1+\varepsilon_i} - h_i d_{i-1+\varepsilon_i} + \alpha_i[u_i y_0 + (x_n - x_0)d_0]}{x_n - x_0}. \quad (6)$$

After some basic computation, we deduce the condition on u_i as

$$u_i = \frac{h_i(d_{i-\varepsilon_i} + d_{i-1+\varepsilon_i}) - \alpha_i(x_n - x_0)(d_0 + d_n)}{y_{i-\varepsilon_i} - y_{i-1+\varepsilon_i} - \alpha_i(y_n - y_0)},$$

where α_i is chosen such that $y_{i-\varepsilon_i} - y_{i-1+\varepsilon_i} - \alpha_i(y_n - y_0) \neq 0$. The condition $u_i > -2$ is required to avoid the singularity.

Example 1 We give graphical examples to the flexibility and diversity of zipper rational FIF over the traditional rational FIF available in the literature. Consider the interpolation data $\{(0, 3), (1, 0.05), (2, 5), (3, 8)\}$. We have considered shape parameters as $u = (100, 100, 100)$ for Fig. 1a–e. The classical quadratic rational interpolant is generated in Fig. 1a by taking the signature $\varepsilon = (0, 0, 0)$ and $\alpha = (0, 0, 0)$. Next the zipper rational FIFs are generated in Fig. 1b–d by taking the signature as $(0, 0, 0)$, $(1, 0, 0)$ and $(1, 1, 1)$, respectively, with a fixed scaling vector $\alpha = (0.2, 0.34, 0.1)$. If we compare Fig. 1b with Fig. 1c and d, it can be noted that the changes in zipper rational FIF is occurring due to changes in the component of the signature. Again by taking signature same as $\varepsilon = (1, 1, 1)$ and scaling vector as $\alpha = (-0.5, 0.34, -0.45)$ we have generated Fig. 1e. Here we observe that major changes in the shape of the zipper rational FIF is happening because of the selection of different scaling vectors. It can be noted that Fig. 1b–e depict $\mathcal{C}^0$-continuous interpolant. To generate $\mathcal{C}^1$-continuous interpolant, we choose the parameters according to Theorem 3 as $\alpha = (0.2, 0.15, -0.1)$ and $u = (-1.5256, -1.2577, -1.8150)$. The corresponding figure is generated in Fig. 1f. It is worth to note that among all the figure generated from Fig. 1a–f, we can only construct Fig. 1b from the available rational FIF. Hence, the zipper rational FIF studied in the present article is more general than the rational FIF studied in the literature [9].

4 Convergence

This section devotes to deduce an upper bound between the zipper rational quadratic FIF and an original function $\Psi \in \mathcal{C}^1(I)$ which generates the data. The following theorem will be useful for finding the upper bound with respect to the uniform norm.

Theorem 4 ([9]) *Let $\Psi \in \mathcal{C}^1(I)$ be the original function which generates the interpolation data $\{(x_i, y_i) : i = 0, 1, 2, \ldots, n\}$. Let R^α be rational quadratic FIF*

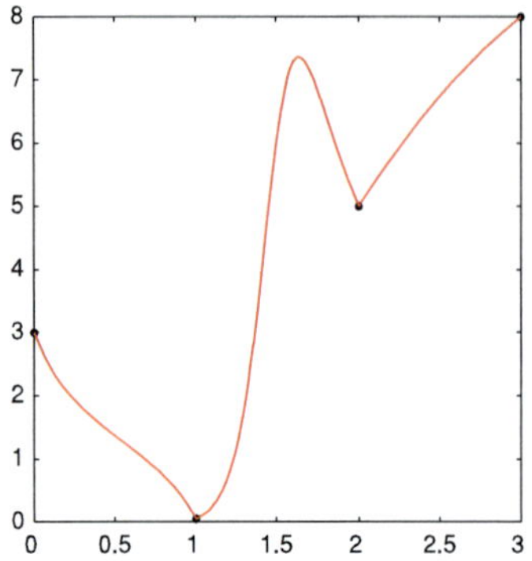

(a): Classical rational quadratic interpolant with $\varepsilon = (0, 0, 0)$ and $\alpha = (0, 0, 0)$.

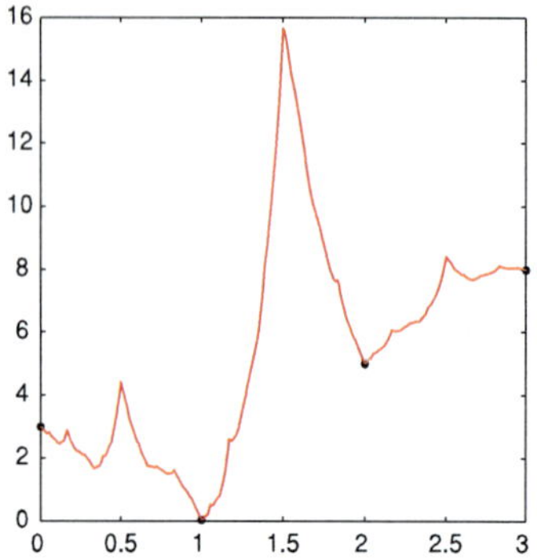

(b): Rational FIF with $\varepsilon = (0, 0, 0)$ and $\alpha = (0.2, 0.34, 0.1)$.

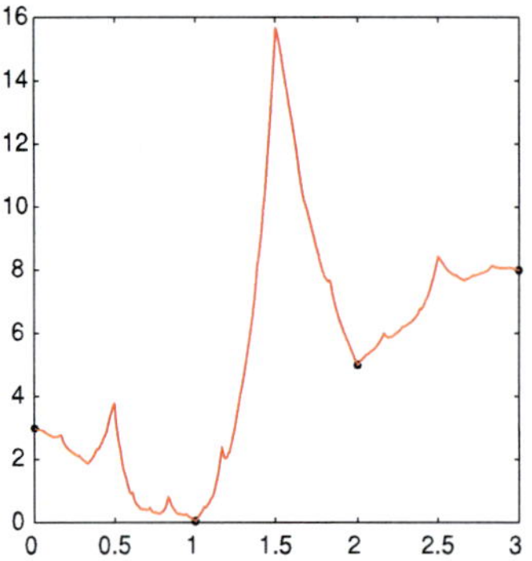

(c): Zipper rational FIF with $\varepsilon = (1, 0, 0)$ and $\alpha = (0.2, 0.34, 0.1)$.

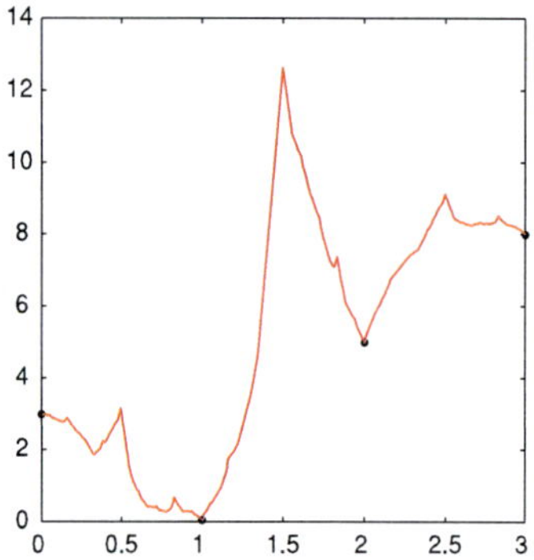

(d): Zipper rational FIF with $\varepsilon = (1, 1, 1)$ and $\alpha = (0.2, 0.34, 0.1)$.

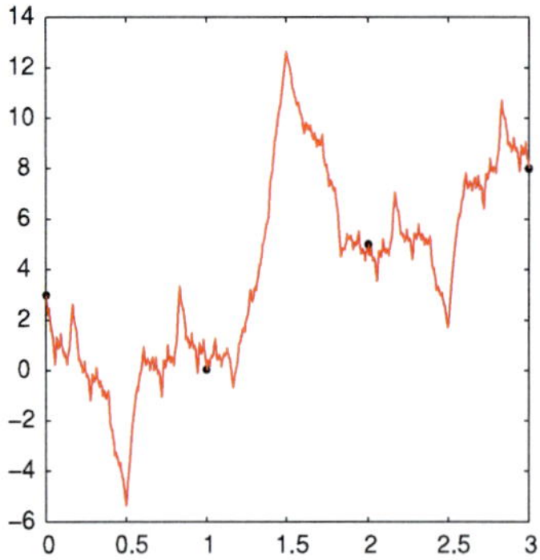

(e): Zipper rational FIF with $\varepsilon = (1, 1, 1)$ and $\alpha = (0.5, 0.34, -0.45)$.

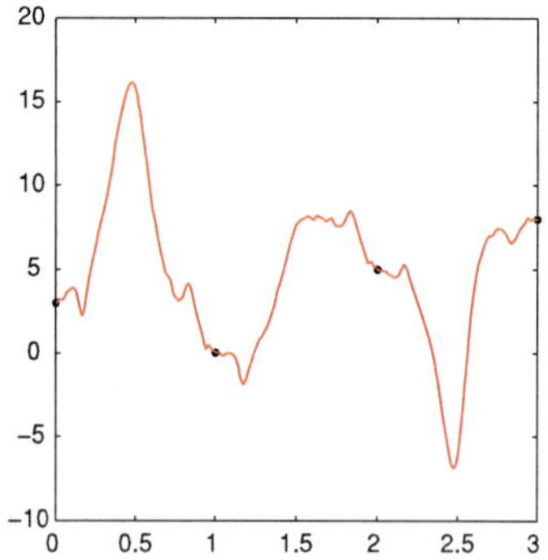

(f): $\mathcal{C}^1$-Zipper rational FIF with $\varepsilon = (1, 1, 1)$ and $\alpha = (0.2, 0.15, -0.1)$.

Fig. 1 Zipper rational quadratic FIFs R_{ε}^{α}

described in (4) *and* S *be the classical rational interpolation function. Suppose* r_i *satisfies* $\left|\frac{\partial r_i(\tau_i,u_i)}{\partial \alpha_i}\right| \leq D$ *for all* $i \in \mathbb{N}_n$*. Then,*

$$\|\Psi - R^\alpha\|_\infty \leq \frac{1}{4\beta}\{(|u|_\infty + 8)\omega(f,h) + h|d|_\infty\} + \frac{|\alpha|_\infty}{1-|\alpha|_\infty}(|S|_\infty + D),$$

where $\beta = \min\{|\beta_i| : i \in \mathbb{N}_n\}$, $h = \max\{h_i : i \in \mathbb{N}_n\}$, $|u|_\infty = \max\{|u_i| : i \in \mathbb{N}_n\}$, $|d|_\infty = \max\{|d_i| : 1 \leq i \leq n\}$ *and* $\omega(f,h) = \sup\limits_{|x_1-x_2|\leq h} \{\Psi(x_1) - \Psi(x_2) : x_1, x_2 \in I\}$.

Theorem 5 *Let* R^α_ε *and* R^α *be the zipper rational quadratic FIF and rational quadratic FIF, respectively, of the given interpolation data* $\{(x_i, y_i) : i = 0, 1, 2, \ldots, n\}$ *with scaling factors* $|\alpha_i| < a_i$. *Then for fixed signature* $\varepsilon \in \{0,1\}^n$,

$$\|R^\alpha_\varepsilon - R^\alpha\|_\infty \leq \frac{|\varepsilon|_\infty}{\lambda(1-|\alpha|_\infty)}\left(\omega(\Psi,h)(2+|u|_\infty) + h\omega(\Psi',h)\right),$$

where $\lambda_i = \min\{q_{i,\varepsilon}(\theta) : 0 \leq \theta \leq 1\}$, $\lambda = \min\{\lambda_i : i \in \mathbb{N}_n\}$ *and*

$$|\varepsilon|_\infty = \begin{cases} 0, & \textit{if } \varepsilon_i = 0, \\ 1, & \textit{if } \varepsilon_i \neq 0. \end{cases}$$

Proof Using the self-referential equation of R^α_ε and R^α, for all $x \in I_i$, we have

$$\begin{aligned}
|R^\alpha_\varepsilon(L_i(x)) - R^\alpha(L_i(x))| &= \Big|\alpha_i(R^\alpha_\varepsilon(x) - R^\alpha(x)) + \frac{(y_{i-1+\varepsilon_i} - y_{i-1})}{q_{i,\varepsilon}(\theta)}(1-\theta)^2 + \frac{(y_{i-\varepsilon_i} - y_i)}{q_{i,\varepsilon}(\theta)}\theta^2 \\
&\quad + \frac{(u_i y_{i-1+\varepsilon_i} + h_i d_{i-1+\varepsilon_i} - u_i y_{i-1} - h_i d_i)}{q_{i,\varepsilon}(\theta)}\theta(1-\theta)\Big| \\
&\leq |\alpha|\|R^\alpha_\varepsilon - R^\alpha\|_\infty + \frac{|y_{i-1+\varepsilon_i} - y_{i-1}|}{|q_{i,\varepsilon}(\theta)|} + \frac{|y_{i-\varepsilon_i} - y_i|}{|q_{i,\varepsilon}(\theta)|} \\
&\quad + \frac{|u_i(y_{i-1+\varepsilon_i} - y_{i-1})| + |h_i(d_{i-1+\varepsilon_i} - d_{i-1})|}{|q_{i,\varepsilon}(\theta)|} \\
&\leq |\alpha|_\infty\|R^\alpha_\varepsilon - R^\alpha\|_\infty + |\varepsilon_i|\frac{|y_i - y_{i-1}|}{\lambda} + |\varepsilon_i|\frac{|y_i - y_{i-1}|}{\lambda} \\
&\quad + |\varepsilon_i|\frac{|u|_\infty|y_i - y_{i-1}| + h|d_i - d_{i-1}|}{\lambda} \\
&\leq |\alpha|_\infty\|R^\alpha_\varepsilon - R^\alpha\|_\infty + |\varepsilon|_\infty\frac{2\omega(\Psi,h) + |u|_\infty\omega(\Psi,h) + h\omega(\Psi',h)}{\lambda},
\end{aligned}$$

which deduced the result.

Theorem 6 *Let* R^α_ε *be the zipper rational quadratic FIF for a given data-generating function* $\Psi \in C^1(I)$ *with fixed signature* $\varepsilon \in \{0,1\}^n$. *Then*

$$\begin{aligned}
\|\Psi - R^\alpha_\varepsilon\|_\infty &\leq \frac{1}{4\beta}\{(|u|_\infty + 8)\omega(f,h) + h|d|_\infty\} + \frac{|\alpha|_\infty}{1-|\alpha|_\infty}(|S|_\infty + D) \\
&\quad + \frac{|\varepsilon|_\infty}{\lambda(1-|\alpha|_\infty)}\left(\omega(\Psi,h)(2+|u|_\infty) + h\omega(\Psi',h)\right).
\end{aligned}$$

Proof Using Theorems 4 and 5 in the following

$$\|\Psi - R^\alpha_\varepsilon\|_\infty \le \|\Psi - R^\alpha\| + \|R^\alpha_\varepsilon - R^\alpha\|_\infty,$$

we obtain the desired result.

5 Positive Zipper Quadratic Rational FIF

The zipper rational quadratic FIF described in (4) may not always generate positive ZRQFIF for a given positive data set. In this section, we deduce the conditions on the scaling factor and shape parameters to construct a positive zipper fractal curve. The following result is required to find the positivity conditions for quadratic polynomials.

Proposition 1 ([18]) *The polynomial $P(z) = a_i + b_i z + c_i z^2$ is non-negative for all $x \ge 0$ if and only if $c \ge 0$, $a \ge 0$ and $b > -2\sqrt{ac}$.*

Theorem 7 *Consider a set of positive data $\{(x_i, y_i) : i = 0, 1, 2, \dots, n\}$ and let R^α_ε be the zipper rational quadratic FIF described in (4). Then, R^α_ε is also positive if the scaling factors and shape parameters satisfy the following for $i \in \mathbb{N}_n$:*

$$0 \le \alpha_i < \min\left\{\frac{y_{i-1+\varepsilon_i}}{y_0}, \frac{y_{i-\varepsilon_i}}{y_n}\right\},$$

$$u_i > \max\left\{-2, \frac{-h_i d_{i-\varepsilon_i} + \alpha_i d_0(x_n - x_0) - 2\sqrt{(y_{i-1+\varepsilon_i} - \alpha_i y_0)(y_{i-\varepsilon_i} - \alpha_i y_n)}}{y_{i-1+\varepsilon_i} - \alpha_i y_0}\right\}.$$

Proof Here $u_i > -2$ is required for the strict positive denominator of $r_{i,\varepsilon}(x)$. Let $G(R^\alpha_\varepsilon)$ be the graph of the rational quadratic zipper FIF R^α_ε on I, that is, $R(f^\alpha_\varepsilon) = \{(x, R^\alpha_\varepsilon(x)) : x \in I\}$. From the construction of R^α_ε given in Sect. 3, it follows that $G(R^\alpha_\varepsilon)$ is the attractor of the zipper $\{I \times K; (L_i(x), F_i(x, y)) : i = 1, 2, \dots, n\}$, where $F_i(x, y) = \alpha_i y + r_{i,\varepsilon}(x)$ and the maps $r_{i,\varepsilon}$ are given as earlier.

Since the zipper rational FIF R^α_ε is obtained by iterating the zipper, and I is the attractor of the IFS $\{I; L_i : i = 1, 2, \dots, n\}$, we have the following. To prove $R^\alpha_\varepsilon > 0$, it is sufficient to prove that $R^\alpha_\varepsilon(L_i(x)) > 0$ for all $i \in \mathbb{N}_n$, $x \in I$, whenever $R^\alpha_\varepsilon(x) > 0$, $x \in I$. Now assume that $\alpha_i > 0$ for all $i \in \mathbb{N}_n$ and $R^\alpha_\varepsilon(x) > 0$. It can be noted that with all the mentioned hypotheses on the IFS parameters, $R^\alpha_\varepsilon(L_i(x))$ will be positive if $p_{i,\varepsilon}(\theta) > 0$ for all $\theta \in [0, 1]$. Consider the substitution $\theta = \frac{z}{z+1}$. Then the aforementioned conditions on $p_{i,\varepsilon} > 0$ is transformed to finding the condition such that $P(z, \varepsilon) = a_{i,\varepsilon} + b_{i,\varepsilon} z + c_{i,\varepsilon} z^2 > 0$ for all $z \ge 0$. Hence, using Proposition 1, we obtain the following restrictions as

$$y_{i-1+\varepsilon_i} - \alpha_i y_0 > 0,$$

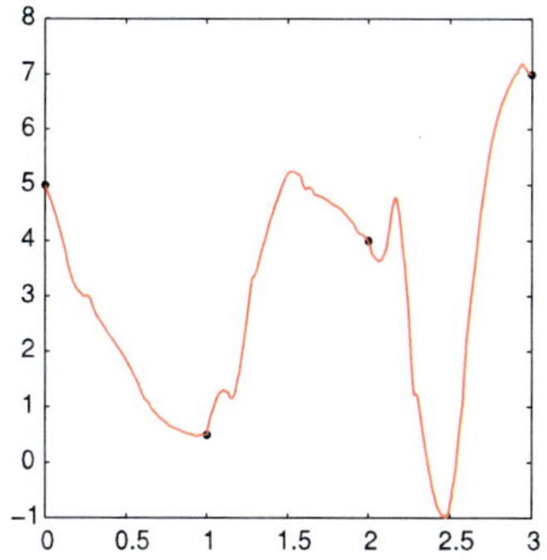

(a): Non-positive rational quadratic interpolant with $\varepsilon = (1, 1, 1)$ and $\alpha = (0.06, 0.16, -0.32), u = (2, 1, -1)$.

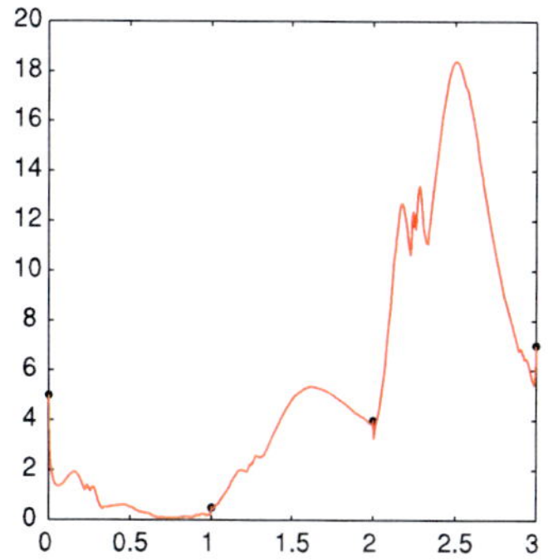

(b): Positive rational quadratic FIF with $\varepsilon = (1, 1, 1)$ and $\alpha = (0.091, 0.06, 0.5), u = (117.6290, -0.5725, -1)$.

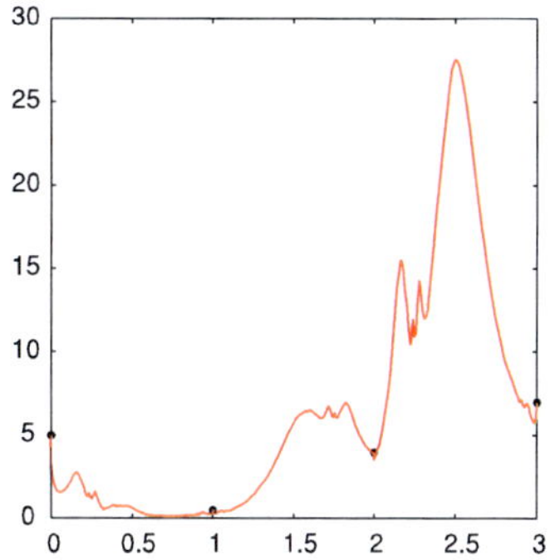

(c): Positive zipper rational quadratic FIF with $\varepsilon = (1, 0, 1)$ and $\alpha = (0.088, 0.09, 0.44), u = (87.1709, -1, -1.5)$.

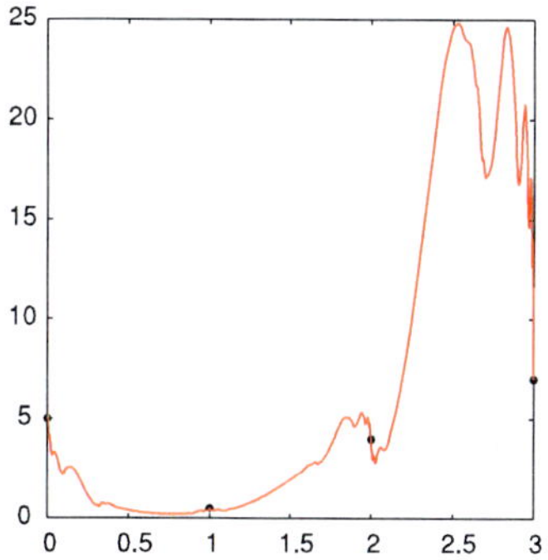

(d): Positive zipper rational quadratic FIF with $\varepsilon = (1, 0, 0)$ and $\alpha = (0.078, 0.08, 0.7), u = (46.4644, 1, -1)$.

Fig. 2 Positive and non-positive zipper rational FIFs R_ε^α

$$y_{i-\varepsilon_i} - \alpha_i y_n > 0,$$

$$u_i(y_{i-1+\varepsilon_i} - \alpha_i y_0) + h_i d_i - \alpha_i d_0(x_n - x_0) > -2\sqrt{(y_{i-1+\varepsilon_i} - \alpha_i y_0)(y_{i-\varepsilon_i} - \alpha_i y_n)}.$$

The required conditions can be obtained from the above restrictions.

We give the following example to implement the above procedure.

Example 2 Consider a set of positive data set $\{(0, 5), (1, 0.5), (2, 4), (3, 7)\}$. We will generate positive rational FIF associated with this positive data by choosing suitable values of parameters according to Theorem 7. We use the arithmetic mean method to compute the derivative values. It can be noted that the proposed interpolant is very sensitive to scale factors, shape parameters and signatures. For $n+1$-data set, we can get maximum 2^n shape-preserving interpolants by varying signature with fixed scale and shape parameters. For an arbitrary choice of $u = (2, 1, -1)$ and $\alpha = (0.06, 0.16, -0.32)$, the corresponding figure is generated in Fig. 2a which is non-positive. To obtain positive zipper interpolant we calculate the parameters using Theorem 7 as $\alpha = (0.091, 0.06, 0.5)$, $u = (117.6290, -0.5725, -1)$ for the signature $\varepsilon = (1, 1, 1)$. The corresponding ZRQFIF is generated in Fig. 2b. Now, we change the signature as $\varepsilon = (1, 0, 1)$ and select the parameters value according to Theorem 7 as $\alpha = (0.088, 0.09, 0.44)$, $u = (87.1709, -1, -1.5)$. The associated ZRQFIF is generated in Fig. 2c. Again if we change the signature as $\epsilon = (1, 0, 0)$, the restrictions of the shape parameters and scale vectors will change accordingly. In this case, we select $\alpha = (0.078, 0.08, 0.7)$, $u = (46.4644, 1, -1)$ using Theorem 7 and generate the ZRQFIF in Fig. 2d. It can be noted that the changes in the fractal interpolant is occurring due to the changes of the component in the signature.

References

1. Ahlberg, J.H., Nilson, E.N., Walsh, J.L.: The Theory of Splines and Their Applications: Mathematics in Science and Engineering: A Series of Monographs and Textbooks, vol. 38. Elsevier, Amsterdam (2016)
2. Aseev, V.V.: On the regularity of self-similar zippers. Ü TThe 6-th Russian-Korean International Symposium on Science and Technology. KORUS-2002 (June 24–30, 2002. Novosibirsk State Technical University, Russia). Part 3 (Ab-stracts) p. 167 (2002)
3. Barnsley, M.F.: Fractal functions and interpolation. Constr. Approx. **2**(4), 303–329 (1986)
4. Barnsley, M.F., Elton, J., Hardin, D., Massopust, P.: Hidden variable fractal interpolation functions. SIAM J. Math. Anal. **20**(5), 1218–1242 (1989)
5. Barnsley, M.F., Harrington, A.N.: The calculus of fractal interpolation functions. J. Approx. Theory **57**(1), 14–34 (1989)
6. Chand, A.K.B., Kapoor, G.P.: Generalized cubic spline fractal interpolation functions. SIAM J. Numer. Anal. **44**(2), 655–676 (2006)
7. Chand, A.K.B., Vijender, N., Navascués, M.A.: Shape preservation of scientific data through rational fractal splines. Calcolo **51**(2), 329–362 (2014)
8. Chand, A.K.B., Vijender, N., Viswanathan, P., Tetenov, A.V.: Affine zipper fractal interpolations. BIT Numer. Math. **60**, 319–344 (2020)
9. Chand, A.K.B., Viswanathan, P., Navascués, M.A.: A new class of rational quadratic fractal functions with positive shape preservation. Fractals, Wavelets, and their Applications, pp. 283–301. Springer, Berlin (2014)
10. Chand, A., Viswanathan, P.: Shape preserving rational cubic spline fractal interpolation (2015). arXiv:1503.02458
11. Hussain, M.Z., Ayub, N., Irshad, M.: Visualization of 2d data by rational quadratic functions. J. Inf. Comput. Sci. **2**(1), 17–26 (2007)
12. Hutchinson, J.E.: Fractals and self-similarity. Indiana Univ. Math. J. **30**(5), 713–747 (1981)
13. Massopust, P.R.: Fractal Functions, Fractal Surfaces, and Wavelets. Academic, San Diego (1994)
14. Massopust, P.R.: On local fractal functions in Besov and Triebel-Lizorkin spaces. J. Math. Anal. Appl. **436**(1), 393–407 (2016)
15. Navascués, M.A.: Fractal polynomial interpolation. Z. Anal. Anwendungen **24**(2), 401–418 (2005)

16. Navascués, M.A., Jha, S., Chand, A.K.B., Sebastián, M.V.: Fractal approximation of Jackson type for periodic phenomena. Fractals **26**(5), 1850079, 14 (2018)
17. Reddy, K.M.: Some aspects of fractal functions in geometric modeling. Department of Mathematics, IIT Madras (2018)
18. Schmidt, J.W., Hess, W.: Positivity of cubic polynomials on intervals and positive spline interpolation. BIT Numer. Math. **28**(2), 340–352 (1988)
19. Tetenov, A.V.: On self-similar Jordan arcs on a plane. Sib. Zh. Ind. Mat. **7**(3), 148–155 (2004)
20. Tetenov, A.V.: Self-similar jordan arcs and the graph directed systems of similarities. Siberian Math. J. **47**(5), 940–949 (2006)

Double Lacunary Statistical Convergence of Weight ν via Ideal in Topological Groups

Ekrem Savaş

Abstract In (Savaş in Commun Math Appl 9(3):265–275, 2018 [34]), Savas has defined double statistical convergence of order α in topological groups. In this paper, we study the concepts of $\mathcal{I}$-double statistical convergence of weight ν and $\mathcal{I}_\theta$-double statistical convergence of weight ν. Furthermore, we establish the relations between these concepts.

Keywords Double lacunary sequence · Double statistical convergence of weight ν · Topological groups

Mathematics Subject Classification (2000) Primary 42B15 · Secondary 40C05

1 Introduction

The idea of statistical convergence was introduced by Fast [9] as follows: A sequence y is said to be statistically convergent to the number ψ if for every $\varphi > 0$

$$\lim_n \frac{1}{n}|i \le n : |y_i - \psi| \ge \varphi\}| = 0,$$

where by $i \le n$ we mean that $i = 0, 1, 2, \ldots, n$ and the vertical bars indicate the number of elements in the enclosed set. In this case, we write $s - \lim y = \psi$ or $y_i \to \psi(s)$.

Nowadays, the concept of statistical convergence has become one of the most active area of research in the field of summability. Cakalli and Savaş [2] studied the statistical convergence of double sequences in topological groups and also in [25], Savas introduced lacunary statistical convergence of double sequences in topological

E. Savaş (✉)
Department of Mathematics, Uşak University, Uşak, Turkey
e-mail: ekremsavas@yahoo.com

D. Giri et al. (eds.), *Proceedings of the Fifth International Conference on Mathematics and Computing*, Advances in Intelligent Systems and Computing 1170, https://doi.org/10.1007/978-981-15-5411-7_19

groups. Recently in [33], Savaş presented double lacunary statistical convergence in topological groups via ideal.
It should be noted that the notion of statistical convergence of order α, $0 < \alpha < 1$ was introduced in [3].

We now present some definitions and preliminaries:

By a lacunary sequence, we mean an increasing sequence $\theta = (k_u)$ of positive integers such that $k_0 = 0$ and $h_u : k_u - k_{u-1} \to \infty$ as $u \to \infty$. Throughout this paper, the intervals determined by θ will be denoted by $I_u = (k_{u-1}, k_u]$ and the ratio $(k_u)(k_{u-1})^{-1}$ will be abbreviated by q_u.

A new type of convergence called lacunary statistical convergence was defined in [11] as follows: A sequence (y_i) of real numbers is said to be lacunary statistically convergent to ψ (or, S_θ-convergent to ψ) if for any $\varphi > 0$,

$$\lim_{u\to\infty} \frac{1}{h_u} |\{i \in I_u : |y_i - \psi| \geq \varphi\}| = 0.$$

The concept of $\mathcal{I}$-convergence was introduced by Kostyrko et al. [12] in a metric space. Later on, it has been discussed by Dems [8] and Das et al. [7]. Several papers deal with this problem, see [5, 6, 18, 22–24, 26–28, 30, 31].

2 Definitions and Notations

We recall some basic definitions.

Definition 2.1 (*See* [4]) A family $\mathcal{I} \subset 2^{\mathbb{N}}$ is said to be an ideal of $\mathbb{N}$ if the following conditions hold:
(a) $R, S \in \mathcal{I}$ implies $R \cup S \in \mathcal{I}$,
(b) $R \in \mathcal{I},\ S \subset R$ implies $S \in \mathcal{I}$,

Definition 2.2 A non-empty family $F \subset 2^{\mathbb{N}}$ is said to be a filter of $\mathbb{N}$ if the following conditions hold:
(a) $\phi \notin F$,
(b) $R, S \in F$ implies $R \cap S \in F$,
(c) $R \in F,\ R \subset S$ implies $S \in F$.

If $\mathcal{I}$ is a proper ideal of $\mathbb{N}$ (i.e. $\mathbb{N} \notin \mathcal{I}$), then the family of sets $F(\mathcal{I}) = \{L \subset \mathbb{N} : exists\ R \in \mathcal{I} : L = \mathbb{N} \setminus R\}$ is a filter of $\mathbb{N}$. It is called the filter associated with the ideal.

Definition 2.3 A proper ideal $\mathcal{I}$ is said to be admissible if $\{n\} \in \mathcal{I}$ for each $n \in \mathbb{N}$.

Definition 2.4 [15]. Let $y = (y_{ij})$ be a double sequence, if for each $\varphi > 0$ there exists $N \in \mathbb{N}$ such that $|y_{ij} - \psi| < \varphi$ whenever $i, j \geq N$, then we say that $y = (y_{ij})$ is convergent in the Pringsheim's sense or $P-$convergent to ψ and ψ is called Pringsheim limit (denoted by $P - \lim y = \psi$).

Mursaleen and Edely has presented the following definition.

Definition 2.5 Let $y = (y_{ij})$ be a double sequence; y is double statistically convergent to ψ provided that for each $\varphi > 0$

$$P - \lim_{m,n} \frac{1}{mn}\{ \text{ number of } (i, j) : i < m \text{ and } j < n, |y_{ij} - \psi| \geq \varphi\} = 0.$$

It is obvious that a convergent double sequence is also st^2-convergent but the inverse is not true in general. Also note that st^2-convergence need not be bounded.

Definition 2.6 Let $y = (y_{ij})$ be double sequence, if for each $\varphi > 0$,

$$L(\varphi) = \left\{(i, j) \in \mathbb{N} \times \mathbb{N} : \left|y_{ij} - \psi\right| \geq \varphi\right\} \in \mathcal{I}.$$

Then we say that the double sequence $y = (y_{ij})$ is convergent to the number ψ with respect to the ideal $\mathcal{I}$. In this case we write $\mathcal{I} - \lim_{i,j} y_{ij} = \psi$.

Let $\nu : \mathbb{N} \to [0, \infty)$ be a function with $\lim\limits_{n\to\infty} \nu(n) = \infty$. The upper density of weight ν was defined in [1] by

$$\overline{d}_\nu(L) = \limsup_{n\to\infty} \frac{L(1, n)}{\nu(n)}$$

for $L \subset \mathbb{N}$. Then the set

$$\mathcal{I}_\nu = \{L \subset \mathbb{N} : \overline{d}_\nu(L) = 0\}$$

forms an ideal. It has been present in [1] that $\mathbb{N} \in \mathcal{I}_\nu$ iff. $\frac{n}{\nu(n)} \to 0$. So we further suppose that $n/\nu(n) \nrightarrow 0$ as $n \to \infty$ so that $\mathbb{N} \notin \mathcal{I}_\nu$ and $\mathcal{I}_\nu$ is a proper admissible ideal of $\mathbb{N}$. The set of all such weight functions ν satisfying the above properties will be denoted by Δ. We now give the following definition.

The double sequence $\theta = \{(k_u, l_v)\}$ is called **double lacunary** if there exists two increasing of integers such that

$$k_0 = 0, h_u = k_u - k_{u-1} \to \infty \text{ as } u \to \infty$$

and

$$l_0 = 0, \bar{h}_v = l_v - l_{v-1} \to \infty \text{ as } v \to \infty.$$

Notations: $k_{u,v} = k_u l_v$, $h_{u,v} = h_u \bar{h}_v$, θ is determined by $I_u = \{(k) : k_{u-1} < k \leq k_u\}$, $I_v = \{(l) : l_{v-1} < l \leq l_v\}$, $I_{u,v} = \{(u, v) : k_{u-1} < k \leq k_u \;\&\; l_{v-1} < l \leq l_v\}$, We will denote the set of all double lacunary sequences by $\mathbf{N}_{\theta_{u,v}}$.

Let $L \subseteq \mathbf{N} \times \mathbf{N}$ has double lacunary density $\delta_2^{\theta}(L)$ if

$$P - \lim_{u,v} \frac{1}{h_{uv}} |\{(i, j) \in I_{uv} : (i, j) \in L\}|$$

exists.

Patterson and Savas [14] presented double lacunary statistically convergence as follows:

Definition 2.7 Let θ be a double lacunary sequence; the double sequence y is st_{θ}^2-convergent to ψ provided that for every $\varphi > 0$,

$$P - \lim_{u,v} \frac{1}{h_{uv}} \left|\{(i, j) \in I_{uv} : |y_{ij} - \psi| \geq \varphi\}\right| = 0.$$

In this case write $st_{\theta}^2 - \lim y = \psi$ or $y_{ij} \to \psi(S_{\theta}^2)$.

More study of double lacunary and double sequences can be found in [15, 17, 19–21, 32].

By E, we will denote an abelian topological Hausdorff group, written additively, which satisfies the first axiom of countability.

Let E be an topological group and $y = (y_{ij})$ be a double sequence in E, if for every neighbourhood V of 0 there exists $N \in \mathbb{N}$ such that $y_{ij} - \psi \in V$ whenever $i, j \geq N$, then we say that $y = (y_{ij})$ is convergent to a point ψ in E in the Pringsheims sense and ψ is called the Pringsheim limit of y.

Throughout $\mathcal{I}_2$ will stand for a proper strongly admissible ideal in $\mathbb{N} \times \mathbb{N}$.

Statistical convergence of double sequences $y = (y_{ij})$ in a topological group was given in [2] as follows.

In a topological group E, double sequence $y = (y_{ij})$ is called statistically convergent to a point ψ of E if for each neighbourhoud V of 0 the set

$$\{(i, j), i \leq n; \text{and}; j \leq m : y_{ij} - \psi \notin V\}$$

has double natural density zero. In this case we write $S^2 - \lim_{i,j} y_{ij} = \psi$ and we denote the set of all statistically convergent double sequences by $S^2(E)$.

Recently the following definition was presented by Savas [25].

Definition 2.8 A sequence $y = (y_{ij})$ is said to be S_{θ}^2-convergent to ψ (or double lacunary statistically convergent to ψ) if for each neighbourhood V of 0,

$$P - lim_{u,v\to\infty}(h_{uv})^{-1} \left|\{(i, j) \in I_{uv} : y_{ij} - \psi \notin V\}\right| = 0.$$

In this case, we write

$$S_{\theta}^2 - \lim_{i,j\to\infty} y_{ij} = \psi \text{ or } y_{ij} \to \psi(S_{\theta}^2).$$

Now we can present our definitions of $\mathcal{I}$- double statistical convergence of weight ν and $\mathcal{I}$- double lacunary statistical convergence of weight ν in topological groups as follows:

Definition 2.9 Let $y = (y_{ij})$ in E be a double sequence, if for each $\eta > 0$ and for each neighbourhood V of 0,

$$\left\{(m, n) \in N \times N : \frac{1}{\nu(mn)} \left|\{i \leq n \text{ and } j \leq m : y_{ij} - \psi \notin V\}\right| \geq \eta\right\} \in \mathcal{I}_2,$$

then we say that $y = (y_{ij})$ is $\mathcal{I}$- double statistically convergent of weight ν to ψ or $S(\mathcal{I}_2)^\nu$-convergent to ψ. In this case, $y_{ij} \to \psi(S(\mathcal{I}_2)^\nu)$. We will denote the set of all $S(\mathcal{I})^\nu$-double statistically convergent sequences by simply $S(\mathcal{I}_2)^\nu(E)$.

Remark 2.10 If $\mathcal{I}_2 = \mathcal{I}_{2fin} = \{P \subset \mathbb{N} \times \mathbb{N}, P \text{ is a finite}\}$, then $S(\mathcal{I}_2)^\nu$-convergence coincides with double statistical convergence of weight ν in E. Furthermore if we take $\nu\,(nm) = (nm)^\alpha$, it reduces to $\mathcal{I}$-double statistical convergence of order α, which is presented by Savas [34].

Definition 2.11 Let $y = (y_{ij})$ in E be a double sequence, if for each neighbourhood V and any $\eta > 0$

$$\left\{(u, v) \in \mathbb{N} \times \mathbb{N} : \frac{1}{\nu(h_{uv})} \left|\{(i, j) \in I_{uv} : y_{ij} - \psi \notin V\}\right| \geq \eta\right\} \in \mathcal{I}_2,$$

then we say that $y = (y_{ij})$ is $\mathcal{I}_\theta$-double statistically convergent of weight ν to ψ or $S_\theta(\mathcal{I}_2)^\nu-$ convergent to ψ. In this case,

$$S_\theta(\mathcal{I}_2)^\nu - \lim_{i,j\to\infty} y_{ij} = \psi \text{ or } y_{ij} \to \psi(S^\theta(\mathcal{I}_2)^\nu)$$

and write

$$S_\theta(\mathcal{I}_2)^\nu(E) = \left\{y = (y_{ij}) : \text{ for some } \psi, S_\theta(\mathcal{I}_2)^\nu - \lim_{i,j\to\infty} y_{ij} = \psi\right\}$$

Remark 2.12 If $\mathcal{I}_2 = \mathcal{I}_{2fin} = \{P \subset \mathbb{N} \times \mathbb{N}, P \text{ is finite}\}$, $\mathcal{I}_\theta-$ double statistical convergence of weight ν becomes lacunary double statistical convergence of weight ν in topological groups. Further, if we take $\mathcal{I} = \mathcal{I}_{fin}$ and $\nu\,(nm) = (nm)^\alpha$, then $\mathcal{I}_\theta-$ double statistical convergence of weight ν, becomes lacunary double convergence of order α in topological groups which is studied by Savas [34].

3 Inclusion Theorems

We have the following theorems.

Theorem 3.1 *Let $\nu_1, \nu_2 \in \Delta$ be such that there exists $K > 0$ and $(n_0, m_0) \in \mathbb{N} \times \mathbb{N}$ such that $\frac{\nu_1(nm)}{\nu_2(nm)} \leq K$ for all $n \geq n_0$ and $m \geq m_0$. Then $S(\mathcal{I}_2)^{\nu_1} \subset S(\mathcal{I}_2)^{\nu_2}$.*

Proof For any neighbourhood V of 0,

$$\frac{\left|\left\{i \leq n \text{ and } j \leq m : y_{ij} - \psi \notin V\right\}\right|}{\nu_2(nm)}$$
$$= \frac{\nu_1(nm)}{\nu_2(nm)} \cdot \frac{\left|\left\{i \leq n \text{ and } j \leq m : y_j - \psi \notin V\right\}\right|}{\nu_1(nm)}$$
$$\leq K \cdot \frac{\left|\left\{i \leq n \text{ and } j \leq m : x_{ij} - \psi \notin V\right\}\right|}{\nu_1(nm)}.$$

for all $n \geq n_0$ and $m \geq m_0$. Further, for any $\eta > 0$ and for each neighbourhood V of 0

$$\left\{(n,m) \in \mathbb{N} \times \mathbb{N} : \frac{\left|\left\{i \leq n \text{ and } j \leq m : y_{ij} - \psi \notin V\right\}\right|}{\nu_2(nm)} \geq \eta\right\}$$
$$\subset \left\{(n,m) \in \mathbb{N} \times \mathbb{N} : \frac{\left|\left\{i \leq n \text{ and } j \leq m : y_{ij} - \psi \notin V\right\}\right|}{\nu_1(nm)} \geq \frac{\eta}{K}\right\}$$
$$\cup(\{1, 2, \ldots, i_0\} \times \{1, 2, \ldots, j_0)\}.$$

So we write that $S(\mathcal{I}_2)^{\nu\nu_1} \subset S(\mathcal{I}_2)^{\nu_2}$. □

By using the same method we can get the following result

Theorem 3.2 *Let $\nu_1, \nu_2 \in \Delta$ be such that there exists $K > 0$ and $(n_0, m_0) \in \mathbb{N} \times \mathbb{N}$ such that $\frac{\nu_1(nm)}{\nu_2(nm)} \leq K$ for all $n \geq n_0$ and $m \geq m_0$. Then*
(i) $S_\theta(\mathcal{I}_2)^{\nu_1}(E) \subset S_\theta(\mathcal{I}_2)^{\nu_2}(E)$.
(ii) In particular $S_\theta(\mathcal{I}_2)^{\nu_1}(E) \subset S_\theta(\mathcal{I}_2)(E)$.

We now record two useful another theorems.

Theorem 3.3 *For any double lacunary sequence θ, $\mathcal{I}_2$-statistical convergence of weight ν implies $\mathcal{I}_2$-lacunary statistical convergence of weight ν if*

$$\liminf_{u,v} \frac{\nu(h_{uv})}{\nu(k_{uv})} > 1.$$

Proof Since $\liminf_{u,v} \frac{\nu(h_{uv})}{\nu(k_{uv})} > 1$, so we get a $H > 1$ such that for sufficiently large u, v we get

$$\frac{\nu\left(h_{uv}\right)}{\nu\left(k_{uv}\right)} \geq H.$$

Since $y_{ij} \to \psi\ (S(\mathcal{I}_2)^{\nu})$, hence for each neighbourhood V of 0 and sufficiently large u, v we have

$$\begin{aligned}
&\frac{1}{\nu\left(k_{uv}\right)}\left|\left\{i \leq n, j \leq m : y_{ij} - \psi \notin V\right\}\right| \\
&\geq \frac{1}{\nu\left(k_{uv}\right)}\left|\left\{(i, j) \in I_{uv} : y_{ij} - \psi \notin V\right\}\right| \\
&\geq H \cdot \frac{1}{\nu\left(h_{uv}\right)}\left|\left\{i \leq n, j \leq m : y_{ij} - \psi \notin V\right\}\right|.
\end{aligned}$$

Then for any $\eta > 0$, and for each neighbourhood V of 0 we get

$$\begin{aligned}
&\left\{(m, n) \in \mathbb{N} \times \mathbb{N} : \frac{1}{\nu\left(h_{uv}\right)}\left|\left\{i \leq n, j \leq m : y_{ij} - \psi \notin V\right\}\right| \geq \eta\right\} \\
&\subseteq \left\{(u, v) \in \mathbb{N} \times \mathbb{N} : \frac{1}{\nu\left(k_{uv}\right)}\left|\left\{(i, j) \in I_{uv} : y_{ij} - \psi \notin V\right\}\right| \geq H\eta\right\} \in \mathcal{I}_2.
\end{aligned}$$

This shows that $y_{ij} \to \psi\ (S_\theta(\mathcal{I}_2)^{\nu})$.

For the coming theorem we suppose that the double lacunary sequence θ satisfies the condition that for any set $\zeta \in F(\mathcal{I}_2)$, $\bigcup\{(n, m) : k_{u-1} < n < k_u$ and $l_{v-1} < m < l_v, (u, v) \in \zeta\} \in F(\mathcal{I}_2)$.

Theorem 3.4 *For double sequence $\theta = \{(k_u, l_v)\}$, satisfying the above condition $S_\theta\left(\mathcal{I}_2\right)^{\nu}(E) \subseteq S\left(\mathcal{I}_2\right)^{\nu}(E)$ if $\sup_{uv} \sum_{i,j=0}^{u-1,v-1} \frac{\nu(h_{i+1,j+1})}{\nu(k_{u-1} l_{v-1})} = B(say) < \infty$.*

Proof Suppose that $S_\theta\left(\mathcal{I}_2\right)^{\nu} - \lim_{i,j\to\infty} y_{ij} = \psi$, say, for any neighbourhood V of 0 and for $\eta, \eta_1 > 0$ define the sets

$$\zeta = \left\{(u, v) \in \mathbb{N} \times \mathbb{N} : \frac{1}{\nu(h_{uv})}|\{(i, j) \in I_{uv} : y_{ij} - \psi \notin V\}| < \eta\right\}$$

and

$$\chi = \left\{(n, m) \in \mathbb{N} \times \mathbb{N} : \frac{1}{\nu(nm)}|\{ki \leq n \text{ and } j \leq m : y_{ij} - \psi \notin V\}| < \eta_1\right\}.$$

It is obvious from our assumption that $\zeta \in F(\mathcal{I}_2)$, the filter associated with the ideal $\mathcal{I}_2$. Further observe that

$$A_{k,l} = \frac{1}{\nu(h_{kl})}|\{(i, j) \in I_{kl} : y_{ij} - L \notin V\}| < \eta$$

for all $(k, l) \in \zeta$. Let m and n be such that $k_{v-1} < m \le k_v$ and $l_{v-1} < n \le l_v$; hence we have the following

$$\frac{1}{\nu(mn)} \left|\left\{i \le m \text{ and } j \le n : y_{ij} - \psi \notin V\right\}\right|$$
$$\le \sup_{(p,q)\in\zeta} A_{p,q}.\sup_{u,v} \sum_{p,q=0}^{u-1,v-1} \frac{\nu(k_{p+1} - k_p)(l_{q+1} - l_q)}{\nu(k_{u-1} l_{v-1})} < B\delta.$$

Taking $\eta_1 = \frac{\delta}{B}$ and in view of the fact that $\bigcup\{(m, n) : k_{u-1} < m < k_u, l_{v-1} < n < l_v, (u, v) \in \zeta\} \subset \chi$ where $\zeta \in F(\mathcal{I}_2)$ it follows from our assumption on θ that the set χ also belongs to $F(\mathcal{I}_2)$.

Finally we prove the following theorem.

Theorem 3.5 *Let $\theta = \{(k_u, l_v)\}$ and $\theta^1 = \{(k_u, l_v)\}$ be two double lacunary sequences such that $I_{uv} \subset J_{uv}$ for all $(u, v) \in \mathbb{N} \times \mathbb{N}$, if*

$$\lim_{u,v\to\infty} \inf \frac{\nu(h_{uv})}{\nu(\ell_{uv})} > 0 \tag{1}$$

then $S_{\theta^1}(\mathcal{I}_2)(E) \subseteq S_\theta(\mathcal{I}_2)(E)$.

Proof Suppose that $I_{uv} \subset J_{uv}$ for all $(u, v) \in \mathbb{N} \times \mathbb{N}$ and let (1) be satisfied. For neighbourhood U of 0, we have

$$\left\{(i, j) \in J_{uv} : y_{ij} - \psi \notin V\right\} \supseteq \left\{(i, j) \in I_{uv} : y_{ij} - \psi \notin V\right\}.$$

Hence we can write

$$\nu(\ell_{uv})^{-1} \left|\{(i, j) \in J_{uv} : y_{ij} - \psi \notin U\}\right|$$
$$\ge \nu(h_{uv}\ell_{uv})^{-1} \nu(h_{uv})^{-1} \left|\{(i, j) \in I_{uv} : y_{ij} - \psi \notin V\}\right|$$

and so for all $(u, v) \in \mathbb{N} \times \mathbb{N}$ we have

$$\left\{(u, v) \in \mathbb{N} \times \mathbb{N} : \nu(h_{uv})^{-1} \left|\{(i, j) \in I_{uv} : y_{ij} - \psi \notin V\}\right| \ge \eta\right\}$$
$$\subseteq \left\{(u, v) \in \mathbb{N} \times \mathbb{N} : \nu(\ell_{uv})^{-1} \left|\{(i, j) \in J_{uv} : y_{ij} - \psi \notin V\}\right|\right.$$
$$\left.\ge \eta \nu(h_{uv}) g(l_{uv})^{-1}\right\} \in \mathcal{I}_2$$

for all $(u, v) \in \mathbb{N} \times \mathbb{N}$ where $I_{uv} = \{(i, j) : k_{u-1} < k \le k_u \ \& \ l_{v-1} < l \le l_v\}$, and $J_{uv} = \{(r, s) : r_{u-1} < r \le r_u \ \& \ s_{v-1} < s \le s_v\}$. Hence $S_\theta^1(\mathcal{I}_2)^\nu(E) \subseteq S_\theta(\mathcal{I}_2)^\nu(E)$.

References

1. Balcerzak, M., Das, P., Filipczak, M., Swaczyna, J.: Generalized kinds of density and the associated ideals. Acta Math. Hungar. **147**(1), 97–115 (2015)
2. Çakalli, H., Savaş, E.: Statistical convergence of double sequence in topological groups. J. Comp. Anal. Appl. **12**(2), 421–426 (2010)
3. Colak, R.: Statistical Convergence of Order α. Modern Methods in Analysis and Its Applications, pp. 121–129. Anamaya Pub., New Delhi, India (2010)
4. Das, P., Savaş, E.: On $\mathcal{I}$-convergence of nets in locally solid Riesz spaces. Filomat **27**(1), 84–89 (2013)
5. Das, P., Savaş, E.: On $\mathcal{I}_\lambda$-statistical convergence in locally solid riesz spaces. Math. Slovaca **65**(6), 1491–1504 (2015)
6. Das, P., Savaş, E.: On $\mathcal{I}$-statistically pre-Cauchy sequences. Taiwan. J. Math. **18**(1), 115–126 (2014)
7. Das, P., Savaş, E., Ghosal, S.K.: On generalizations of certain summability methods using ideals. Appl. Math. Lett. **24**, 1509–1514 (2011)
8. Dems, K.: On $\mathcal{I}$-cauchy sequences. Real Anal. Exch. **30**, 123–128 (2004–2005)
9. Fast, H.: Sur la convergence statistique. Colloq Math. **2**, 241–244 (1951)
10. Fridy, J.A.: On ststistical convergence. Analysis **5**, 301–313 (1985)
11. Fridy, J.A., Orhan, C.: Lacunary statistical convergence. Pacific J. Math. **160**, 43–51 (1993)
12. Kostyrko, P., Šalát, T., Wilczynki, W.: $\mathcal{I}$-convergence. Real Anal. Exch. **26**(2), 669–685 (2000/2001)
13. Mursaleen, M., Edely, O.H.: Statistical convergence of double sequences. J. Math. Anal. Appl. **288**, 223–231 (2003)
14. Patterson, R.F., Savaş, E.: Lacunary statistical convergence of double sequences. Math. Commun. **10**, 55–61 (2005)
15. Pringsheim, A.: Zur theorie der zweifach unendlichen Zahlenfolgen. Math. Annalen **53**, 289–321 (1900)
16. Šalát, T.: On statistically convergent sequences of real numbers. Math. Slovaca **30**, 139–150 (1980)
17. Savaş, E., Patterson, R.F.: Double sequence spaces defined by Orlicz functions. Iran. J. Sci. Technol. Trans. A Sci. **31**(2), 183–188 (2007)
18. Savaş, E., Das, P.: A generalized statistical convergence via ideals. Appl. Math. Lett. **24**, 826–830 (2011)
19. Savaş, E.: On some new double lacunary sequences spaces via Orlicz function. J. Comput. Anal. Appl. **11**(3), 423–430 (2009)
20. Savaş, E., Patterson, R.F.: Some double lacunary sequence spaces defined by Orlicz functions. Southeast Asian Bull. Math. **35**(1), 103–110 (2011)
21. Savaş, E., Patterson, R.F.: Double σ-convergence lacunar statistical sequences. J. Comput. Anal. Appl. **11**(4), 610–615 (2009)
22. Savaş, E.: $\mathcal{I}_\lambda$-double statistical convergence of order α in topological groups. Ukrainian Math. J. **68**(9), 1437–1446 (2017)
23. Savaş, E.: On some new sequence spaces in 2-normed spaces using Ideal convergence and an Orlicz function. J. Ineq. Appl. (2010). https://doi.org/10.1155/2010/482392
24. Savaş, E.: On generalized double statistical convergence via ideals. In: The Fifth Saudi Science Conference, pp. 16–18 (2012)
25. Savaş, E.: Lacunary statistical convergence of double sequences in topological groups. J. Inequal. Appl. (2014)
26. Savaş, E.: On $\mathcal{I}$-lacunary statistical convergence of order α for sequences of sets. Filomat **29**(6), 1223–1229 (2015)
27. Savaş, E.: $\mathcal{I}_\theta$ -statistically convergent sequences in topological groups. Mat. Bilten **39**(2), 19–28 (2015)

28. Savaş, E., Savaş Eren, R.E.: $\mathcal{I}_\theta$-statistical convergence of order α in topological groups. In: Applied Mathematics in Tunisia, Springer Proc. Math. Stat., vol. 131, pp. 141–148. Springer, Cham (2015)
29. Savaş, E.: $\mathcal{I}_\lambda$-double statistically convergent sequences in topological groups. Commun. Comput. Inform. Sci. **655**, 349–357 (2017)
30. Savaş, E.: $\mathcal{I}$-statistically convergent sequences in topological groups. In: International Conference "Kangro-100. Methods of Analysis and Algebra", dedicated to the Centennial of Professor Gunnar Kangro. Tartu, Estonia (2013)
31. Savaş, E., Das, P., Dutta, S.: A note on strong matrix summability via ideals. Appl. Math Lett. **25**(4), 733–738 (2012)
32. Savaş, E., Savaş, R.: Double lacunary statistical convergence of order α. Indian J. Math. **57**(1), 1–15 (2015)
33. Savaş, E.: Double Lacunary Statistical Convergence in Topological Groups Via Ideal presented for the 30th International Conference of The Jangjeon Mathematical Society (ICJMS'2017), Algeria (2017)
34. Savaş, E.: Double lacunary statistical convergence of order α in topological groups via ideal. Commun. Math. Appl. **9**(3), 265–275 (2018)

Monotonicity Preserving Rational Cubic Graph-Directed Fractal Interpolation Functions

A. K. B. Chand and K. M. Reddy

Abstract The idea of graph-directed fractal interpolation function (FIF) is introduced recently to represent several dependent data sets from graph-directed iterated function system (GDIFS). When dependent data sets are generated from $\mathscr{C}^1$-smooth functions with irregular derivatives, it is not ideal to use affine FIF or classical splines in such scenario. Thus, we initiate the use of smooth graph directed FIFs for two or more sets of interpolation data that are not independent, and generated from original smooth functions having fractal characteristics in their derivatives. For this task, we have proposed a new class $\mathscr{C}^1$-rational cubic graph-directed FIFs (RCGDFIFs) using cubic rational function involving two shape parameters in each sub-interval. For applications of the proposed RCGDFIFs in modeling of monotonic data sets, we have deduced sufficient condition based on the restriction of the corresponding rational GDIFS parameters.

Keywords Iterated function system · Rational fractal interpolation functions · Graph-directed fractal functions · Monotonicity

1 Introduction

The concept of fractal interpolation functions (FIFs) has been broached by Barnsley [2] based on the setting of an iterated function system (IFS). For a prescribed set of interpolation data, it is possible to construct an IFS which is contractive in nature. Consequently, the IFS has a fixed point or self-referential set and it matches with the graph of an interpolating curve. This interpolating curve is known as a FIF which

A. K. B. Chand (✉)
Indian Institute of Technology Madras, Chennai 600036, India
e-mail: chand@iitm.ac.in

K. M. Reddy
VIT-AP University, Amaravati 522237, Andhra Pradesh, India
e-mail: mahipalnitw@gmail.com

D. Giri et al. (eds.), *Proceedings of the Fifth International Conference on Mathematics and Computing*, Advances in Intelligent Systems and Computing 1170, https://doi.org/10.1007/978-981-15-5411-7_20

may be smooth or non-smooth depending on the parameters of the IFS. The main differences of a FIF with the traditional interpolation techniques consist (i) in the definition in terms of a functional equation that implies a self-similarity in all scales, (ii) in the constructive way through iterations that are used to compute the interpolant instead of analytic formulae, (iii) in the presence of scaling factors that replace the unicity of the traditional interpolant for a fixed set of interpolation data, (iv) in the fact that the Minkowski dimension of a FIF is often non-integer. The theory of FIFs has emerged beyond its mathematical settings and has become a powerful tool in various branches of science as well as engineering. Barnsley and Harrington [3] proved the existence of a smooth or spline (differentiable) FIF, and developed the fundamental relation between scaling factors and differentiable FIF. Note that certain derivative of a fractal spline possesses the fractal characteristics. But only preserving smoothness is not sufficient for geometric modeling problems, whereas a data set under consideration for interpolation may have some fundamental shapes like irregularity in derivative, positivity, monotonicity, and convexity(concavity). The processes of evaluating interpolant that mimic the shape properties associated with a data set is called shape preserving or iso-geometric interpolation, and the corresponding interpolant is referred to as shape preserving interpolant [4–9, 11–13, 17–19]. In applications like computer-aided geometric design [15], data visualization, image analysis [16], and cartography constriction of interpolating curves with different shapes for a prescribed data plays a pivotal role. Recently, Deniz and Özdemir [10] has proposed graph-directed iterated function system (GDIFS) for finite number of data sets, and proved the existence of graph-directed fractal functions interpolating corresponding data sets with graphs as the attractors of the GDIFS. But this type of interpolants are not suitable to approximate two or more sets of interpolations data that are not independent, and are generated from original smooth functions having fractal characteristics in their derivatives. To overcome this deficiency, we propose a novel class of rational cubic graph-directed FIFs with two shape parameters. Further, we restrict the values of the IFS parameters so that the RCGDFIFs preserve the monotonicity features associated with similar type of monotonic data.

This paper is divided into five parts. In Sect. 2, we review the basics of FIF theory. In Sect. 3, we construct a graph-directed rational cubic FIF of finite number of interpolation data sets. In Sect. 4, we study the theory of monotonic RCGDFIFs by using suitable IFS parameters. The results of Sects. 3, 4 are illustrated with suitable examples in Sect. 5.

2 Fractal Interpolation Function

For construction of FIF and smooth FIF, we have collected the basics tools from the references [1–3]. We denote $\mathbb{N}_k$ as the subset of $\mathbb{N}$ containing first k natural numbers and $\mathbb{N}_N^0 = \{0, 1, \ldots, N\}$. Let $\{(x_i, y_i) : i \in \mathbb{N}_N^0\}$ be the given interpolation data set with $x_0 < x_1 < x_2 < \cdots < x_N$. Denote $I = [x_0, x_N]$ and $I_i = [x_{i-1}, x_i]$, $i \in \mathbb{N}_N$. Suppose $L_i : I \rightarrow I_i$ are contraction homeomorphisms satisfying the join-up conditions in the x-direction as

$$L_i(x_0) = x_{i-1},\ L_i(x_N) = x_i,\ i \in \mathbb{N}_N. \tag{1}$$

Let $F_i : I \times \mathbb{R} \to \mathbb{R}$ be continuous maps satisfying

$$F_i(x_0, y_0) = y_{i-1},\ F_i(x_N, y_N) = y_i,\ i \in \mathbb{N}_N. \tag{2}$$

For $i \in \mathbb{N}_N$, define functions $W_i(x, y) = \big(L_i(x), F_i(x, y)\big)$. Consider the collection $\{I \times \mathbb{R};\ W_i : i \in \mathbb{N}_N\}$ referred to as an iterated function system (IFS). The following is a fundamental result in the theory of fractal functions.

Theorem 1 ([1]) *Let $\mathscr{C}(I)$, the space of all real-valued continuous functions on a compact interval I, be endowed with the supremum norm and consider the subspace*

$$\mathscr{C}_{y_0,y_N}(I) := \{g \in \mathscr{C}(I) :\ \ g(x_0) = y_0,\ \ g(x_N) = y_N\}.$$

The following hold.

1. *The IFS $\{I \times \mathbb{R};\ W_i, i = 1, 2, \ldots, N\}$ has a unique attractor $G(f)$ which is the graph of a continuous function $f : I \to \mathbb{R}$ satisfying $f(x_i) = y_i$ for $i \in \mathbb{N}_N^0$.*
2. *The function f is the fixed point of the Read-Bajraktarević (RB) operator T defined on $\mathscr{C}_{y_0,y_N}(I)$ as*

$$(Tg)(x) = F_i\big(L_i^{-1}(x), g \circ L_i^{-1}(x)\big),\quad x \in I_i,\ i \in \mathbb{N}_N. \tag{3}$$

The above function f appearing in Theorem 1 is knowns as Fractal Interpolation Function (FIF) corresponding to the given data set $\{(x_i, y_i) : i \in \mathbb{N}_N\}$, and it obeys

$$f(x) = S_i f\big(L_i^{-1}(x)\big) + q_i\big(L_i^{-1}(x)\big),\ S_i \in (-1, 1),\ i \in \mathbb{N}_N. \tag{4}$$

The most widely used and popular FIFs are defined by the system of maps

$$L_i(x) = a_i x + b_i,\quad F_i(x, y) = S_i y + q_i(x), \tag{5}$$

where $q_i : I \to \mathbb{R}$ are continuous functions satisfying

$$q_i(x_0) = y_{i-1} - S_i y_0,\quad q_i(x_N) = y_i - S_i y_N.$$

The parameter S_i is referred to as vertical scaling factor of the transformation F_i, and $S = (S_1, S_2, \ldots, S_N) \in (-1, 1)^N$ is called scale vector. For a differentiable FIF, the scaling factors should satisfy certain conditions, and it is given in the following:

Theorem 2 [3] *Suppose $\{(x_i, y_i) : i \in \mathbb{N}_N^0\}$ is a given set of data points taken from a smooth function. Assume that $L_i(x) = a_i x + b_i$ satisfies (1) and $F_i(x, y) = S_i y + q_i(x)$ obeys (2) for $i \in \mathbb{N}_N$. Let $|\alpha_i| < a_i^p$ for some integer $p \geq 0$ and $q_i \in \mathscr{C}^p[x_0, x_N]$, $i \in \mathbb{N}_N$. Let*

$$F_{i,k}(x,y) = \frac{S_i y + q_i^{(k)}(x)}{a_i^k},\ y_{0,k} = \frac{q_1^{(k)}(x_0)}{a_1^k - S_1},\ y_{N,k} = \frac{q_N^{(k)}(x_N)}{a_N^k - S_N},\quad k = 1, 2, \dots, p.$$

If $F_{i-1,k}(x_N, y_{N,k}) = F_{i,k}(x_0, y_{0,k})$ for $i \in \mathbb{N}_N$ and $k = 1, 2, \dots, p$, then the IFS $\{I \times \mathbb{R}; (L_i(x), F_i(x, y)), i \in \mathbb{N}_N\}$ determines a FIF $f \in \mathscr{C}^p[x_0, x_N]$, and $f^{(k)}$ is the FIF determined by $\{I \times \mathbb{R}; (L_i(x), F_{i,k}(x, y)), n \in \mathbb{N}_N\}$ for $k = 1, 2, \dots, p$.

Since most of the classical smooth splines can be generalized to fractal splines by using the above theorem, fractal interpolation constitutes an advance in the technique of approximation.

3 Construction of Rational Cubic Graph Directed Fractal Interpolation Functions

In order to interpolate several shaped data generating from $\mathscr{C}^1$-smooth functions simultaneously, we develop a new type graph-directed FIFs that are attractors of graph-directed IFSs. Let $\mathscr{D}^\alpha := \{(x_j^\alpha, F_j^\alpha, d_{F_j}^\alpha) : j = 0, 1, \dots, N_\alpha\}$ be Hermite data sets, where F_j^α and $d_{F_j}^\alpha$ are the functional and derivative values of data generating function Φ^F with $N_\alpha \geq 2$ for $\alpha = 1, 2, \dots, n$ satisfying

$$\frac{x_j^\alpha - x_{j-1}^\alpha}{x_{N_\beta}^\beta - x_0^\beta} < 1, \text{ for all } \alpha \neq \beta, \alpha, \beta = 1, 2, ..., n,\ j = 1, 2, \dots, N_\alpha. \tag{6}$$

Let $D_G = (V, E)$ be a directed graph, where V is the set of vertices and E is the set of edges. Throughout the section, we follow the following notations:

1. $k^{\alpha\beta}$ is the number of elements in $E^{\alpha\beta}$
2. $E^{\alpha\beta}$ represents the set of edges from α to β with elements $e_i^{\alpha\beta}$, $i = 1, 2, \dots k^{\alpha\beta}$.

We assume that $E^\alpha := \bigcup_{\beta \in V} E^{\alpha\beta} \neq \emptyset$.

Let $\{X^\alpha \mid \alpha \in V\}$ be a finite collection of complete metric spaces and $w_i^{\alpha\beta} : X^\beta \to X^\alpha$ be a contraction mapping corresponding to the edge e_i^{rs} (but in the opposite direction of e_i^{rs}) for $i = 1, 2, \dots, K^{\alpha\beta}$ and $r, s \in V$. It can be shown that there exists a unique family of nonempty compact sets $A^\alpha \subset X^\alpha$ such that

$$A^\alpha = \bigcup_{\beta \in V} \bigcup_{i=1}^{K^{\alpha\beta}} \omega_i^{\alpha\beta}(A^\beta).$$

The system $\left\{X^\alpha; w_i^{\alpha\beta}\right\}$ is called a graph-directed iterated function system (GDIFS) realizing the graph D_G and A^α's ($\alpha \in V$) are called the attractors of the system (see [14] for more details).

Proposition 1 [10] *Let $\mathscr{D}^\alpha := \{(x_j^\alpha, F_j^\alpha) : j = 0, 1, \ldots, N_\alpha\}$ be data sets in $\mathbb{R}^2$ with $N_\alpha \geq 2$ for $\alpha = 1, 2, \ldots, n$ satisfying (6). Let $\{\mathbb{R}^2; \omega_i^{\alpha\beta}\}$ be graph-directed IFSs associated with the data sets $\mathscr{D}^\alpha$ with attractors A^α ($\alpha = 1, 2, \ldots, n$), such that A^α is the graph of a function which interpolates the data set $\mathscr{D}^\alpha$ for each α.*

For simplicity, consider two Hermite data sets as $\mathscr{D}^1 = \{(x_0, F_0, d_{F_0}), (x_1, F_1, d_{F_1}), \ldots, (x_N, F_N, d_{F_N})\}$ and $\mathscr{D}^2 = \{(x_0, G_0, d_{G_0}), (x_1, G_1, d_{G_1}), \ldots, (x_M, G_M, d_{G_M})\}$ with $n = 2$. Denote $I = [x_0, x_N]$, $|I| = |x_N - x_0|$, $|I_i| = |x_{i-1} - x_i|$, $i \in \mathbb{N}_N$ and $J = [y_0, y_M]$, $|J| = |y_M - y_0|$, $|J_j| = |y_{j-1} - y_j|$, $j \in \mathbb{N}_M$. Let $D_G = (V, E)$ be the directed graph with $V = \{1, 2\}$ such that $k^{11} + k^{12} = N$, $k^{21} + k^{22} = M$ and $\frac{x_i - x_{i-1}}{|J|} < 1$, $\frac{y_j - y_{j-1}}{|I|} < 1$.

Define the functions $w_i^{\alpha\beta} : \mathbb{R}^2 \to \mathbb{R}^2$ as

$$w_i^{\alpha\beta} = (L_i^{\alpha\beta}(x), U_i^{\alpha\beta}(x, y)), i = 1, 2, \ldots, k^{\alpha\beta}, \ \alpha, \beta = \{1, 2\}$$

such that

- for $i = 1, 2, \ldots, k^{11}$,

$$\begin{aligned} &w_i^{11}(x_0, F_0) = (x_{i-1}, F_{i-1}), \ w_i^{11}(x_N, F_N) = (x_i, F_i), \\ &L_i^{11}(x_0) = x_{i-1}, \ L_i^{11}(x_N) = x_i, \\ &U_i^{11}(x_0, F_0) = F_{i-1}, U_i^{11}(x_N, F_N) = F_i, \\ &(U_i^{11})^{(1)}(x_0, d_{F_0}) = d_{F_{i-1}}, (U_i^{11})^{(1)}(x_N, d_{F_N}) = d_{F_i}, \end{aligned} \tag{7}$$

- for $i = 1, 2, \ldots, k^{12}$,

$$\begin{aligned} &w_i^{12}(y_0, G_0) = (x_{i-1}, F_{i-1}), \ w_i^{12}(y_M, G_M) = (x_i, F_i), \\ &L_i^{12}(y_0) = x_{i-1}, \ L_i^{12}(y_M) = x_i, \\ &U_i^{12}(y_0, G_0) = F_{i-1}, U_i^{12}(y_M, G_M) = F_i, \\ &(U_i^{12})^{(1)}(y_0, d_{G_0}) = d_{F_{i-1}}, (U_i^{12})^{(1)}(y_M, d_{G_M}) = d_{F_i}, \end{aligned} \tag{8}$$

- for $i = 1, 2, \ldots, k^{21}$,

$$\begin{aligned} &w_i^{21}(x_0, F_0) = (y_{i-1}, G_{i-1}), \ w_i^{21}(x_N, F_N) = (y_i, G_i), \\ &L_i^{21}(x_0) = y_{i-1}, \ L_i^{21}(x_N) = y_i, \\ &U_i^{21}(x_0, F_0) = G_{i-1}, U_i^{21}(x_N, F_N) = G_i, \\ &(U_i^{21})^{(1)}(x_0, d_{F_0}) = d_{G_{i-1}}, (U_i^{21})^{(1)}(x_N, d_{F_N}) = d_{G_i}, \end{aligned} \tag{9}$$

- for $i = 1, 2, \ldots, k^{22}$,

$$\begin{aligned} &w_i^{22}(y_0, G_0) = (y_{i-1}, G_{i-1}),\ w_i^{22}(y_M, G_M) = (y_i, G_i),\\ &L_i^{22}(y_0) = y_{i-1},\ L_i^{22}(y_M) = y_i,\\ &U_i^{22}(y_0, G_0) = G_{i-1}, U_i^{22}(y_M, G_M) = G_i,\\ &(U_i^{22})^{(1)}(y_0, d_{G_0}) = d_{G_{i-1}}, (U_i^{22})^{(1)}(y_M, d_{G_M}) = d_{G_i}. \end{aligned} \tag{10}$$

From each of the above join-up conditions, one can easily observe that

$$\begin{aligned} a_i^{11} &= \frac{x_i - x_{i-1}}{|I|},\ e_i^{11} = \frac{x_N x_{i-1} - x_0 x_i}{|I|} \quad \text{for } i = 1, 2, \ldots, k^{11},\\ a_i^{12} &= \frac{x_i - x_{i-1}}{|J|},\ e_i^{12} = \frac{y_M x_{i-1} - y_0 x_i}{|J|} \quad \text{for } i = 1, 2, \ldots, k^{12},\\ a_i^{21} &= \frac{y_i - y_{i-1}}{|I|},\ e_i^{21} = \frac{x_N y_{i-1} - x_0 y_i}{|J|} \quad \text{for } i = 1, 2, \ldots, k^{21},\\ a_i^{22} &= \frac{y_i - y_{i-1}}{|J|},\ e_i^{21} = \frac{y_M y_{i-1} - y_0 y_i}{|J|} \quad \text{for } i = 1, 2, \ldots, k^{22}. \end{aligned} \tag{11}$$

Here we use rational cubic FIF with two shape parameters in our construction. For construction of GDIFS, consider

$$U_i^{\alpha\beta}(x, y) = S_i^{\alpha\beta} y + R_i^{\alpha\beta}\big((L_i^{\alpha\beta})^{-1}(x)\big),\ \ R_i^{\alpha\beta}\big((L_i^{\alpha\beta})^{-1}(x)\big) = \frac{P_i^{\alpha\beta}(\theta^{\alpha\beta})}{Q_i^{\alpha\beta}(\theta^{\alpha\beta})}, \tag{12}$$

where

$$\begin{aligned} P_i^{\alpha\beta}(\theta^{\alpha\beta}) &= A_i^{\alpha\beta}(1-\theta^{\alpha\beta})^3 + B_i^{\alpha\beta}\theta^{\alpha\beta}(1-\theta^{\alpha\beta})^2\\ &\quad + C_i^{\alpha\beta}(\theta^{\alpha\beta})^2(1-\theta^{\alpha\beta}) + D_i^{\alpha\beta}(\theta^{\alpha\beta})^3,\\ Q_i^{\alpha\beta}(\theta^{\alpha\beta}) &= r_i^{\alpha\beta}(1-\theta^{\alpha\beta}) + t_i^{\alpha\beta}\theta^{\alpha\beta}, \end{aligned}$$

where $r_i^{\alpha\beta}$, $t_i^{\alpha\beta}$ are positive real parameters. For $\mathscr{C}^1$-smooth GDFIF, by using Theorem 2, it is assumed that $|S_i^{\alpha\beta}| < a_i^{\alpha\beta}$ for all possible values of α and β. The unknowns $A_i^{\alpha\beta}$, $B_i^{\alpha\beta}$, $C_i^{\alpha\beta}$, $D_i^{\alpha\beta}$ are evaluated using (7)–(10). Then, we have found

- for $i = 1, 2, \ldots, k^{11}$,

$$\begin{aligned} A_i^{11} &= r_i^{11}(F_{i-1} - S_i^{11}F_0),\ D_i^{11} = t_i^{11}(F_i - S_i^{11}F_N),\\ B_i^{11} &= r_i^{11}(d_{F_{i-1}}a_i^{11} - S_i^{11}d_{F_0})|I| + (2r_i^{11} + t_i^{11})(F_{i-1} - S_i^{11}F_0),\\ C_i^{11} &= -t_i^{11}(d_{F_i}a_i^{11} - S_i^{11}d_{F_N})|I| + (r_i^{11} + 2t_i^{11})(F_i - S_i^{11}F_N), \end{aligned} \tag{13}$$

- for $i = 1, 2, \ldots, k^{12}$,

$$
\begin{aligned}
A_i^{12} &= r_i^{12}(F_{i-1} - S_i^{12} G_0),\ \ D_i^{12} = t_i^{12}(F_i - S_i^{21} G_M),\\
B_i^{12} &= r_i^{12}(d_{F_{i-1}} a_i^{12} - S_i^{12} d_{G_0})|J| + (2r_i^{12} + t_i^{12})(F_{i-1} - S_i^{12} G_0),\\
C_i^{12} &= -t_i^{12}(d_{F_i} a_i^{12} - S_i^{12} d_{G_M})|I| + (r_i^{12} + 2t_i^{12})(F_i - S_i^{12} F_N),
\end{aligned} \tag{14}
$$

– for $i = 1, 2, \dots, k^{21}$,

$$
\begin{aligned}
A_i^{21} &= r_i^{21}(G_{i-1} - S_i^{21} F_0),\ \ D_i^{21} = t_i^{21}(G_i - S_i^{21} F_N),\\
B_i^{21} &= r_i^{21}(d_{G_{i-1}} a_i^{21} - S_i^{21} d_{F_0})|J| + (2r_i^{21} + t_i^{21})(G_{i-1} - S_i^{21} F_0),\\
C_i^{21} &= -t_i^{21}(d_{G_i} a_i^{21} - S_i^{21} d_{F_N})|I| + (r_i^{21} + 2t_i^{21})(G_i - S_i^{21} F_N),
\end{aligned} \tag{15}
$$

– for $i = 1, 2, \dots, k^{22}$,

$$
\begin{aligned}
A_i^{22} &= r_i^{22}(G_{i-1} - S_i^{22} G_0),\ \ D_i^{22} = t_i^{22}(G_i - S_i^{22} G_M),\\
B_i^{22} &= r_i^{22}(d_{G_{i-1}} a_i^{22} - S_i^{22} d_{G_0})|J| + (2r_i^{22} + t_i^{22})(G_{i-1} - S_i^{22} G_0),\\
C_i^{22} &= -t_i^{22}(d_{G_i} a_i^{22} - S_i^{22} d_{G_M})|J| + (r_i^{22} + 2t_i^{22})(G_i - S_i^{22} G_M).
\end{aligned} \tag{16}
$$

Substituting the values of parameters in (13)–(16) the functional equations corresponding to (12), we obtain the desired graph-directed fractal interpolations Φ^f and Φ^g as

$$
\begin{aligned}
\phi^f(L_i^{11}(x)) &= S_i^{11}\phi^f(x) + R_i^{11}(x, S_i^{11}) \ \text{ for } \ i = 1, 2, \dots, k^{11},\\
\phi^f(L_i^{12}(y)) &= S_i^{12}\phi^f(x) + R_i^{12}(x, S_i^{12}) \ \text{ for } \ i = 1, 2, \dots, k^{12},\\
\phi^g(L_i^{21}(x)) &= S_i^{21}\phi^g(x) + R_i^{21}(x, S_i^{21}) \ \text{ for } \ i = 1, 2, \dots, k^{21},\\
\phi^g(L_i^{22}(y)) &= S_i^{22}\phi^g(x) + R_i^{22}(x, S_i^{22}) \ \text{ for } \ i = 1, 2, \dots, k^{22}.
\end{aligned} \tag{17}
$$

4 Monotonicity Preserving RCGDFIF

In this section, we restrict the GDIFS parameters based on the sufficient conditions so that the developed RCGDFIFs preserve the monotonicity for monotonic data sets.

Theorem 1 *Let* $\{(x_0, F_0), (x_1, F_1), \dots, (x_N, F_N)\}$ *and* $\{(y_0, G_0), (y_1, G_1), \dots, (y_M, G_M)\}$, *where* $N, M \geq 2$ *are two strictly monotonic data sets, i.e.,* $F_i > F_{i-1}$, $i \in \mathbb{N}_N$ *and* $G_j > G_{j-1}$, $j \in \mathbb{N}_M$. *For* $i = 1, 2, \dots, k^{\alpha\beta}$, $\alpha, \beta \in V := \{1, 2\}$, *if*

(i) *the scaling factors* $S^{\alpha\beta}$ *are chosen such that*

$$\left.\begin{array}{l} 0 \le S_i^{11} < \min\left\{a_i^{11}, \frac{a_i^{11} d_{F_{i-1}}}{d_{F_0}}, \frac{a_i^{11} d_{F_i}}{d_{F_N}}, \frac{F_i - F_{i-1}}{x_N - x_1}\right\} \\ 0 \le S_i^{12} < \min\left\{a_i^{12}, \frac{a_i^{12} d_{F_{i-1}}}{d_{G_0}}, \frac{a_i^{12} d_{F_i}}{d_{G_M}}, \frac{F_i - F_{i-1}}{y_M - y_1}\right\} \\ 0 \le S_i^{21} < \min\left\{a_i^{21}, \frac{a_i^{21} d_{G_{i-1}}}{d_{F_0}}, \frac{a_i^{21} d_{G_i}}{d_{F_N}}, \frac{G_i - G_{i-1}}{x_N - x_1}\right\} \\ 0 \le S_i^{22} < \min\left\{a_i^{22}, \frac{a_i^{22} d_{G_{i-1}}}{d_{G_0}}, \frac{a_i^{22} d_{G_i}}{d_{G_M}}, \frac{G_i - G_{i-1}}{y_M - y_1}\right\} \end{array}\right\}, \tag{18}$$

(ii) the shape parameters are chosen such that $r_i^{\alpha\beta} > 0$, $t_i^{\alpha\beta} > 0$ *and*

$$\begin{gathered} r_i^{11} > \max\left\{0, \frac{t_i^{11}(d_{F_i} h_i - S_i^{11} d_{F_N}|I| - [F_i - F_{i-1} - S_i^{11}(F_N - F_0)])}{F_i - F_{i-1} - S_i^{11}(F_N - F_0) - [d_{F_{i-1}} h_i - S_i^{11} d_{F_0}|I|]}\right\}, \\ r_i^{12} > \max\left\{0, \frac{t_i^{12}(d_{F_i} h_i - S_i^{12} d_{G_M}|J| - [F_i - F_{i-1} - S_i^{12}(G_M - G_0)])}{F_i - F_{i-1} - S_i^{12}(G_M - G_0) - [d_{F_{i-1}} h_i - S_i^{12} d_{G_0}|J|}\right\}, \\ r_i^{21} > \max\left\{0, \frac{t_i^{21}(d_{G_i} h_i^* - S_i^{21} d_{F_N}|I| - [G_i - G_{i-1} - S_i^{21}(F_N - F_0)])}{G_i - G_{i-1} - S_i^{21}(F_N - F_0) - [d_{G_{i-1}} h_i^* - S_i^{21} d_{F_0}|I|]}\right\}, \\ r_i^{22} > \max\left\{0, \frac{t_i^{22}(d_{G_i} h_i^* - S_i^{22} d_{G_M}|J| - [G_i - G_{i-1} - S_i^{22}(G_N - G_0)])}{G_i - G_{i-1} - S_i^{22}(G_N - G_0) - [d_{G_{i-1}} h_i^* - S_i^{22} d_{G_0}|J|]}\right\}, \\ h_i = x_i - x_{i-1}, h_i^* = y_i - y_{i-1}, i \in \{1, \dots, k^{\alpha\beta}\}, \end{gathered} \tag{19}$$

then for a fixed set of positive real numbers $r_i^{\alpha\beta}$ *and* $t_i^{\alpha\beta}$, $i \in \{1, \dots, k^{\alpha\beta}\}$, *the* $\mathscr{C}^1$-*RCGDFIFs preserve the monotonicity feature of the given data sets.*

Proof Let the monotonic data sets $\{(x_0, F_0), (x_1, F_1), \dots, (x_N, F_N)\}$ and $\{(y_0, G_0), (y_1, G_1), \dots, (y_M, G_M)\}$ be generated from some smooth monotonic functions f and g, respectively. Suppose ϕ^f and ϕ^g are RCGDFIFs associated with the RGDIFS constructed from the above data sets. Differentiating (12), we have

$$(\phi^f)'(x) = \begin{cases} S_i^{11}(\phi^f)'((L_i^{11})^{-1}(x)) + \frac{\mathscr{N}_i^{11}(\theta^{11})}{a_i^{11}(q_i^{11}(\theta^{11}))^2}, & \text{for } i = 1, 2, \dots, k^{11}, \\ S_i^{12}(\phi^g)'((L_i^{12})^{-1}(x)) + \frac{\mathscr{N}_i^{12}(\theta^{12})}{a_i^{12}(q_i^{12}(\theta^{12}))^2}, & \text{for } i = 1, 2, \dots, k^{12}, \end{cases} \tag{20}$$

where $\mathscr{N}_i^{11}(\theta^{11}) = M_{1,i}^{11}(1-\theta^{11})^3 + M_{2,i}^{11}\theta^{11}(1-\theta^{11})^2 + M_{3,i}^{11}(\theta^{11})(1-\theta^{11})^2 + M_{4,i}^{11}(\theta^{11})^3$, $\mathscr{N}_i^{12}(\theta^{12}) = M_{1,i}^{12}(1-\theta^{12})^2 + M_{2,i}^{12}\theta^{12}(1-\theta^{12})^2 + M_{3,i}^{12}(\theta^{12})(1-\theta^{12})^2 + M_{4,i}^{12}(\theta^{12})^3$,

$$\begin{aligned} M_{1,i}^{11} &= (r_i^{11})^2\left(d_{F_{i-1}} - \frac{S_i^{11}}{a_i^{11}} d_{F_0}\right), \; M_{4,i}^{11} = (t_i^{11})^2\left(d_{F_i} - \frac{S_i^{11}}{a_i^{11}} d_{F_N}\right), \\ M_{2,i}^{11} &= \frac{(2(r_i^{11})^2 + 4r_i^{11}t_i^{11})}{h_i}\left(F_i - F_{i-1} - S_i^{11}[F_N - F_0]\right) \\ &\quad -(r_i^{11})^2\left[d_{F_{i-1}} - \frac{S_i^{11}}{a_i^{11}} d_{F_0}\right] - 2r_i^{11}t_i^{11}\left[d_{F_i} - \frac{S_i^{11}}{a_i^{11}} d_{F_N}\right], \\ M_{3,i}^{11} &= \frac{(2(t_i^{11})^2 + 4r_i^{11}t_i^{11})}{h_i}\left(F_i - F_{i-1} - S_i^{11}[F_N - F_0]\right) \end{aligned}$$

$$-(t_i^{11})^2\left[d_{F_i}-\frac{S_i^{11}}{a_i^{11}}d_{F_N}\right]-2r_i^{11}t_i^{11}\left[d_{F_{i-1}}-\frac{S_i^{11}}{a_i^{11}}d_{F_0}\right],$$

and

$$M_{1,i}^{12}=(r_i^{12})^2\left(d_{F_{i-1}}-\frac{S_i^{12}}{a_i^{12}}d_{G_0}\right),\ M_{4,i}^{12}=(t_i^{12})^2\left(d_{F_i}-\frac{S_i^{12}}{a_i^{12}}d_{G_M}\right),$$

$$M_{2,i}^{12}=\frac{\left(2(r_i^{12})^2+4r_i^{12}t_i^{12}\right)}{h_i}\left(F_i-F_{i-1}-\alpha_i^{12}[G_M-G_0]\right)$$
$$-(r_i^{12})^2\left[d_{F_{i-1}}-\frac{S_i^{12}}{a_i^{12}}d_{G_0}\right]-2r_i^{12}t_i^{12}\left[d_{F_i}-\frac{S_i^{12}}{a_i^{12}}d_{G_M}\right],$$

$$M_{3,i}^{12}=\frac{\left(2(t_i^{12})^2+4r_i^{11}t_i^{12}\right)}{h_i}\left(F_i-F_{i-1}-S_i^{12}[G_M-G_0]\right)$$
$$-(t_i^{12})^2\left[d_{F_i}-\frac{S_i^{12}}{a_i^{11}}d_{G_M}\right]-2r_i^{12}t_i^{12}\left[d_{F_{i-1}}-\frac{S_i^{12}}{a_i^{12}}d_{G_0}\right].$$

Similarly, consider the functional equation of ϕ^g for $x\in[x_{i-1},x_i]$ as

$$\phi^g(x)=\begin{cases}S_i^{21}\phi^f((L_i^{21})^{-1}(x))+M_i^{21}((L_i^{21})^{-1}(x),S_i^{21}), & \text{for } i=1,2,\dots,k^{21},\\ S_i^{22}\Phi^g((L_i^{22})^{-1}(x))+M_i^{22}((L_i^{22})^{-1}(x),S_i^{22}), & \text{for } i=1,2,\dots,k^{22},\end{cases}\tag{21}$$

where $M_i^{2\beta}((L_i^{2\beta})^{-1}(x),\alpha_i^{2\beta})=\frac{p_i^{2\beta}(\theta^{2\beta})}{q_i^{2\beta}(\theta^{2\beta})}$, $\beta\in\{1,2\}$ and $p_i^{2\beta}(\theta^{2\beta})$ and $q_i^{2\beta}(\theta^{2\beta})$ are defined using (15)–(16). Differentiating (21), we obtain the following:

$$(\phi^g)'(x)=\begin{cases}S_i^{21}(\Phi^f)'((L_i^{21})^{-1}(x))+\frac{\mathcal{N}_i^{21}(\theta^{21})}{a_i^{21}(q_i^{21}(\theta^{21}))^2}, & \text{for } i=1,2,\dots,k^{21},\\ S_i^{22}(\Phi^g)'((L_i^{22})^{-1}(x))+\frac{\mathcal{N}_i^{22}(\theta^{22})}{a_i^{22}(q_i^{22}(\theta^{22}))^2}, & \text{for } i=1,2,\dots,k^{22},\end{cases}$$

where $\mathcal{N}_i^{21}(\theta^{21})=M_{1,i}^{21}(1-\theta^{21})^3+M_{2,i}^{21}\theta^{21}(1-\theta^{21})^2+M_{3,i}^{21}(\theta^{21})^2(1-\theta^{21})+M_{4,i}^{21}(\theta^{21})^3$, $\mathcal{N}_i^{22}(\theta^{22})=M_{1,i}^{22}(1-\theta^{22})^3+M_{2,i}^{22}\theta^{22}(1-\theta^{22})^2+M_{3,i}^{22}(\theta^{22})^2(1-\theta^{22})+M_{4,i}^{22}(\theta^{22})^3$,

$$M_{1,i}^{21}=(r_i^{21})^2\left(d_{G_{i-1}}-\frac{S_i^{21}}{a_i^{21}}d_{F_0}\right),\ M_{4,i}^{21}=(t_i^{21})^2\left(d_{G_i}-\frac{S_i^{21}}{a_i^{21}}d_{F_N}\right),$$

$$M_{2,i}^{21}=\frac{\left(2(r_i^{21})^2+4r_i^{21}t_i^{21}\right)}{h_i^*}\left(G_i-G_{i-1}-S_i^{21}[F_N-F_0]\right)$$
$$-(r_i^{21})^2\left[d_{G_{i-1}}-\frac{S_i^{21}}{a_i^{21}}d_{F_0}\right]-2r_i^{21}t_i^{21}\left[d_{G_i}-\frac{S_i^{21}}{a_i^{21}}d_{F_N}\right],$$

$$M_{3,i}^{21}=\frac{\left(2(t_i^{21})^2+4r_i^{21}t_i^{21}\right)}{h_i^*}\left(G_i-G_{i-1}-S_i^{21}[F_N-F_0]\right)$$

$$-(t_i^{21})^2\left[d_{G_i}-\frac{S_i^{21}}{a_i^{21}}d_{F_N}\right]-2r_i^{21}t_i^{21}\left[d_{G_{i-1}}-\frac{S_i^{21}}{a_i^{21}}d_{F_0}\right],$$

and

$$M_{1,i}^{22}=(r_i^{22})^2\left(d_{G_{i-1}}-\frac{S_i^{22}}{a_i^{22}}d_{G_0}\right),\ M_{4,i}^{22}=(t_i^{22})^2\left(d_{G_i}-\frac{S_i^{22}}{a_i^{21}}d_{G_M}\right),$$

$$M_{2,i}^{22}=\frac{(2(r_i^{22})^2+4r_i^{22}t_i^{22})}{h_i^*}(G_i-G_{i-1}-S_i^{22}[G_M-G_0])$$

$$-(r_i^{22})^2\left[d_{G_{i-1}}-\frac{S_i^{22}}{a_i^{22}}d_{G_0}\right]-2r_i^{22}t_i^{22}\left[d_{G_i}-\frac{S_i^{22}}{a_i^{22}}d_{G_M}\right],$$

$$M_{3,i}^{22}=\frac{(2(t_i^{22})^2+4r_i^{21}t_i^{22})}{h_i^*}(G_i-G_{i-1}-S_i^{22}[G_M-G_0])$$

$$-(t_i^{22})^2\left[d_{G_i}-\frac{S_i^{21}}{a_i^{22}}d_{G_M}\right]-2r_i^{22}t_i^{22}\left[d_{G_{i-1}}-\frac{S_i^{22}}{a_i^{22}}d_{G_0}\right].$$

Without loss of generality, assume that the data sets are strictly increasing. Then, the necessary conditions $(\phi^f)'(x_j)>0,\ j\in\mathbb{N}_N\cup\{0\}$ and $(\phi^g)'(y_j)>0,\ j\in\mathbb{N}_M\cup\{0\}$. Assume that the scaling factors S_i^{11} and S_i^{12} are positive. It is clear that from (20) and (4), the RCGDFIF ϕ^f and Φ^g are monotonic if $(\phi^f)'$ and $(\phi^g)'$ are positive over I and J, respectively. Since the shape parameters $r_i^{\alpha\beta}$ and $t_i^{\alpha\beta}, \alpha,\beta\in V:=\{1,2\}$ are positive, the positivity of $(\phi^f)'$ and $(\Phi^g)'$ depends on the positivity of $A_i^{\alpha\beta}(\theta^{\alpha\beta}),\ \alpha,\beta\in V$. It is sufficient to find the positivity conditions for coefficients $M_{k,i}^{\alpha\beta},\ \alpha,\beta\in V,\ k\in\mathbb{N}_4$ in order to get the positivity of $A_i^{\alpha\beta}(\theta^{\alpha\beta}),\ \alpha,\ \beta\in V$. We will deduce the range restricted conditions on the IFS parameters $S_i^{11},\ r_i^{11}$ so that $M_{K,i}^{11}>0,\ K\in\mathbb{N}_4,\ i=1,2,\ldots,k^{11}$.
Consider

$$M_{1,i}^{11}=(r_i^{11})^2\left(d_{F_{i-1}}-\frac{S_i^{11}}{a_i^{11}}d_{F_0}\right).$$

Now $M_{1,i}>0$ if $S_i^{11}<\frac{a_i^{11}d_{F_{i-1}}}{d_{F_0}}$.
Similarly, consider

$$M_{4,i}^{11}=(t_i^{21})^2\left(d_{F_i}-\frac{S_i^{11}}{a_i^{21}}d_{F_N}\right).$$

With the selection of $S_i^{11}<\frac{a_i^{11}d_{F_i}}{d_{F_N}}$, we obtain the positivity of $M_{4,i}^{11}>0$.

Now consider

$$M_{2,i}^{11} = \frac{\left(2(r_i^{11})^2 + 4r_i^{11}t_i^{11}\right)}{h_i}\left(F_i - F_{i-1} - S_i^{11}[F_N - F_0]\right) - (r_i^{11})^2\left[d_{F_{i-1}} - \frac{S_i^{11}}{a_i^{11}} d_{F_0}\right] - 2r_i^{11}t_i^{11}\left[d_{F_i} - \frac{S_i^{11}}{a_i^{11}} d_{F_N}\right],$$

$$M_{3,i}^{11} = \frac{\left(2(r_i^{11})^2 + 4r_i^{11}t_i^{11}\right)}{h_i}\left(F_i - F_{i-1} - S_i^{11}[F_N - F_0]\right) - (t_i^{11})^2\left[d_{F_i} - \frac{S_i^{11}}{a_i^{11}} d_{F_N}\right] - 2r_i^{11}t_i^{11}\left[d_{F_{i-1}} - \frac{S_i^{11}}{a_i^{11}} d_{F_0}\right].$$

From simple calculation, we infer that $M_{2,i}^{11}$ and $M_{3,i}^{11}$ are non negative if $r_i^{11} > \frac{t_i^{11}\left(d_{F_i}h_i - S_i^{11}d_{F_N}|I| - [F_i - F_{i-1} - S_i^{11}(F_N - F_0)]\right)}{F_i - F_{i-1} - S_i^{11}(F_N - F_0)] - (d_{F_{i-1}}h_i - S_i^{11}d_{F_0}|I|)}$ Combining all these conditions, we obtain the desired restrictions on IFS parameters for strictly increasing data sets. Applying the same arguments, we get condition for $\mathscr{N}_i^{12}(\theta^{12})$ is positive for $i = 1, 2, \ldots, k^{12}$, and hence Φ^f is strictly increasing. Similarly, selecting the IFS parameters listed in last two inequalities of (18)–(19), we obtain the monotonicity of ϕ^g. Similarly, for strictly decreasing data sets, we have $D_{F_j} < 0,\ j \in \mathbb{N}_N \cup \{0\}$ and $D_{G_j} < 0,\ j \in \mathbb{N}_M \cup \{0\}$ and we have to prove $(\phi^f)'(x) < 0$ and $(\phi^g)'(x) < 0$. Take

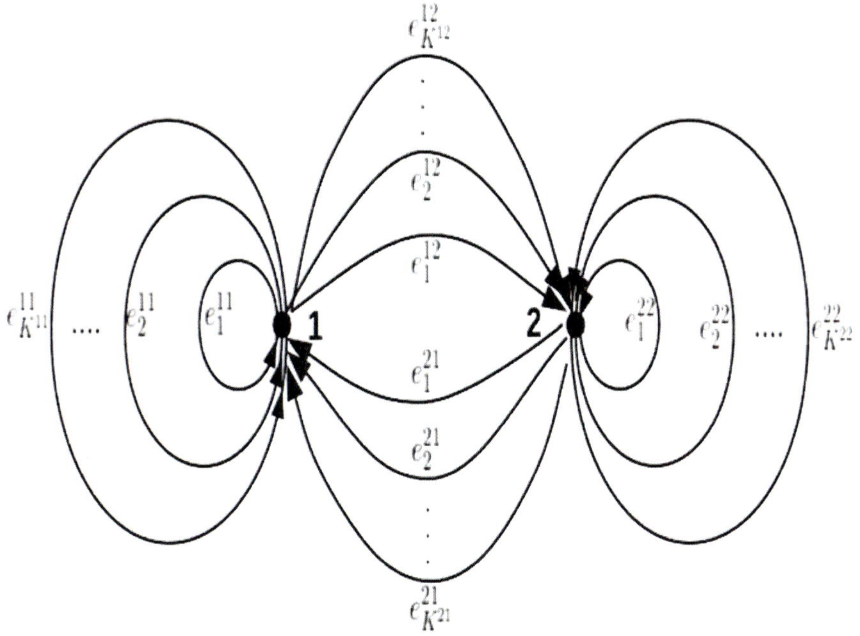

Fig. 1 Directed graph between two sets of interpolation data

$S_i^{\alpha\beta} \geq 0,\ \alpha, \beta \in V$, we have to find conditions on IFS parameters for which $A_i^{\alpha\beta}(\theta^{\alpha\beta}) < 0,\ \alpha,\ \beta \in V$, respectively. It is found that the same restrictions (18)–(19) are also sufficient in this case.

5 Numerical Examples

In this section, we have generated examples of monotonic RCGDFIFs for two different dependent monotonic data sets. Consider two monotonically increasing data sets as $\{(0, 0.1), (1, 0.4), (2, 0.9), (3, 1.4), (4, 1.5), (5, 1.6)\}$ and $\{(0, 0.01), (1, 0.25), (2, 1), (3, 2.5), (4, 3.5)\}$ realizing the graph with $k^{11} = 3,\ k^{12} = 2, k^{21} = 1, k^{22} = 3$ as in the directed graph shown in Fig. 1. Let us denote the graph-directed FIFs are ϕ^f and ϕ^g associated with the given data sets, respectively, for fixed choice of IFS parameters.

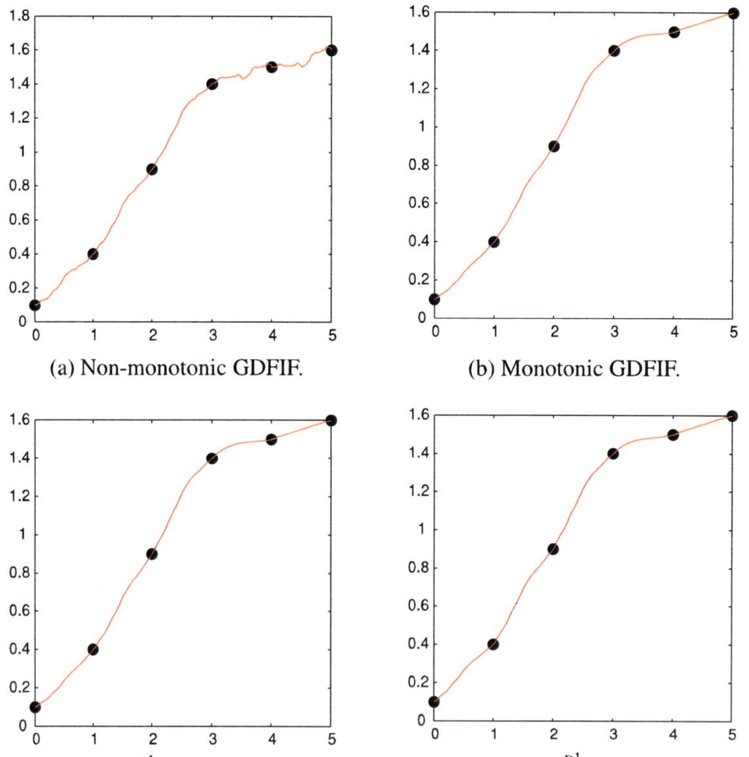

(a) Non-monotonic GDFIF.

(b) Monotonic GDFIF.

(c) Effects of $S^{D1} = (0, 0, 0, 0, 0)$ and r in Fig. 2 (b)

(d) Effects of S^{D1} and r in Fig. 2 (b).

Fig. 2 Rational cubic graph-directed FIFs ϕ^f for monotonic data set

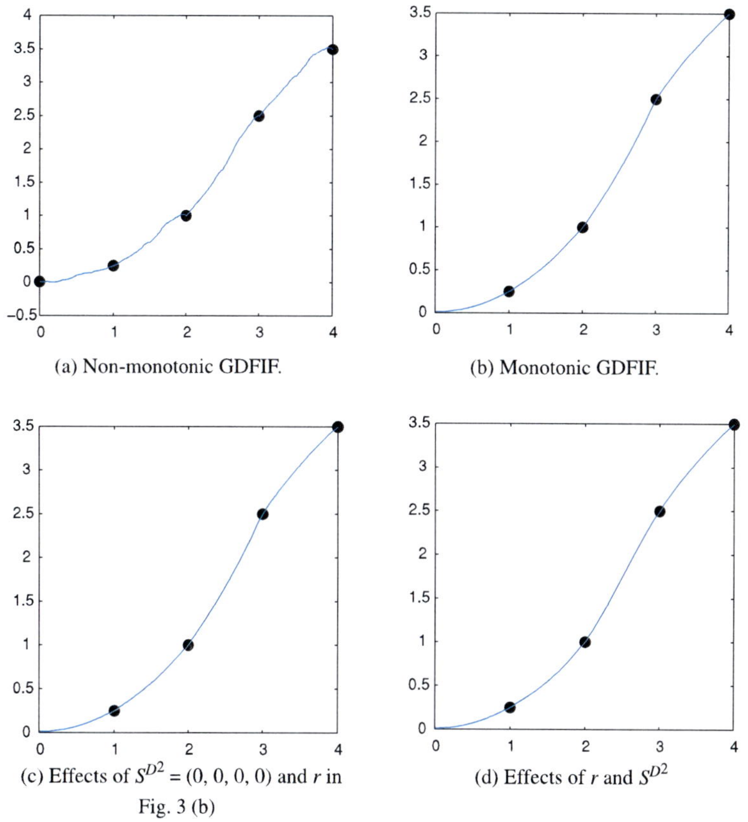

Fig. 3 Rational cubic graph-directed FIFs ϕ^g for monotonic data set

For an arbitrary choice of GDIFS parameter in (17), we obtain non-monotonicity graph-directed FIFs ϕ^f and ϕ^g in the Figs. 2a and 3a respectively. When the scaling and shape parameters are elected from the desired ranges as prescribed in Theorem 1 (see Table 1)), the monotonicity graph-directed FIFs are constructed in Figs. 2, 3b–d. We have modified the scaling factor $S_i^{D^1} = (0, 0, 0, 0, 0)$ and shape parameter r with respect to the Fig. 2b to generate RCGDFIF in Fig. 2c. We have constructed another graph-directed FIF Fig. 2d by changing the IFS parameters r and α with respect to Fig. 2b. Similarly, by changing the IFS parameters in Fig. 3a, we have generated different shape of graph-directed rational FIFs in Fig. 3b–d.

Table 1 Scaling vectors and shape parameters used in the construction of monotonic RCGDFIFs in Figs. 2, 3

Figures	Scaling vectors S	Shape parameter vectors r, t
2a	(0.19, 0.19, 0.19, 0.19, 0.19)	(10, 10, 10, 10, 30, 30), (50, 50, 50, 20, 20)
2b	(0.075, 0.133, 0.125, 0, 0)	(120.32, 96.87, 81.25, 50, 50), (50, 50, 50, 20, 20)
2c	(0, 0, 0, 0, 0)	(10, 10, 10, 30, 30), (50, 50, 50, 20, 20)
2d	(0.075, 0.125, 0.125, 0, 0)	10^5 (1, 1, 1, 1, 1), (50, 50, 50, 20, 20)
3a	(0.23, 0.23, 0.23, 0.23)	(10, 30, 30, 30), (100, 0.05, 0.05, 0.05)
3b	(0.001, 0.001, 0.001, 0.001)	(50, 99.95, 99.95, 99.95), (100, 0.05, 0.05, 0.05)
3c	(0, 0, 0, 0)	(10, 30, 30, 30), (100, 0.05, 0.05, 0.05)
3d	(0.001, 0.001, 0.001, 0.001)	10^5 (1, 1, 1, 1), (100, 0.05, 0.05, 0.05)

6 Conclusion

In this paper, we have proposed a graph-directed rational cubic FIFs for two or more dependent data sets. The sufficient conditions are derived to preserve monotone nature of interpolation data sets. The theoretical results are verified through examples and effects of GDIFS parameters are illustrated. The proposed model can be used to modeling of several monotonic dependent data sets in various science and engineering problems.

Acknowledgements The first author is thankful for the MATRICS Project: MTR/2017/000574 from the Science and Engineering Research Board (SERB), Government of India.

References

1. Barnsley, M.F.: Fractal functions and interpolation. Constr. Approx. **2**(4), 303–329 (1986)
2. Barnsley, M.F.: Fractals Everywhere. Academic Press, Orlando, Florida (1988)
3. Barnsley, M.F., Harrington, A.N.: The calculus of fractal functions. J. Approx. Theory **57**(1), 14–34 (1989)
4. Chand, A.K.B., Reddy, K.M.: Constrained fractal interpolation functions with variable scaling. Sib. Élektron. Mat. Izv. **15**, 60–73 (2018)
5. Chand, A.K.B., Vijender, N., Navascués, M.A.: Shape preservation of scientific data through rational fractal splines. Calcolo **51**, 329–362 (2014)
6. Chand, A.K.B., Vijender, N., Agarwal, R.P.: Rational iterated function system for positive/monotonic shape preservation. Adv. Differ. Equ. **30**, 1–19 (2014)

7. Chand, A.K.B., Viswanathan, P., Reddy, K.M.: Towards a more general type of univariate constrained interpolation with fractal splines. Fractals **23**(4), 12 (2015)
8. Chand, A.K.B., Viswanathan, P., Reddy, K.M.: A novel approach to surface interpolation: marriage of Coons technique and univariate fractal functions. In: Mathematical Analysis and Its Applications, Springer Proc. Math. Stat., vol. 143, pp. 577–592. Springer, New Delhi (2015)
9. Chand, A.K.B., Tyada, K.R.: Constrained shape preserving rational cubic fractal interpolation functions. Rocky Mountain J. Math. **48**(1), 75–105 (2018)
10. Deniz, A., Özdemir, Y.: Graph-directed fractal interpolation functions. Turk. J. Math. **41**, 829–840 (2017)
11. Delbourgo, R., Gregory, J.A.: $\mathscr{C}^2$ rational quadratic spline interpolation to monotonic data. IMA. J. Numer. Anal. **3**, 141–152 (1983)
12. Delbourgo, R., Gregory, J.A.: Shape preserving piecewise rational interpolation. SIAM J. Sci. Stat. Comp. **6**(1), 967–976 (1985)
13. Duan, Q., Xu, G., Liu, A., Wang, X., Cheng, F.: Constrained interpolation using rational Cubic spline with linear denominators. Korean J. Comput. Appl. Math. **6**, 203–215 (1999)
14. Edgar, G.: Measure, Topology and Fractal Geometry. Springer, New York (2008)
15. Farin, G.: Curves and Surfaces for Computer Aided Geometric Design: A Practical Guide, Computer Science and Scientific Computing. Academic Press, San Diego, Calif, USA (1990)
16. Fisher, Y.: Fractal Image Compression. Springer, New York, USA (1995)
17. Katiyar, S.K., Reddy, K.M. , Chand, A.K.B.: Constrained data visualization using rational bi-cubic fractal functions. In: Mathematics and Computing, Commun. Comput. Inf. Sci., vol. 655, pp. 265–277. Springer, Singapore (2017)
18. Katiyar, S.K., Chand, A.K.B., Saravana Kumar, G.: A new class of rational cubic spline fractal interpolation function and its constrained aspects. Appl. Math. Comput. **346**, 319–335 (2019)
19. Reddy, K.M., Chand, A.K.B., Viswanathan, P.: Data visualization by rational fractal function based on function values. J. Anal. **28**(1), 261–277 (2020)

Author Index

D. Giri et al. (eds.), *Proceedings of the Fifth International Conference on Mathematics and Computing*, Advances in Intelligent Systems and Computing 1170, https://doi.org/10.1007/978-981-15-5411-7